SPLINES AND VARIATIONAL METHODS

P. M. Prenter
Colorado State University

DOVER PUBLICATIONS, INC.
Mineola, New York

Bibliographical Note

This Dover edition, first published in 2008, is an unabridged republication of the work originally published in 1975 by John Wiley & Sons, Inc., New York.

Library of Congress Cataloging-in-Publication Data

Prenter, P. M., 1934-
Splines and variational methods / P.M. Prenter. — Dover ed.
p. cm.
Originally published: New York : Wiley, 1975.
Includes index.
ISBN-13: 978-0-486-46902-7
ISBN-10: 0-486-46902-6
1. Boundary value problems—Numerical solutions. 2. Finite element method. 3. Collocation methods. 4. Spline theory. I. Title.

QA379.P74 2008
511'.422—dc22

2008029636

Manufactured in the United States of America
Dover Publications, Inc., 31 East 2nd Street, Mineola, N.Y. 11501

PREFACE

This book introduces the numerical solution of boundary value problems by variational methods. Special emphasis is placed on the finite element and collocation methods. We have deliberately attempted to present the majority of the material in a very elementary manner, to make it accessible to serious students of engineering and mathematics who have no more mathematical background than elementary applied linear algebra and very basic advanced calculus or their equivalents. Hilbert space and more abstract functional analytic concepts have been introduced only when needed to put the theory on a firm mathematical footing or to unify a body of material and point out new directions of thinking about boundary value problems. Some will undoubtedly find this point of view distasteful. However, the approach seems to be in keeping with the development of a vast engineering literature on the subject. Engineers are unquestionably the pioneers in the development and research on finite elements, and a large number of their early successes were produced without benefit of abstract functional analysis; basic calculus and a strong physical sense of what will work were their only tools. (This is not to say that powerful mathematics is unnecessary to establish further successes and point out serious failures.) We have tried to reach the audience described, to introduce them gradually to the mathematician's way of thinking about such problems. In this sense, the book is no more than an introduction to more advanced works, as well as an access route to the enormous literature and great number of open research questions in the area.

A secondary direction we have tried to point out, hopefully with success, is the powerful application of approximation theoretic notions to very applied problems. It seems unfortunate that so few approximation theorists are interested in the dynamic application of their art to very difficult and important physical problems. We hope that this is a changing situation and that more of these individuals will turn their attentions and talents to some of the numerical problems facing the engineer. Although our emphasis is on splines as a fine approximating tool, it is clear that splines, as we know them are not a panacea for all problems of numerical approximation. We feel that many fruitful new ideas are in the winds and that this is an exciting time for approximation theory. Perhaps the splines theorists, with their penchant for functional catholicism, will claim these stars of the unborn. Be that as it may, it is a fruitful era.

No attempt has been made to present the material from a physical viewpoint. This is recognized as a serious omission, since the womb or birthplace of the Rayleigh-Ritz-Galerkin method is the idea of minimizing the energies of physical systems over finite dimensional approximating spaces. Moreover, this outlook is still the most natural physical way of approaching many applied problems. Many excellent mathematics and physics texts take precisely this point of view, however; thus perhaps we can be forgiven our transgressions.

The book has been successfully used over a 4-year period to teach mixed audiences of first-year graduate students in engineering and mathematics, as well as industrial scientists. It has also been used for short expository lectures here and overseas. By giving reading assignments and covering only the pragmatic high points of Chapters 1 through 5 we have been able to go through the entire text in one quarter. This has entailed an attempt to present sufficient proofs to keep the mathematicians happy and to delete sufficient proofs to keep the engineers from becoming miserable. An integral part of the course has always included numerous large- and small-scale computational exercises and a lot of programming.

The author is grateful for help and encouragement from a number of individuals. Among these are Bob Rice and Dr. Ron Guenther of the Marathon Oil Corporation and Oregon State University, who thought a book on the subject that could be read by ordinary mortals would be advisable; Drs. Bob Russell and Neall Strand, true friends in need, who proofread the entire original version of the manuscript; my department chairman Dr. E. R. Deal, for providing understanding and typing assistance; Professor C. Jacobsz of the Council for Scientific and Industrial Research (CSIR) of the Republic of South Africa and to the CSIR for providing an appointment as a senior research scientist and bringing me to South Africa to deliver an expository lecture series and to work on my book and my research; to Professor Karl Nickel and the Technical University of Karlsruhe for bringing me to Germany to lecture from and work on the book and my research; to Dr. Erik Thompson of the Civil Engineering Department at Colorado State University for many helpful discussions and for introducing me to the engineering literature; to my students Dr. Tom Dence, Tim Simpson, Don De Gryse, A. Sato, and C. Chen, who pointed out errors in many portions of the manuscript, as did Dr. Don Jones of the University of Michigan; and finally to my many other students here and overseas whose interest and intolerance for obscuring simple mathematical concepts with unnecessary abstract mathematical tools has tempered my outlook. The author cannot ignore her debt to the influence of Dr. I. J. Schoenberg of the Mathematics Research Center, University of Wisconsin, whose personal kindness and beautiful lectures first stimulated the author's interest in splines, and of Professor Mihklin of the USSR, whom the author has never met but whose mathematical outlook has obviously been most influential. Final thanks are also due to a sequence (rather long, but thankfully finite) of impeccable typists who have worked on various phases of the manuscript. Among these are Mrs. Evelyn Anderson and Mrs. Beth Murphy of CSU and Mrs. S. Van Wyk of the CSIR. It is hoped that there are not too many serious mathematical errors; for any that may appear, the author assumes full responsibility.

PADDY PRENTER

Pretoria
Karlsruhe
Fort Collins
March 1975

CONTENTS

1

INTRODUCTORY IDEAS

1.1 A SIMPLY STATED PROBLEM

The need for good techniques for the approximation of functions arises in many settings; one of these is the numerical solution of differential equations. For example, suppose you are given the differential equation

$$\frac{d^2x}{dt^2}+a(t)\frac{dx}{dt}+b(t)x(t)=f(t), \qquad a\leqslant t\leqslant b \tag{1}$$

subject to the *boundary conditions*

$$x(a)=\alpha \qquad \text{and} \qquad x(b)=\beta,$$

where α and β are constant and $a(t)$, $b(t)$, and $f(t)$ are functions of t defined on the interval $[a,b]$. Moreover, suppose you know that this equation, subject to the boundary conditions, has a *unique* solution $x(t)$ which you would like to find. Our first problem then is

PROBLEM 1

How do we find $x(t)$?

As those who have worked to any extent in ordinary differential equations know, the answer to this query is decidedly gloomy. In particular

Answer. For most choices of $a(t)$, $b(t)$, and $f(t)$ we cannot find $x(t)$ exactly.

This being the case, we compromise. Since we cannot find $x(t)$ exactly, we can try to find it approximately. This leads us to a new problem.

PROBLEM 2

How do we find a good approximation $\tilde{x}(t)$ to the solution $x(t)$ of our differential equation?

This problem is far more tractable, and there are many ways of answering it. All possible solutions depend in some way on the answer to yet another problem.

PROBLEM 3

Given a function $x(t)$, what kind of functions $\tilde{x}(t)$ make good approximations to $x(t)$?

and to the companion problems

PROBLEM 4

What is meant by a *good approximation*?

and

PROBLEM 5

How does one compute a good approximation $\tilde{x}(t)$ to a given function $x(t)$?

Providing some answers to these simple questions and their two-dimensional analogs is precisely what this book is all about. Since *linear spaces*, *subspaces*, *norms*, and *basis* are notions fundamental to all we do in the sequel, we start with a minor digression to define these entities.

1.2 LINEAR SPACES

A *real linear space* X is simply a set of mathematical objects called *vectors* which add according to the usual laws of arithmetic and can be multiplied by real numbers in accord with the usual laws of arithmetic. Specifically to qualify as a real linear space, elements of X must satisfy the following conditions or *axioms*.

For all x, y, and z, in X and for all real numbers α and β, αx is in X, $x+y$ is in X, $x+y=y+x$, $(x+y)+z=x+(y+z)$, $1\cdot x=x$, $(\alpha+\beta)x=\alpha x+\beta x$, $\alpha(\beta x)=(\alpha\beta)x$, and $\alpha(x+y)=\alpha x+\alpha y$. There exists a *zero vector* $\odot$ in X with the property that $x+\odot=x$ for all x. Finally, for each x there is a unique vector $-x$, called the *inverse of* x, such that $-x+x=\odot$.

If all these axioms are satisfied when multiplication is multiplication by complex numbers, we say X is a *complex linear space*. We usually deal with the real linear spaces.

Examples abound. Among these are the set E^2 of vectors in the plane with addition defined as coordinatewise addition and the set $C[a,b]$ of functions $f(t)$ continuous on the closed interval $[a,b]$. In each case we must first define addition and scalar multiplication.

EXAMPLE 1 THE REAL PLANE E^2

The set $X=E^2$ is simply the set

$$E^2=\{(x_1,x_2)\colon x_1 \text{ and } x_2 \text{ are real numbers}\}$$

of all ordered pairs of real numbers or vectors in a plane. Addition is coordinatewise addition, and multiplication is coordinatewise multiplication. In particular, given a real number α and vectors $x=(x_1,x_2)$ and $y=(y_1,y_2)$, we define

$$x+y=(x_1+y_1,x_2+y_2)$$

$$\alpha x=(\alpha x_1,\alpha x_2).$$

The reader can easily verify that all the axioms of a linear space are satisfied where $\odot=(0,0)$ is the zero vector (see Figure 1.1).

The real linear space $C[a,b]$ is much more interesting.

EXAMPLE 2 THE SPACE $C[a,b]$

The space $C[a,b]$ is simply the set of all functions continuous on $[a,b]$. To define addition and scalar multiplication, let α be any real number and let $f(t)$ and $g(t)$ be two continuous functions from $C[a,b]$. We define $f+g$ and αf in the usual way. That is,

$$(f+g)(t)=f(t)+g(t), \qquad a\leqslant t\leqslant b$$

and

$$(\alpha f)(t)=\alpha\cdot f(t), \qquad a\leqslant t\leqslant b.$$

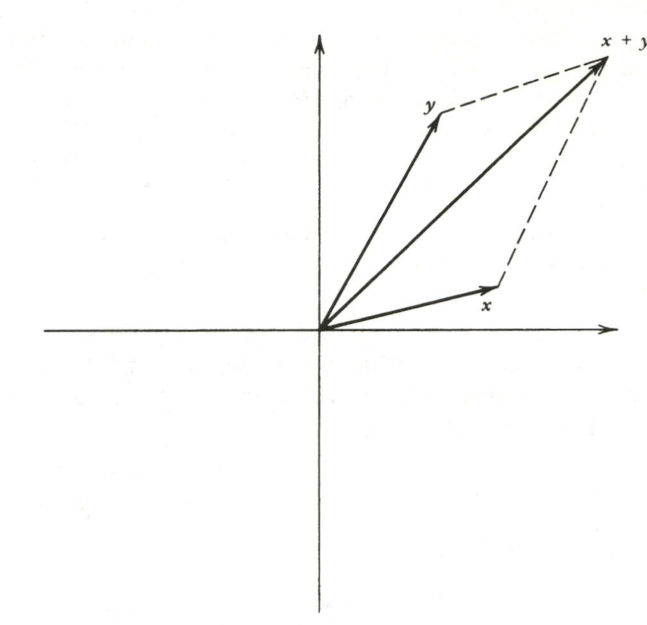

Figure 1.1 Addition of vectors in the plane E^2.

Noting that the sum of any two continuous functions is a continuous function and that a constant times a continuous function is again continuous, it is easy to verify that $C[a,b]$ forms a real linear space. The *zero vector* of this space is the function that vanishes identically on the interval $[a,b]$.

A space to which we shall take frequent recourse is

EXAMPLE 3 $P_n[a,b]$, POLYNOMIALS OF DEGREE n.

A real polynomial $p(t)$ of exact degree n or less in one variable t is a function

$$p(t) = a_n t^n + a_{n-1} t^{n-1} + \cdots + a_1 t + a_0,$$

where $a_0, a_1, \ldots, a_n$ are given real numbers and t is a real variable. The polynomial $p(t)$ is said to have *exact degree* n, if and only if $a_n \neq 0$. Thus $2t^3 + 5t - 1$ is of degree 4 but is of exact degree 3. Let

$$P_n[a,b] = \left\{ a_n t^n + \cdots + a_1 t + a_0 : \quad \begin{matrix} \text{each } a_i \text{ is a real constant} \\ \text{and } a \leqslant t \leqslant b \end{matrix} \right\}.$$

Thus $P_n[a,b]$ is the set of all polynomials of exact degree n or less, defined on the interval $[a,b]$. Of course adding two polynomials of degree n or less produces a polynomial of degree n or less, whereas multiplying a polynomial of degree n or less by a constant gives a polynomial of degree n or less. Knowing this, it is easily checked that $P_n[a,b]$ is a real linear space. The zero in this space is again the zero function, which is the same as the zero polynomial. Note too that since every polynomial is continuous, $P_n[a,b] \subset C[a,b]$ (read $P_n[a,b]$ is a *subset* of $C[a,b]$).

We are interested in special subsets of linear spaces, which are known as *subspaces*. A subset M of a linear space X is called a *linear subspace* (or simply a *subspace*) of X if

> M is a subset of X and for each x and y in M and for any pair of scalars α and β, $\alpha x + \beta y$ belongs to M.

For example, a real line passing through the origin is a linear subspace of the plane E^2. Also, the set $P_n[a,b]$ is a subspace of $C[a,b]$. We use these concepts repeatedly in the sequel.

EXERCISES

1. Prove that $C[a,b]$ is a linear space.
2. Let P.C.$[a,b]$ denote the set of all functions that are piecewise continuous on $[a,b]$. In particular, a function $f(t)$ belongs to P.C.$[a,b]$ if and only if it has at most a finite number of discontinuities on $[a,b]$ and $\int_a^b [f(t)]^2\,dt < \infty$. Is P.C.$[a,b]$ a linear space? Prove your answer. If $f \in$ P.C.$[a,b]$, does $f'(t) \in$ P.C.$[a,b]$? Why? If $f(t)$ is a simple jump function, such as

$$f(t) = \begin{cases} 0 & \text{when} \quad a \leqslant t < \dfrac{b-a}{2} \\ 1 & \text{when} \quad \dfrac{b-a}{2} \leqslant t \leqslant b, \end{cases}$$

   does either $f(t)$ or $f'(t)$ belong to $C[a,b]$? Explain.
3. Let $X = \{(x_0, y_0) + (x, y)$: x and y are real and x_0 and y_0 are constants$\}$. Is X a linear subspace of E^2? Explain.
4. Let $X = \text{P.C.}^p[a,b] = \{f(t)\colon f^{(p-1)} \in C[a,b] \text{ and } f^{(p)} \in \text{P.C.}[a,b]\}$. Is X a linear space? Why? If $f \in X$, does the fundamental theorem of calculus hold? In particular, does

$$f^{(p-1)} - f^{(p-1)}(t) = \int_{t_0}^{t} f^{(p)}(s)\,ds$$

   for all t, t_0 in $[a,b]$? Explain.

1.3 NORMED LINEAR SPACES

Actually, we are interested in more than just linear spaces. We want linear spaces in which we can assign a notion of *length* $\|x\|$ to each vector x in X. Such linear spaces are known as *normed linear spaces*, and the real number $\|x\|$ is referred to as the *norm of* x. In particular, a *norm* is a real valued function defined on a linear space X having the following properties for each real number α and each pair of vectors x and y from X:

1. $\|x\|>0$ unless $x=0$
2. $\|\alpha x\|=|\alpha|\cdot\|x\|$
3. $\|x+y\|\leqslant\|x\|+\|y\|$ (triangle inequality).

The reader can check that these properties of a norm do coincide with usual geometric properties of length. Each of the linear spaces we have cited in our examples is easily made a normed linear space through an appropriate choice for $\|\ \|$. For example, if $X=E^2$, the usual choice for the length or norm $\|\ \|_2$ of a vector $x=(x_1,x_2)$ in the real plane is

$$\|x\|_2=\sqrt{x_1^2+x_2^2}\ .$$

But there are many other choices available. For example, the function $\|\ \|_1$ defined by

$$\|x\|_1=|x_1|+|x_2|$$

is a norm as is the function $\|\ \|_\infty$, defined by

$$\|x\|_\infty=\max\{|x_1|,|x_2|\}.$$

To prove that each of these function is a norm, you must verify that each one satisfies each of the conditions 1, 2, and 3 required of a norm. For example, to see that $\|\ \|_1$ satisfies condition 1, note that $\|x\|_1=0$ implies $|x_1|+|x_2|=0$. But this is possible if and only if $x_1=x_2=0$. Thus $\|x\|_1=0$ if and only if $x=(0,0)$, the zero vector. Moreover, since $\|x\|_1$ is clearly nonnegative for all x, we see that condition 1 is satisfied. Conditions 2 and 3 follow readily from elementary properties of absolute values of real numbers.

We are especially interested in the *Tchebycheff* or *uniform norm* on the space $C[a,b]$.

DEFINITION TCHEBYCHEFF NORM

Let $f(t)\in C[a,b]$. The real-valued function $\|\ \|$ defined by

$$\|f\|=\max_{a\leqslant t\leqslant b}|f(t)|$$

is known as the *Tchebycheff norm of f*.

One must, of course, prove that this function actually is a norm. This is easily accomplished because of elementary properties of absolute value of real numbers. In particular, $\|f\| \geq 0$, since $|f(t)| \geq 0$ for all t in $[a,b]$. Also $\|f\| = 0$ only if $|f(t)| = 0$ for all t in $[a,b]$. Thus $\|f\| > 0$ unless $f(t) \equiv 0$ on $[a,b]$. Moreover, $|\alpha f(t)| = |\alpha| \cdot |f(t)|$ implies

$$\max_{a \leq t \leq b} |\alpha f(t)| = \max_{a \leq t \leq b} |\alpha| \cdot |f(t)| = |\alpha| \max_{a \leq t \leq b} |f(t)| = |\alpha| \cdot \|f\|.$$

Thus 1 and 2 are clearly true. Condition 3 follows from the triangle inequality $|f(t) + g(t)| \leq |f(t)| + |g(t)|$ for absolute values of real numbers. Thus the Tchebycheff norm is indeed a norm.

A second norm on $C[a,b]$ which is also very important to us is the so-called L_2 *norm* $\|\ \|_2$. In particular

DEFINITION L_2 NORM

Let $f(t) \in C[a,b]$. The real valued function $\|\ \|_2$ defined by

$$\|f\|_2 = \sqrt{\int_a^b [f(t)]^2 \, dt}$$

is known as the L_2 *norm of f*.

Verification that $\|\ \|_2$ is a norm on $C[a,b]$ is left as an exercise (see Exercise 2).

Defining a norm on a linear space X introduces the companion notion of the *distance* $\|x - y\|$ between two points x and y belonging to X. For example, if $X = E^2$, $x = (1,3)$ and $y = (-2,4)$, then

$$\|x - y\| = \sqrt{9 + 1} = \sqrt{10}$$

$$\|x - y\|_1 = 3 + 1 = 4$$

$$\|x - y\|_\infty = \max\{3, 1\} = 3.$$

On the other hand, if $X = C[a,b]$ with the Tchebycheff norm, and $f(t)$ and $g(t)$ are any two functions belonging to X,

$$\|f - g\| = \max_{a \leq t \leq b} |f(t) - g(t)|.$$

To take a specific case, let $f(t) = \cos \pi t$, $g(t) = t^2$ and $X = C[0,1]$. Then

$$\|f-g\| = 2$$

$$\|f\| = \|f-0\| = 1$$

$$\|g\| = \|g-0\| = 1$$

(see Figure 1.2).

Given a continuous function $f(t)$ from $C[a,b]$, it is helpful to consider the set of all functions $g(t)$ from $C[a,b]$ for which $\|f-g\| < \epsilon$. Such collections of functions, known as ϵ-neighborhoods of f, are nice things—we can draw pictures of them. Suppose, for example, $f(t) = \sin 2t + 2$, where $0 \leqslant t \leqslant \pi$, and $X = C[0,\pi]$. Then $\{g \in C[0,\pi] : \|\sin 2t + 2 - g\| < \epsilon\}$ is the family of all continuous functions $g(t)$ from $C[0,\pi]$ which thread through a tube of width 2ϵ, symmetric about the graph of $\sin 2t + 2$ (see Figure 1.3). It is clear that $\|f-g\| < \epsilon$ if and only if the graph of $g(t)$ threads through the shaded region about the graph of $f(t)$. If $g(t)$ were a function approximating $f(t)$ and $\|f-g\| < \epsilon$, it appears g would be a good approximation to f if ϵ were small. We borrow this simple geometry to give *an initial definition of a good approximation.*

DEFINITION

Let $f \in C[a,b]$ with the Tchebycheff norm. A function g belonging to $C[a,b]$ is a *good approximation* to f provided $\|f-g\| < \epsilon$ for a sufficiently small ϵ.

In some situations this definition may not be a sufficient measure of goodness of approximation. To see this, consider the function $f(t) = \frac{1}{2}\epsilon \cos 4\pi t$ on the interval $[0,1]$, where ϵ is a very small constant. Then the function $g(t) \equiv 0$ on $[0,1]$ (the zero vector in $C[0,1]$) is a "good approximation" to f by our definition, since

$$\|f-g\| = \max_{0 \leqslant t \leqslant 1} \left|\frac{\epsilon}{2} \cos 4\pi t\right| = \frac{\epsilon}{2} < \epsilon.$$

However, note that $y'(t) \equiv 0$, and $f'(t) = (-4\pi\epsilon/2)\sin 4\pi t$. Thus $\|f' - g'\| = \|f'\| = 2\pi\epsilon$. Similarly $\|f'' - g''\| = 8\pi^2\epsilon$, $\|f''' - g'''\| = 32\pi^3\epsilon$, $f^{(n)} - g^{(n)}\| = (4\pi)^n\epsilon/2$, and so forth. Thus if you require a "good approximation" g to a given function f to be one that makes each of the quantities $\|f-g\|$, $\|f' - g'\|, \ldots, \|f^{(n)} - g^{(n)}\|$ small, the definition just given simply will not do, since $(4\pi)^n\epsilon/2$ could be quite a large number even though ϵ was quite a small number. In this case, the way out of the difficulty is to retain the essential character of the definition of a good approximation, namely,

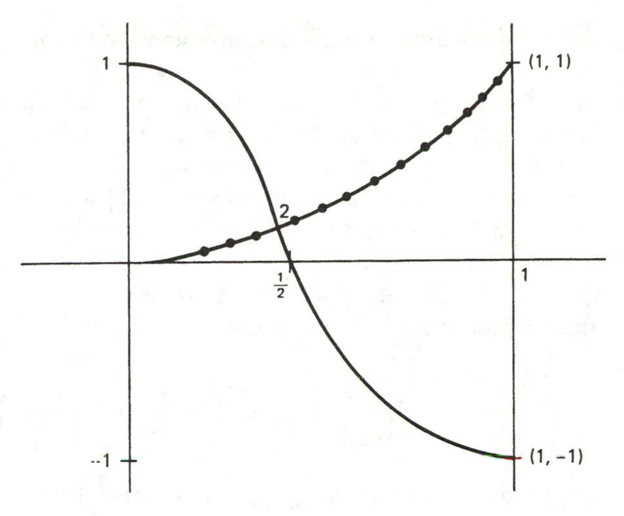

Figure 1.2 Dotted line: $g(t)=t^2$; solid line: $f(t)=\cos \pi t$. $\|f-g\|=2$. Graph of f and g.

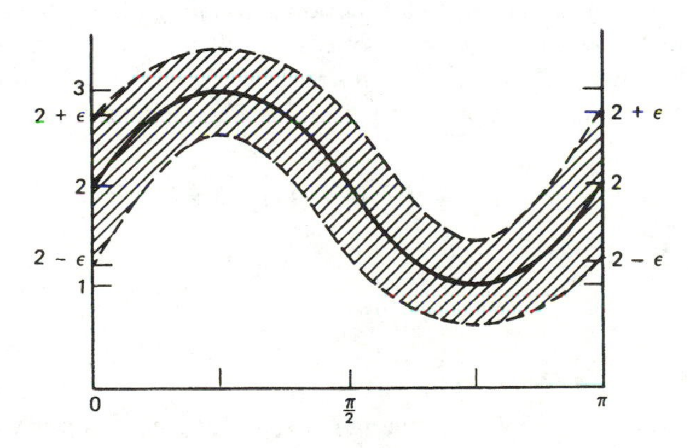

Figure 1.3 Graph of a "tube" of width 2ϵ symmetric about $\sin 2t+2$.

$\|f-g\|$ small, but change the linear space and the norm. For the example we have cited, you could take $X=C^n[a,b]$, the space of n times continuously differentiable functions defined on $[a,b]$. For each $f\in C^n[a,b]$ define the norm $\| \ \|_n$ of f as

$$\|f\|_n=\|f\|+\|f'\|+\|f''\|+\cdots+\|f^{(n)}\|.$$

You must, of course, check that this function really is a norm and that $C^n[a,b]$ with addition and scalar multiplication defined as in $C[a,b]$ really is a linear space. Then g would be a good approximation to f if $\|f-g\|_n<\epsilon$ for some small ϵ.

The Tchebycheff norm is not always available, and we sometimes have to use the L_2 norm on $C[a,b]$ (or on some larger space containing $C[a,b]$). The contrast in the geometry of these two spaces is well worth noting. For example, if f and g are any two functions from $C[a,b]$ with the L_2 norm, the distance $\|f-g\|_2$ between f and g has a geometric interpretation that is quite different from the distance between two continuous or bounded functions with the Tchebycheff norm. In particular, every student of elementary calculus knows that for continuous f and g

$$\left(\frac{1}{\sqrt{b-a}}\right)\|f-g\|_2=\left(\int_a^b[f(t)-g(t)]^2\,dt\right)^{1/2}\left(\frac{1}{\sqrt{b-a}}\right)$$

is *the "average distance" from f to g on the interval* $[a,b]$. The same interpretation applies when f and g are any two functions from $L_2[a,b]$ (see next section). In the sequel we are sometime forced to consider a function f to be a good approximation to a function g if $\|f-g\|_2<\epsilon$ for some small ϵ. To show just how much *weaker* a measure of good approximation this is than the Tchebycheff or uniform, norm $\|f-g\|$, let $f(t)\equiv 0$ on $[-1,1]$ and let $g^2(t)$ be the roof function graphed in Figure 1.4. Then

$$\|f-g\|_2^2=\int_{-1}^{1}[g(t)]^2\,dt=\epsilon.$$

Thus

$$\|f-g\|_2=\sqrt{\epsilon}\ .$$

However, no matter how small we make ϵ, so long as it is positive we have

$$\|f-g\|=\max_{0<t<1}|f(t)|=1.$$

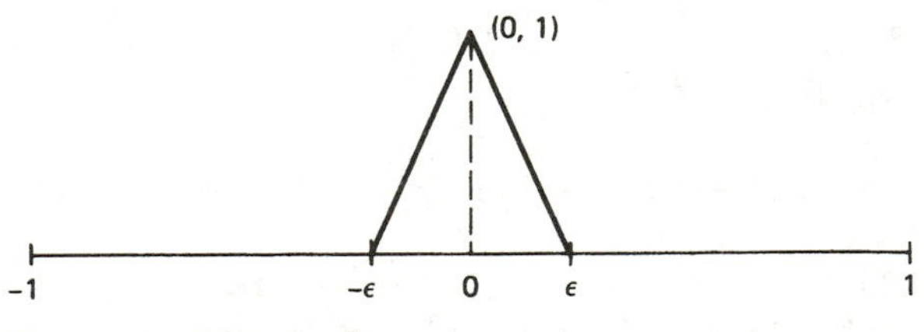

Figure 1.4 Graph of $g^2(t)$.

Thus f can be a very good approximate to g in the L_2 norm but a perfectly dreadful approximation to g in the Tchebycheff norm. At times we simply have to live with this condition. In the theory of Fourier series, the problem has evoked some very beautiful theorems which are outside the scope of this book, including the well-known Féjer-Cesaro summation theorems.

EXERCISES

1. Let $X=E^n=\{(x_1,x_2,\ldots,x_n)$: x is real$\}$. For each $x\in X$ let

$$\|x\|_2=\sqrt{x_1^2+x_2^2+\cdots+x_n^2}\,,$$

$$\|x\|_1=|x_1|+|x_2|+\cdots+|x_n|,$$

$$\|x\|_\infty=\max_{1\leqslant i\leqslant n}|x_i|.$$

   Prove that each of these quantities is a norm. When $i=1$, 2, 3, graph $\{x\colon \|x-a\|_i<1\}$, where a is some fixed point of X.

2. Let $X=C[a,b]$, let $f(t)\in C[a,b]$, and define $\|f\|_2=(\int_a^b f^2(t)\,dt)^{1/2}$. Is $\|\ \|_2$ a norm for X? Prove it. Suppose $(f_n:n=1,2,\ldots)$ is a sequence of functions in X such that $\lim f_n(t)=f(t)$ exists for all t in $[a,b]$. Then we say f_n converges pointwise to f. Must $f(t)\in X$? Does $\|f-f_n\|_2\to 0$ as $n\to\infty$? Explain.

3. Let $X=$ P.C.$[a,b]$, the piecewise continuous functions $f(t)$ on $[a,b]$ for which $\int_a^b[f(t)]^2<\infty$. Let $(f_n:n=1,2,\ldots)$ be a sequence of functions from X for which $\int_a^b[f_n(t)-f_m(t)]^2\to 0$ as $m,n\to\infty$. Does there exist a function $f(t)\in X$ such that
   (a) $(f_n(t):n=1,2,\ldots)$ converges pointwise to $f(t)$?
   (b) $\int_a^b[f(t)-f_n(t)]^2\to 0$ as $n\to\infty$?
   Explain.

4. Let x_0 be a fixed vector in some normed linear space X. Let $d>0$ be a real number and let $K=\{x:\|x-x_0\|\leqslant d\}$. Prove that K is closed and bounded.

5. Let $\|\ \|$ be a norm for a linear space X. Prove that $g(x)=\|x\|$ is a continuous function on X.

1.4 THE SPACE $L_2[a,b]$

There is a normed linear space of which mathematicians are especially enamoured (with good reason); many engineers are not formally familiar with it, however, even though they frequently solve many of their problems in precisely this setting. This is the space $L_2[a,b]$ of functions $f(t)$, which are Lebesgue square integrable on $[a,b]$. To obtain this space, first define what is known as a set of measure zero (e.g., the rational numbers, any finite set of real numbers, and the integers) and a measureable function.

A subset E of $[a,b]$ is of measure zero if for each $\epsilon > 0$ there exists a countable number of open intervals whose total length is less that ϵ which contains E. For example, the rational numbers $E = \{1/2, 1/3, \ldots, 1/n, \ldots\}$ have measure zero, since

$$E \subset \bigcup_{n=2}^{\infty} \left(\frac{1}{n} - \frac{\epsilon}{2^n}, \frac{1}{n} + \frac{\epsilon}{2^n} \right)$$

and

$$\sum_{n=2}^{\infty} \frac{\epsilon}{2^n} < \epsilon.$$

The irrational numbers in $[0,1]$, on the other hand, have measure 1.

A function $f(t)$ is measurable if it is the limit pointwise everywhere except on a set of measure zero of an increasing sequence $f_n(t)$ of step functions (such convergence is called "convergence almost everywhere"). Moreover, two functions $f(t)$ and $g(t)$ are considered to be the same function if $f(t) = g(t)$ for all t in $[a,b]$ except a set of measure zero. For example, $L_2[a,b]$ does not recognize any difference between the functions

$$f(t) = \left\{ \begin{array}{ll} 0 & \text{if } t \text{ is a rational number} \\ 1 & \text{otherwise} \end{array} \right\}$$

$$g(t) \equiv 1 \qquad \text{on } [a,b]$$

$$h(t) = \left\{ \begin{array}{ll} 0 & \text{if } t = \frac{1}{2} \\ 1 & \text{otherwise} \end{array} \right\}.$$

Having accomplished all this and having proved a few theorems, we define a new kind of integration known as *Lebesgue integration*, which differs somewhat from *Riemann integration* (the kind taught in first-quarter calculus). In particular, if $f_n(t)$ is an increasing sequence of step functions that converges almost everywhere to $f(t)$, the *Lebesgue integral* $\int_a^b f(t)\,dt$ of $f(t)$ on $[a,b]$ is defined by

$$\int_a^b f(t)\,dt = \lim_n \int_a^b f_n(t)\,dt,$$

where the Riemann and Lebesgue integrals of step functions are the same. Formally then, we have

DEFINITION $L_2[a,b]$

The set $L_2[a,b]$ is the collection of all measurable functions $f(t)$ defined on $[a,b]$ for which

$$\int_a^b [f(t)]^2\,dt < \infty,$$

where integration is Lebesgue integration.

One can now prove that *the Riemann integral of a continuous function and of a function that is continuous except at a finite number of points* (*i.e.*, *a jump function*) *is exactly the same as the Lebesgue integral of that function* (*i.e.*, $C[a,b]$ *is a linear subspace of* $L_2[a,b]$). What, if anything, have we gained by the introduction of the space $L_2[a,b]$? A first answer to this question is that going to $L_2[a,b]$ enables us to integrate many more functions than would have been possible with the Riemann integral. For example, our function $f(t)$

$$f(t) = \left\{ \begin{array}{ll} 0 & \text{if } t \text{ is a rational number} \\ 1 & \text{otherwise} \end{array} \right\}$$

does not have a Riemann integral. However, it does have a Lebesgue integral and, in fact,

$$\int_0^1 f(t)\,dt = 1.$$

The particular answer is not always altogether satisfying to very applied people. However, it is important and is linked to an even more compelling reason for turning to $L_2[a,b]$ rather than, say, $C[a,b]$. This answer depends on the norm for $L_2[a,b]$. Before defining a norm on $L_2[a,b]$ one must ascertain that it is indeed a linear space. Defining addition and scalar multiplication of functions from $L_2[a,b]$ exactly as we did in $C[a,b]$, it follows from the Lebesgue theory of integration that $L_2[a,b]$ *is a linear space*. In particular, one can prove that the sum of any two square integrable functions and a constant times a square integrable function is again square integrable. Put more concisely, if $f,g \in L_2[a,b]$ and α is any real number,

$$f+g \in L_2[a,b] \qquad \text{and} \qquad \alpha f \in L_2[a,b].$$

The remaining axioms for a linear space then follow quite simply. The norm $\| \ \|_2$ for this linear space is defined as an integral. In particular

DEFINITION L_2 NORM

Let $f(t) \in L_2[a,b]$. The real valued function $\| \ \|_2$ defined by

$$\|f\|_2 = \left(\int_a^b [f(t)]^2 \, dt \right)^{1/2}$$

is known as *the L_2 norm of f*.

Proof that this function actually forms a norm depends on the Lebesgue theory. The student who has not formally studied this theory can get some feeling for the space by thinking of, say, functions f that are continuous (piecewise continuous); then try to prove which of the properties of a norm hold on this smaller space (you will find that all of them do). That is,

THEOREM

The real-valued function $\| \ \|_2$ defined by

$$\|f\|_2 = \left(\int_a^b [f(t)]^2 \, dt \right)^{1/2}$$

is a norm for the linear space $C[a,b]$ as well as for the linear space $L_2[a,b]$.

Our comments on the distance $\|f-g\|_2$ between two functions f and g from $L_2[a,b]$ are much the same as those given in Section 1.3 for functions from $C[a,b]$ with the L_2 norm. In particular, the weakness of the L_2 norm as a measure of approximation in contrast with the Tchebycheff norm still obtains. However, it is only fair to point out that there are many simple functions possessing an L_2 norm which have no Tchebycheff norm. An instance of this is given by the function

$$f(t) = t^{-1/3}, \qquad 0 \leqslant t \leqslant 1.$$

We now return to our original question, "Why bother with $L_2[a,b]$?" For example, why not simply use $C[a,b]$, since $\| \ \|_2$ is a norm for this space as well. One reason is that there are sequences $(f_n(t))$ of continuous functions for which $\|f_n - f_m\|_2 \to 0$ as m and $n \to \infty$ but for which there is no continuous function $f(t)$ for which $\|f_n - f\|_2 \to 0$ as $n \to \infty$. The technical description of this phenomenon (see Chapter 6) is that $C[a,b]$ *is not* *complete in the norm* $\| \; \|_2$.* What makes $L_2[a,b]$ desirable is that it is complete in the norm $\| \; \|_2$. Put somewhat differently, if $(f_n(t))$ is a sequence of function in $L_2[a,b]$ for which $\|f_n - f_m\| \to 0$ as $m,n \to \infty$, there exists a function $f(t)$ in $L_2[a,b]$ such that $\|f - f_n\|_2 \to 0$ as $n \to \infty$. This property can be most important when one is constructing approximate solutions to a differential equations by certain variational methods.†

For us, however, the single most compelling reason for turning to $L_2[a,b]$ is its "catchall catholicism." It simply has more functions in it than $C[a,b]$ or P.C.$[a,b]$ or even $R[a,b]$, the space of functions that are Riemann integrable on $[a,b]$. Moreover, since each of these spaces can carry the L_2 norm on its own, it is convenient to start in the roomier space and pick up the others as subspaces when it is convenient. There are times in applied settings when even $L_2[a,b]$ is not sufficiently catholic, and one must turn to the even more open-minded fraternity of distributions. However, in this book, we seldom need anything much more complicated than the P.C.$[a,b]$ spaces.

DEFINITION P.C.$[a,b]$

A function $f(t)$ is in P.C.$[a,b]$ if it is continuous at all but a finite number of points of $[a,b]$ and if $\int_a^b [f(t)]^2 dt < \infty$.

EXERCISES

1. Is P.C.$[a,b]$ contained in $L_2[a,b]$? Why? Is it complete? Prove your answer.

1.5 BASIS FOR A LINEAR SPACE

Suppose X is a linear space and $\{x_1, x_2, \ldots, x_n\} = B_n$ is a set of n vectors from X. The set B_n is said to be *linearly independent* if and only if the equation

$$a_1 x_1 + a_2 x_2 + \cdots + a_n x_n = \odot$$

*For that matter, neither is $R[a,b]$, the space of functions which are Riemann integrable on $[a,b]$.

†Moreover, L_2 integration is extremely important to the proof of the existence of solutions of any but the most simple of partial differential equations, many of the natural physical properties of such equations being expressed as integrals. In fact, the greater part of the existence theorems for solutions of partial differential equations and systems of such equations is couched precisely in the language of strong convergence in $L_2[\Omega]$ (i.e., $\|f - f_n\|_2 \to 0$ as $n \to \infty$) or of weak convergence in $L_2[\Omega]$ (i.e., $\int_\Omega f_n g \to \int_\Omega fg$ as $n \to \infty$ for all g in L_2), where Ω is some closed, bounded subset of R^n.

implies $a_1 = a_2 = \cdots = a_n = 0$. If B_n is not linearly independent, it is said to be *linearly dependent*. By way of example, consider the set $B_n = \{1, t, t^2, \ldots, t^n\}$. Each function t^k, $0 \leqslant k \leqslant n$, is a polynomial of exact degree k. Thus if we restrict t to the interval $[a,b]$, we see that B_n is a subset of $P_n[a,b]$. We prove

THEOREM

The set $B_n = \{1, t, t^2, \ldots, t^n\}$ is a linearly independent subset of the linear space $P_n[a,b]$ of all polynomials of exact degree n or less.

Proof. Suppose $p(t) = a_n t^n + \cdots + a_1 t + a_0 = \odot$. Then $p(t) \equiv 0$ for all t in $[a,b]$. Thus $p(t)$ is a polynomial with more than n roots, and by the fundamental theorem of algebra $a_n = a_{n-1} = \cdots = a_0 = 0$. Thus B_n is independent. ■

It is helpful to look at a somewhat different example. This time let $X = C[a,b]$, and let $f(t) = |t|$ and $g(t) = t$ (see Figure 1.5).

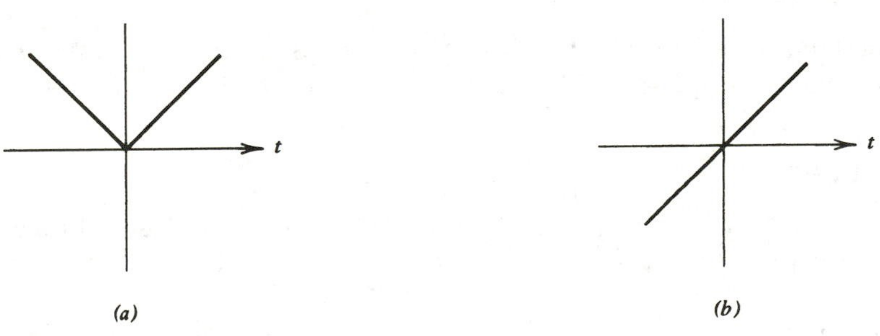

Figure 1.5 (*a*) Graph of $f(t) = |t|$. (*b*) Graph of $g(t) = t$.

We have seen that $C[a,b]$ is a linear space whose zero vector is the zero function. If we let $[a,b] = [0,1]$, it is clear that $f(t) = g(t)$ for all $t \epsilon [0,1]$. Thus $a_1 f + a_2 g = \odot$ means

$$a_1 f(t) + a_2 g(t) = 0 \qquad \text{for all } t \text{ in } [0,1],$$

and this does *not* imply $a_1 = a_2 = 0$. In particular we can take $a_1 = 1$ and $a_2 = -1$. Thus the functions f and g are *linearly dependent* on $[0,1]$. On the other hand, if we change the interval, the situation alters considerably. For example, if $X = C[-1,1]$, then $\{|t|, t\}$ is *linearly independent* in X. To see this, suppose

$$H(t) = a_1 t + a_2 |t| = 0 \qquad \text{for all } t \text{ in } [-1,1].$$

Then

$$H(1)=a_1+a_2=0$$

$$H(-1)=a_1-a_2=0.$$

It follows simply that $a_1=a_2=0$.

The notion of linear independence in a linear space X leads to that of the dimension of X. A linear space X is said to be *n-dimensional* if it contains a set of n linearly independent vectors $\{x_1,x_2,\ldots,x_n\}$ and every set of $n+1$ vectors is linearly dependent. A set of linearly independent vectors $\{x_1,x_2,\ldots,x_n\}$ from X is said to form a *basis* for X if every vector x in X can be written as a linear combination

$$x=c_1x_1+c_2x_2+\cdots+c_nx_n$$

of the vectors x_i, $1\leqslant i\leqslant n$. The c_i's are necessarily unique (see exercises). Proofs are available in any good linear algebra book that every n-dimensional linear space has a basis and every basis for such a space consists of exactly n vectors.

EXAMPLE 1 LET $X=E^2$ AND LET $B_2=\{(1,0), (0,1)\}$.

The set B_2 is independent (why?) and forms a basis for the set of all vectors in the plane.

EXAMPLE 2 LET $X=P_n[a,b]$ AND LET $B_n=\{1,t,\ldots,t^n\}$.

Each polynomial $p(t)$ from X is a unique linear combination

$$p(t)=a_nt^n+\cdots+a_1t+a_0$$

of the vectors of B_n. Moreover, the set B_n is linearly independent, thus forms a basis for $P_n[a,b]$. It follows that $P_n[a,b]$ is an $n+1$-dimensional linear space. There are, incidentally, many other bases for this space, some of them *much better* for computational purposes than B_n (see Chapter 2).

EXAMPLE 3 LET $X=C[-1,1]$ AND LET $B_2=\{t,|t|\}$.

Although B_2 is independent, it does not form a basis for all of $C[-1,1]$, since one cannot find constants a_1 and a_2 such that

$$\sin t=a_1t+a_2|t|.$$

The space $C[a,b]$ is actually infinite dimensional, and we shall have occasion to work with a number of infinite dimensional spaces. It is interesting to ask yourself whether, given any function $f(t)$ from $C[a,b]$, the family $\{1,t,t^2,\ldots\}$ is such that the equation $f(t)=\sum_{k=0}^{\infty}c_kt^k$ has a (unique) solution $c_0,c_1,c_2,\ldots$.

A simpler and more crucial matter to note is that if X is a linear space that is either finite or infinite dimensional and $B_n = \{\phi_1, \phi_2, \ldots, \phi_n\}$ is a linearly independent set of vectors from X, the *span of* B_n is an n-dimensional *linear subspace of* X. The *span* of a set of vectors $\{\psi_1, \psi_2, \ldots, \psi_n\}$ is the set of all possible linear combinations

$$a_1\psi_1 + a_2\psi_2 + \cdots + a_n\psi_n.$$

The ψ_i's need not be independent to talk about their span. For example, the set of functions $\{t, |t|\}$ from the set of functions $C[0,1]$ is not linearly independent in $C[0,1]$, but the span of $\{t, |t|\}$ exists and consists of the set of all straight line segments

$$y = mt, \qquad m \text{ real}, \qquad 0 \leqslant t \leqslant 1$$

passing through the origin of the y–t plane. On the other hand, the set of vectors $\{(1,1), (-2,1)\}$ is linearly independent in E^2 and spans all of E^2. Put another way, the set of vectors shown in Figure 1.6 forms a basis for E^2.

As yet another example, note that the set of polynomials $\{1, t, t^2, \ldots, t^n\}$ is linearly independent, forming a basis for $P_n[a,b]$. But each polynomial from $P_n[a,b]$ is a continuous function and, as such, is also a member of $C[a,b]$. Thus the span of $\{1, t, t^2, \ldots, t^n\}$ is an n-dimensional linear subspace of $C[a,b]$. We shall take frequent recourse to these ideas.

EXERCISES

1. Prove that $\{1, \sin t, \sin 2t, \sin 3t, \ldots, \sin nt\}$ is *linearly independent* on $[0, 2\pi]$.
2. Let $B = \{1, t, t^2, \ldots, t^n\}$. Find a set of polynomials $\mathcal{O} = \{p_0(t), p_1(t), \ldots, p_n(t)\}$ such that

$$\int_a^b p_i(t) p_j(t)\, dt = \begin{cases} 1 & \text{if } i = j \\ 0 & \text{if } i \neq j. \end{cases}$$

3. Prove that $\mathcal{O}$ is also a *basis* for $P_n[a,b]$. The collection $\mathcal{O}$ is called the set of *ortohgonal polynomials* and $p_j(t)$ is called the jth Legendre polynomial when $[a,b] = [-1,1]$.
4. Find a *basis* other than B or $\mathcal{O}$ for $P_n[a,b]$ (see Problem 2).
5. Let $a = t_0 < t_1 < t_2 < \cdots < t_n = b$ be $n+1$ distinct points of $[a,b]$. Let $\phi_i(t)$ be the function graphed in Figure 1.7. Prove that $\{\phi_0, \phi_1, \ldots, \phi_n\}$ is linearly independent on $[a,b]$.

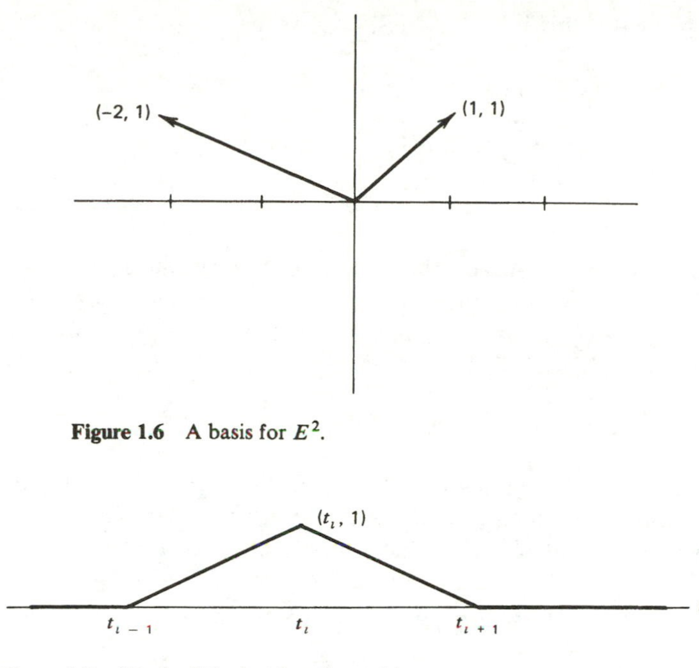

Figure 1.6 A basis for E^2.

Figure 1.7 Graph of the hat function $\phi_i(t)$.

1.6 APPROXIMATING FROM FINITE DIMENSIONAL SUBSPACES

We now turn to the problem of finding good approximations to a given function of one variable. One of the oldest approximation techniques (dating back at least to Fourier in 1822 and to Euler and Lagrange in the eighteenth century) is that of approximating a given function $f(t)$ by a finite sum

$$\tilde{f}(t) = c_1\phi_1(t) + c_2\phi_2(t) + \cdots + c_n\phi_n(t)$$

of simpler functions $\phi_i(t)$ about which a good deal is known and which are fairly simple to compute. The coefficients c_k are constants to be determined by constraints that we shall impose on $\tilde{f}(t)$. The notion of a Fourier series derives from this simple but fruitful idea. In this case one attempts to write $f(t)$ as an infinite sum

$$\sum_{k=0}^{\infty} [c_k \sin akt + b_k \cos akt],$$

where a is a given constant. When using Fourier series to compute, one usually truncates this infinite series, approximating it by the finite sum

$$\tilde{f}(t) = c_0 + c_1 \sin at + b_1 \cos at + c_2 \sin 2at$$
$$+ b_2 \cos 2at + \cdots + c_n \sin ant + b_n \cos ant,$$

where the constants c_i and b_i, $1 \leqslant i \leqslant n$, are again determined by constraints on $\tilde{f}(t)$.

Continuing this trend of thought, one has recourse to even simpler approximating functions than trigonometric functions. For example, we could let the $\phi_i(t)$'s be the polynomials or piecewise polynomials. In particular, if $\phi_k(t) = t^{k-1}$, $1 \leqslant k \leqslant n+1$, then

$$\tilde{f}(t) = c_n t^n + c_{n-1} t^{n-1} + \cdots + c_1 t + c_0$$

is a polynomial of exact degree n or less and the c_k's are again constants to be determined by some particular set of constraints on $\tilde{f}(t)$.

Suppose now that $f(t)$, the function to be approximated, is continuous, belonging to $C[a,b]$ for some given interval $[a,b]$. Then if you also choose your functions $\phi_1(t)$, $\phi_2(t), \ldots, \phi_n(t)$ from $C[a,b]$, the approximating function

$$\tilde{f}(t) = c_1\phi_1(t) + c_2\phi_2(t) + \cdots + c_n\phi_n(t)$$

belongs to the span of $\{\phi_1, \ldots, \phi_n\}$, which is a finite dimensional subspace of $C[a,b]$. If $\{\phi_1, \phi_2, \ldots, \phi_n\}$ is a linearly independent set of functions, then $X_n = \text{span}\{\phi_1, \phi_2, \ldots, \phi_n\}$ is an n-dimensional subspace of $C[a,b]$.

If we select this technique of approximating $f(t)$ by a function $\tilde{f}(t) = c_1\phi_1(t) + c_2\phi_2(t) + \cdots + c_n\phi_n(t)$, we must decide what we mean by "$\tilde{f}(t)$ is a good approximation to $f(t)$." Remaining alert to the fallacies in this decision pointed out at the end of Section 1.2, let us initially decide that $\tilde{f}$ is a good approximation to f, provided $\|f - \tilde{f}\|$ is small where $\|\ \|$ is the Tchebycheff norm on $C[a,b]$. *Our problem, once we have chosen the ϕ_i's is to choose the constants c_i in the definition of $\tilde{f}(t)$ in such a way as to make*

$$\|f - (c_1\phi_1 + c_2\phi_2 + \cdots + c_n\phi_n)\|$$

small and hopefully to force a unique solution for these c_i's.

Since we have n unknown constants to determine, we must impose at least n *constraints* or conditions on $\tilde{f}(t)$ to determine these constants. Moreover, we must choose these constraints in such a way that forces

$\|f-\tilde{f}\|$ to be small. There are a great number of choices among constraints on $\tilde{f}(t)$. We might, for example, insist on

INTERPOLATORY CONSTRAINTS

$\tilde{f}(t_i)=f(t_i)$ at some n distince values t_i, $1 \leqslant i \leqslant n$, of t in the interval $[a,b]$;

or a

PURE VARIATIONAL CONSTRAINT

The constants $c_1, c_2, \ldots, c_n$ are chosen to minimize $\|f-\tilde{f}\|$among all functions from X_n. That is, insist that $\tilde{f}$ solve the minimization problem

$$\|f-\tilde{f}\|=\min\left\{\|f-\bar{f}\|:\bar{f}\in X_n\right\}.$$

MIXTURE OF INTERPOLATING AND SMOOTHNESS CONSTRAINTS

1. $\tilde{f}(t_i)=f(t_i)$ at some k distinct values t_i, $1 \leqslant i \leqslant k$, of t in the interval $[a,b]$,
2. $\tilde{f}'(t_1)=f'(t_1)$ and $\tilde{f}'(t_k)=f'(t_k)$, and
3. $\tilde{f}(t)$ is twice continuously differentiable.

or on

ORTHOGONALITY CONSTRAINTS

The constants $c_1, c_2, \ldots, c_n$ are chosen so as to force $f(t)-\tilde{f}(t)$ to be orthogonal (perpendicular) to n given functions $\psi_1(t), \psi_2(t), \ldots, \psi_n(t)$. That is, we insist that

$$\left(f-\tilde{f}, \psi_i\right)=\int_a^b\left[f(t)-\tilde{f}(t)\right]\psi_i(t)\,dt=0, \qquad 1 \leqslant i \leqslant n.$$

Usually one chooses $\psi_i(t)=\phi_i(t)$, although this is not essential. Such constraints with $\psi_i=\phi_i$ are another means of arriving at truncated Fourier series; thus they are equivalent to the pure variational constraints just given if $\|f\|$ replaced by $\|f\|_2=(\int_a^b f^2(t)\,dt)^{1/2}$.

If your functions $\{\phi_1, \phi_2, \ldots, \phi_n\}$ are such that the interpolatory constraints problem stated previously always has a unique solution $\tilde{f}(t)$ for any choice of points $t_1, t_2, \ldots, t_n$, the set $\{\phi_1, \phi_2, \ldots, \phi_n\}$ is said to form a *Haar system* for $C[a,b]$, and the functions $\tilde{f}(t)$ obtained from solving the resulting system

$$\phi_1(t_1)c_1+\phi_2(t_1)c_2+\cdots+\phi_n(t_1)c_n=f(t_1)$$

$$\phi_1(t_2)c_1+\phi_2(t_2)c_2+\cdots+\phi_n(t_2)c_n=f(t_2)$$

$$\vdots$$

$$\phi_1(t_n)c_1+\phi_2(t_n)c_2+\cdots+\phi_n(t_n)c_n=f(t_n)$$

are referred to as a *generalized polynomial interpolate to* $f(t)$ with knots at $t_1,\ldots,t_n$. On the other hand, if the minimization problem given under the "Pure Variational Constraints" has a solution $\tilde{f}(t)$, the function $\tilde{f}(t)$ is called a *best approximation* or *minimax fit* to $f(t)$ from $X_n=\text{span}\ \{\phi_1,\phi_2,\ldots,\phi_n\}$. The latter is quite difficult to compute because of the particular norm we have chosen. We address ourselves to a variety of other types of best approximates later in the book.

There are many other ways of constraining $\tilde{f}(t)$ to force a unique solution for the c_i's. We begin our search very modestly, resting content with polynomial and piecewise polynomial choices for $\tilde{f}(t)$ that satisfy a variety of *interpolating and smoothness constraints* of which "Interpolatory Constraints" is but a single example.

Before proceeding to these matters, it is appropriate to comment that as we learn in Chapters 6, 7, and 8, each of the methods of constraint mentioned previously has an analog in solving operator equations (e.g., differential and integral equations) by variational methods. The analog of interpolatory constraints is the *method of collocation*, that of pure variational constraints and mixtures of interpolatory and smoothness constraints is the *Rayleigh-Ritz* (finite element) method or the *method of least squares*, and the operator analog of orthogonality constraints is the *Galerkin* method.

EXERCISES

1. Let $\{\phi_1,\phi_2,\ldots,\phi_n\}=\{1,t,t^2,\ldots,t^n\}$. Let $f(t)$ be any function defined on $[a,b]$ and let $t_1<t_2<\cdots<t_n$ be any n distinct points of $[a,b]$. Let $\tilde{f}(t)=c_1\phi_1(t)+c_2\phi_2(t)+\cdots+c_n\phi_n(t)$. Prove that $\tilde{f}(t_i)=f(t_i)$, $1\leqslant i\leqslant n$, always has a unique solution.
2. Let $\{\phi_0(t),\phi_1(t),\ldots,\phi_n(t)\}$ be the hat functions graphed in Figure 1.8. Let $a\leqslant t_0<t_1<\cdots<t_n\leqslant b$ be any $n+1$ distinct points of $[a,b]$ and let

$f(t) = c_0\phi_0(t) + c_1\phi_1(t) + \cdots + c_n\phi_n(t)$. Does the system $\tilde{f}(t_i) = f(t_i)$, $0 \leqslant i \leqslant n$, where $f(t)$ is any function defined on $[a,b]$, always have a solution? Explain.

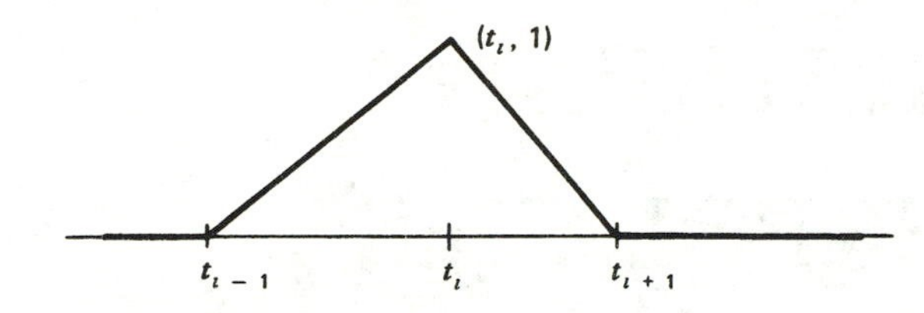

Figure 1.8 Graph of $\phi_i(t)$.

2

LAGRANGIAN INTERPOLATES

2.1 INTRODUCTION

We begin our study of one-dimensional approximates with functions

$$\tilde{f}(t) = c_0\phi_0(t) + c_1\phi_1(t) + \cdots + c_N\phi_N(t), \qquad a \leqslant t \leqslant b,$$

interpolating a given set of data. In the parlance of Chapter 1, we seek functions $\tilde{f}(t)$ subject to a set of *pure interpolatory constraints*

$$\tilde{f}(t_i) = f(t_i), \qquad 0 \leqslant i \leqslant N,$$

where $t_0 < t_1 < \cdots < t_N$ are $N+1$ distinct points of $[a,b]$. Many applied problems give rise to a need for such approximates, and the simplest example is *data fitting*. In this case we are given any $N+1$ real numbers $y_0, y_1, \ldots, y_N$ and we seek $\tilde{f}(t)$, solving $\tilde{f}(t_i) = y_i$, $0 \leqslant i \leqslant N$. There are infinitely many curves passing through $\{(t_i, y_i) : 0 \leqslant i \leqslant N\}$, hence infinitely many solutions to this problem. A popular, nonstatistical, coarse first approximate to such a curve is that obtained by constructing the polygonal line passing through the data as illustrated in Figure 2.1. Let this curve be the graph of $\tilde{f}(t)$. Such an $\tilde{f}(t)$ is technically known as a piecewise Lagrange polynomial of degree 1. It can be quite a good approximate (as we shall see) if the object is simply to make $\|f-\tilde{f}\|$ small, where $\|f-\tilde{f}\|$ is the Tchebycheff norm of $f-\tilde{f}$ on $C[a,b]$ and f is the function we are approximating. However, $\tilde{f}(t)$ is not smooth, since it cannot be differentiated at the knots $t_0, t_1, \ldots, t_N$. In fact, the derivative $\tilde{f}'(t)$ is constant on each of the open intervals (t_i, t_{i+1}) with jump discontinuities at each of the knots, $t_1, t_2, \ldots, t_{N-1}$ (see Figure 2.2).

To illustrate this problem by an example, suppose you are out for a Sunday afternoon drive and $f(t)$ is the distance of your car at time t from its starting point. Let $f(t_i) = y_i$ for each i, and suppose y_i is given exactly. Then $f'(t)$ is a velocity of your car which, from the physics of the situation, well might be a very smooth function. If you accept $\tilde{f}(t)$ as a "good approximate" to $f(t)$, you may tacitly be agreeing to assume that $\tilde{f}'(t)$ is

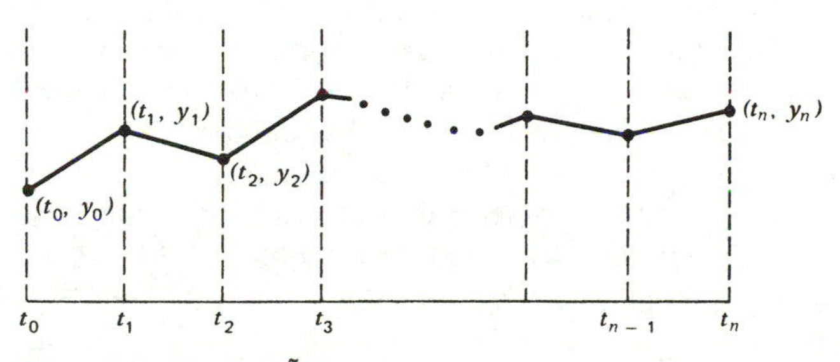

Figure 2.1 Graph of $\tilde{f}'(t)$.

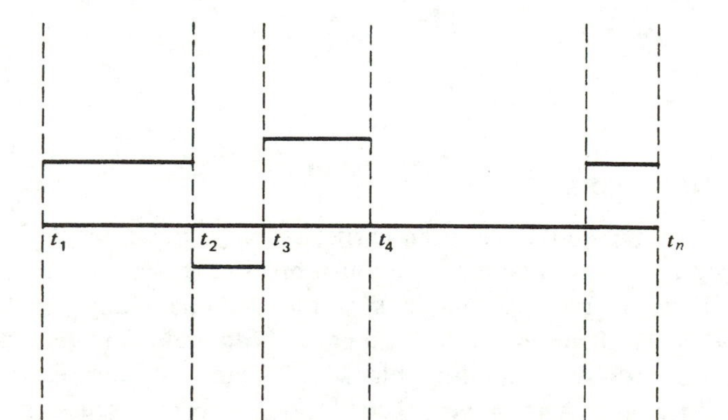

Figure 2.2 A polygonal approximating function $\tilde{f}(t)$.

also a good approximate to $f'(t)$. If your lust for smoothness is such that the polygonal model is unacceptable in that you wish to smooth the approximate velocity $\tilde{f}'(t)$, you must find a differentiable function interpolating your data. There are also infinitely many of these functions which are computable. The simplest of these are polynomials, piecewise Hermite polynomials, and cubic splines. Each of these approximating families will be discussed in some detail.

The previous example was a problem in data fitting. In addition to data fitting problems, there are a large class of problems to which the approximates of this and subsequent chapters apply. Among these are the approximate solution of differential and integral equations by variational methods, numerical quadrature (integration) formulas, numerical differentiation formulas, and various classical numerical techniques for solving first-order nonlinear differential equations, including Adams-Bashford, Adams-Moulton, and predictor corrector techniques to mention a few. Particular attention is focused on variational methods in the later chapters.

We begin our discussion very classically by studying full interpolating polynomials. Such functions are not used as often today as a few years ago (and we shall see when they should be avoided); nonetheless, we treat them in some detail because they are the building blocks or "grandparents" of polynomial splines. Moreover, we encounter in our analysis of these simple functions two problems that concern us throughout the book. These are the estimation of error and the finding of simple algorithms for the numerical computation of our approximating functions $\tilde{f}(t)$. Thus the polynomials serve as a simple introduction to a way of thinking about numerical computation. Unless otherwise stated, our measure of error throughout the next three chapters is the Tchebycheff norm

$$\|f\| = \max_{a < t < b} |f(t)|$$

on $C[a,b]$, where $f \in C[a,b]$.

2.2 ON POLYNOMIALS

Volumes can be and have been written about polynomials. A list of mathematicians who have made serious contributions to our basic knowledge of these marvelous functions would include many of the better known names of mathematics. Since polynomials, or generalizations thereof, are the basic building blocks of many of our approximating functions, we give here a very brief and heuristic discussion of some important properties of these functions.

A *polynomial p(x)* of degree n in one variable is a function

$$p(x) = a_n x^n + a_{n-1}x^{n-1} + \cdots + a_1 x + a_0,$$

where $a_0, a_1, \ldots, a_n$ are real numbers. The polynomial $p(x)$ is said to have exact degree n if and only if $a_n \neq 0$. Thus the polynomial $2x^3 + 5x - 1$ is of degree 4 but is of exact degree 3. Polynomials have marvelous approximation properties, and it is precisely these properties that most concern us. By way of illustration, let us state a well-known theorem of Weierstrass.

THE WEIERSTRASS APPROXIMATION THEOREM

Given any interval $[a,b]$, any real number $\varepsilon > 0$, and any real-valued function continuous on $[a,b]$, there exists a polynomial $p(x)$ such that $\|f - p\| < \varepsilon$.

A constructive proof of this theorem due to Bernstein can be found in the book of Davis (1963). As a simple application, let $f(x) = \sin x$ and suppose $\epsilon = 0.01$. The Weierstrass theorem guarantees the existence of a polynomial $p(x)$ whose graph lies in the band of width 0.02 symmetric about the graph of $\sin x$ (Figure 2.3). Evaluation of such a polynomial at points x in $[0, 2\pi]$ would yield a table of cosines accurate to two decimal places. There are, in fact, infinitely many such polynomials. The famous polynomials

$$P_n(x) = \sum_{k=0}^{n} \binom{n}{k} f\left(\left|\frac{k}{n}\right|\right) x^k (1-x)^{n-k}$$

due to Bernstein give us an algorithm for computing some of them. In addition, it can be shown that if $f(x)$ has k continuous derivatives, there actually exists a polynomial $p(x)$ such that $|f^{(j)}(x) - p^{(j)}(x)| < \varepsilon$ for all x in $[a,b]$ and all $j = 0, 1, \ldots, k$. One can again use Bernstein polynomials. However, since the uniform convergence of the nth Bernstein polynomial to $f(x)$ as n goes to infinity is very slow, these polynomials are not very useful for computational purposes. Proofs of these statements can be found in book by Davis (1963). The paper of Deutch (1966) is also an interesting generalization of the Weierstrass theory.

We are especially concerned with so-called interpolating polynomials. To this end let f be a continuous function on $[a,b]$ and let $a = x_0 < x_1 < \cdots < x_n = b$ be a partition (subdivision) of the interval $[a,b]$. The numbers $x_0, x_1, \ldots, x_n$ are frequently called *knots* by those working in spline theory. We adopt a slight aberration of their nomenclature (see Section 2.7).

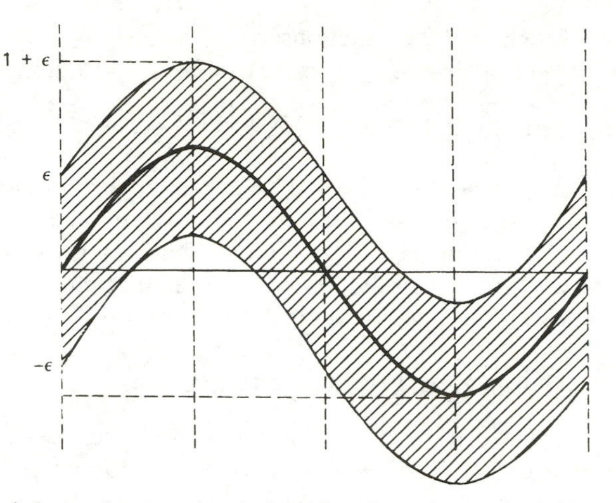

Figure 2.3 Shaded area denotes a band of width ϵ about $\sin x$ or ϵ-neighborhood of $\sin x$ in the Tchebycheff norm.

DEFINITION KNOTS

A set of real numbers $x_0, x_1, \ldots, x_n$ at which we constrain the behavior of an approximating function $\tilde{f}(t)$ is called the *knots* of $\tilde{f}(t)$.

The constraints of this chapter are all pure interpolating constraints. A function p defined on $[a,b]$ is said to *interpolate f at the knots* $x_0, x_1, \ldots, x_n$, provided $f(x_i) = p(x_i)$ for each $i = 0, 1, 2, \ldots, n$. One of the most simply stated problems of approximation theory is to find a polynomial

$$p(x) = a_n x^n + a_{n-1} x^{n-1} + \cdots + a_1 x + a_0$$

that interpolates f at the knots $x_0, x_1, \ldots, x_n$ and to estimate the error $\|f - p\|$ committed in using p as an approximate to f. This problem is deeper than it appears at first glance if one wishes to make $\|f - p\|$ small. Obviously one can vary the degree N of $p(x)$ as well as the number n of knots. In addition, one can move the knots about. Each of these mutations affects $\|f - p\|$. The *optimal placing of knots* is especially important, since it can often lead to high-accuracy approximations by considerably smaller order systems than crude equispacing of knots. The reader can draw some simple pictures to convince himself. This is particularly true in deriving quadrature rules from polynomials as we shall see. We shall look at this problem in some special cases but cannot give it the depth of discussion it deserves. We ask the erudite reader to forgive us our transgressions so that we can focus on the less sophisticated goal of finding functions $s(x)$, which are fairly simple to compute and make $\|f - s\|$ as small as we wish for a large class of functions.

2.3 LAGRANGE INTERPOLATION

The Lagrange polynomials are among the simplest and most practical of the interpolating polynomials. Given $n+1$ real numbers y_i, $0 \leqslant i \leqslant n$, and $n+1$ distinct real numbers $x_0 < x_1 < \cdots < x_n$, *the Lagrange polynomial of degree n* associated with $\{x_i\}$ and $\{y_i\}$ is a polynomial $p(x)$ of degree n solving the interpolation problem

$$p(x_i) = y_i, \qquad 0 \leqslant i \leqslant n.$$

Such polynomials always exist and are unique. This is easy to see when $n=1$, for we know that given a pair of points (x_0, y_0) and (x_1, y_1) there exists a unique straight line

$$y = mx + b$$

passing through these two points, and the equation of a straight line is a polynomial of degree 1 (see Figure 2.4).

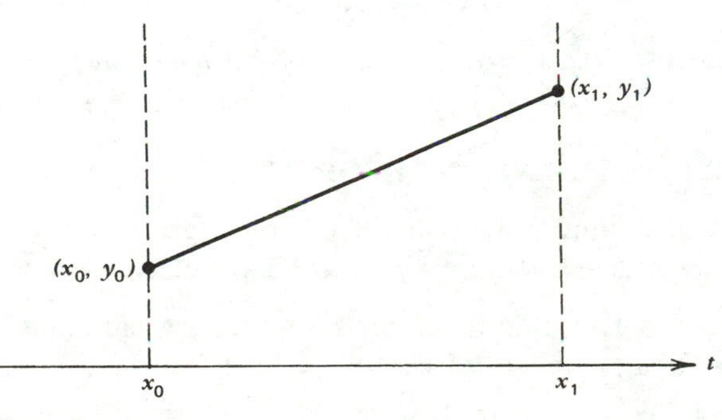

Figure 2.4 Lagrange polynomial of degree 1.

It is only natural to try to generalize this simple observation to the interpolation of more than two points by asking if there exists a unique quadratic

$$p_2(x) = a_0 + a_1 x + a_2 x^2$$

interpolating three distinct points (x_0, y_0), (x_1, x_1), (x_2, y_2); a unique cubic

$$p_3(x) = b_0 + b_1 x + b_2 x^2 + b_3 x^3$$

interpolating four distinct points (x_0, y_0), (x_1, y_1), (x_2, y_2), (x_3, y_3); and so forth (see Figure 2.5).The answer is again yes. The resultant polynomial $p_n(x)$ is, of course, precisely the *Lagrange polynomial of degree n* interpolat-

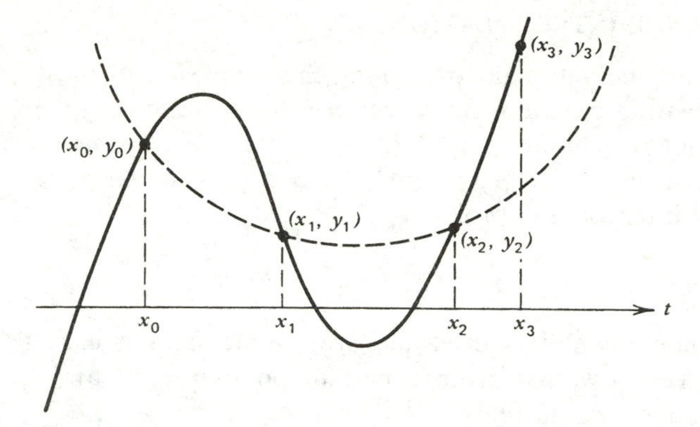

Figure 2.5 Lagrange interpolates of degrees 2 and 3 to $f(x)$. Dotted line: $p_2(x)$; solid line: $p_3(x)$; $y_i = f(x_i)$.

ing our data. If there exists a function $f(x)$ defined at $x_0, x_1, \ldots, x_n$ such that

$$f(x_i) = y_i, \qquad 0 \leqslant i \leqslant n,$$

$p_n(x)$ is known as *the Lagrange interpolate of degree n to f(x) with knots at* $x_0 < x_1 < \cdots < x_n$. The content of these remarks is the following theorem.

LAGRANGE INTERPOLATION THEOREM

There exists a unique polynomial $p_n(x)$ of degree n which assumes prescribed values at $n+1$ distinct real numbers $x_0 < x_1 < \cdots < x_n$.

Proof. The proof of this theorem is simple; we can simply write down $p_n(x)$. In particular, if we let $w(x) = (x - x_0)\ (x - x_1) \cdots (x - x_n)$, $p_n(x)$ is given by

$$\boxed{p_n(x) = \sum_{i=0}^{n} f(x_i) l_i(x),} \tag{1}$$

where

$$\begin{aligned} l_i(x) &= \frac{w(x)}{(x - x_i) w'(x_i)} \\ &= \frac{(x - x_0) \cdots (x - x_{i-1})(x - x_{i+1}) \cdots (x - x_n)}{(x_i - x_0) \cdots (x_i - x_{i-1})(x_i - x_{i+1}) \cdots (x_i - x_n)}. \end{aligned} \tag{2}$$

Note that $l_i(x_j) = 0$ if $i = j$ and $l_i(x_i) = 1$. Thus

$$l_i(x_j) = \delta_{ij}, \qquad 0 \leqslant i, j \leqslant n,$$

where δ_{ij} is Kronecker's delta

$$\delta_{ij} = \left\{ \begin{array}{ll} 1 & \text{if} \quad i=j \\ 0 & \text{if} \quad i \neq j \end{array} \right\}.$$

Furthermore, each $l_i(x)$ is a polynomial of degree n. Thus $p_n(x)$ given by (1) is a polynomial of the required kind, and existence is evident. Uniqueness is also simply shown: suppose $q(x)$ is also a polynomial of degree n interpolating $f(x)$ at $x_0 < x_1 < \cdots < x_n$. Let

$$r(x) = p_n(x) - q(x).$$

Then $r(x)$ is a polynomial of degree n with $n+1$ zeros at $x_0, x_1, \ldots, x_n$. It follows from the fundamental theorem of algebra that $r(x)$ is identically equal to zero (i.e., the only polynomial of degree n with more than n roots is the zero polynomial). Thus $p_n(x)$ is unique, as was to be proved. ■

It is important in the sequel to view this problem and its solution in terms of the approximation theoretic language of Chapter 1. In particular we have proved that if

$$\phi_k(x) = x^k$$

and $f(x_i) = y_i$, $0 \leqslant i \leqslant n$, there exists a unique function

$$\tilde{f}(x) = c_1\phi_0(x) + c_2\phi_1(x) + \cdots + c_n\phi_n(x)$$

approximating $f(x)$, subject to the *pure interpolatory constraints*

$$\tilde{f}(x_i) = f(x_i), \qquad 0 \leqslant i \leqslant n.$$

We have already noted that a polynomial $p(x)$ of degree n in one variable is any function

$$p(x) = a_n x^n + a_{n-1} x^{n-1} + \cdots + a_0$$

belonging to the span of $\{1, x, \ldots, x^n\}$ on the real line, where $a_n, a_{n-1}, \ldots, a_0$ are *real* numbers, and $p(x)$ has *exact degree n* if and only if $a_n \neq 0$. Keeping this in mind we can see that a Lagrange interpolating polynomial $p_n(x)$ of degree n could well be of *exact* degree 1 or even of exact degree 0. To illustrate this in a simple case, let $x_i = i$, $0 \leqslant i \leqslant 5$, and let $y_i = 2$ for all i, $0 < i < 5$. Then

$$p_5(x) \equiv 2$$

is the Lagrange interpolate of degree 5 (and exact degree 0) to $f(x) \equiv 2$. It follows that

$$\sum_{i=0}^{5} 2l_i(t) = 2\sum_{i=0}^{5} l_i(t) \equiv 2.$$

We now want to compute Lagrange polynomials.

2.4 COMPUTATION AND CHOICE OF BASIS

The computation of Lagrange polynomials is a fairly simple matter, but we devote a full section to it *to illustrate in a very elementary setting the importance of choice of basis to numerical computations*. With this in mind, let us suppose $x_0 < x_1 < \cdots < x_n$ are n distinct real numbers and $f(x)$ is a real-valued function defined at each of the x_i's, $0 \leqslant i \leqslant n$. Let $p_n(x)$ be the Lagrange polynomial of degree n interpolating $f(x)$ at $x_0 < x_1 < \cdots < x_n$. We shall give three different methods for computing $p_n(x)$, commenting on the merits and demerits of each. The main pedagogic point, however, is that *these three distinct methods are no more than three distinct algorithms for computing $p_n(x)$ with respect to three distinct basis for the linear space P_n of polynomials of degree n.*

METHOD I (MATRIX METHOD)

Since $p_n(x)$ is a polynomial of degree n, we know $p_n(x)$ belongs to span $\{1, x, \ldots, x^n\}$ and, as such,

$$p_n(x) = c_0 + c_1 + \cdots + c_n x^n \tag{1}$$

is a unique linear combination of the polynomials $1, x, \ldots, x^n$. But $p_n(x)$ solves the interpolation problem

$$p_n(x_i) = f(x_i), \qquad 0 \leqslant i \leqslant n. \tag{2}$$

Thus the system of $n+1$ linear equations

$$c_0 + c_1 x_i + c_2 x_i^2 + \cdots + c_n x_i^n = f(x_i), \qquad 0 \leqslant i \leqslant n,$$

has a unique solution $(c_0, c_1, \ldots, c_n)$. In matrix form, we have

$$Ac = b,$$

where A is the *Vandermonde matrix*

$$A = \begin{bmatrix} x_0^n & x_0^{n-1} & \cdots & x_0 & 1 \\ x_1^n & x_1^{n-1} & \cdots & x_1 & 1 \\ & \vdots & & & \\ x_n^n & x_n^{n-1} & \cdots & x_n & 1 \end{bmatrix},$$

$b=(f(x_0),f(x_1),\ldots,f(x_n))^T$, and $c=(c_0,c_1,\ldots,x_n)^T$. To find $p_n(x)$ by this method, invert the linear system $Ac=b$. Note that the matrix A is *full*; thus for large n it could be ill conditioned. For these reasons, this method is the least desirable for both machine and hand computation. However, it does give the nice theoretical by-product that *Vandermonde's matrix is nonsingular*.

METHOD II (CANONICAL BASIS METHOD)

We see from our proof of existence and uniqueness of Lagrange interpolating polynomials that every Lagrange polynomial $p_n(x)$ of degree n is of the form

$$p_n(x)=c_0l_0(x)+c_1l_1(x)+\cdots+c_nl_n(x), \tag{3}$$

and that

$$p_n(x)=f(x_0)l_0(x)+f(x_1)l_1(x)+\cdots+f(x_n)l_n(x) \tag{4}$$

if $p_n(x_i)=f(x_i)$, $0\leqslant i\leqslant n$. In this case, since we know the $l_i(x)$'s, we can write down $p_n(x)$ directly from (4) and *there is no system of linear equations to invert*. If we did write out the linear system resulting from combining (2) with (4), the coefficient matrix would be the identity matrix. Primarily we want to point out that every polynomial of degree n can be expressed in the form of (3). Thus $\{l_0(x),l_1(x),\ldots,l_n(x)\}$ is yet another basis for the linear space P_n. Moreover, *we see that computing $p_n(x)$ in terms of the basis $\{l_0(x),l_1(x),\ldots,l_n(x)\}$ is much simpler than computing $p_n(x)$ in terms of the basis $\{1,x,\ldots,x^n\}$.*

Choosing a basis for ease of computation is an important factor that must be reckoned with throughout the text and the optimal choice of basis depends on exactly what we are trying to compute. For example, although the basis $\{l_0,l_1,\ldots,l_n\}$ is good for computing the Lagrange interpolate $p_n(x)$, Chapter 8 indicates that it is not the best basis to use when computing polynomials of degree n which approximate the solution to a differential equation by the method of collocation.

Before the advent of high-speed electronic computers, much effort was directed (understandably) to finding the easiest way to compute certain well-known approximating functions by hand or with a desk calculator. The last method we mention is precisely such a method for Lagrange polynomials.

METHOD III (DIVIDED DIFFERENCE METHOD)

The method involves computing $p_n(x)$ subject to the interpolation conditions (2) after letting

$$p_n(x)=a_0\phi_0(x)+a_1\phi_1(x)+\cdots+a_n\phi_n(x),$$

where $\{\phi_0(x),\phi_1(x),\ldots,\phi_n(x)\}$ is the basis

$$\begin{aligned}
\phi_0(x)&=1\\
\phi_1(x)&=(x-x_0)\\
\phi_2(x)&=(x-x_0)(x-x_1)\\
&\vdots\\
\phi_n(x)&=(x-x_0)(x-x_1)\cdots(x-x_{n-1})
\end{aligned}$$

for P_n. In this case, it can be proved that

$$p_n(x)=f(x_0)+f[x_0,x_1](x-x_0)+f[x_0,x_1,x_2](x-x_0)(x-x_1)$$
$$+\cdots+f[x_0,x_1,\ldots,x_n](x-x_0)(x-x_1)\cdots(x-x_{n-1}), \tag{5}$$

where the kth divided difference $f[x_0,x_1,\ldots,x_k]$ is defined recursively through the equations

$$f[x_i,x_{i+1}]=\frac{f(x_{i+1})-f(x_i)}{x_{i+1}-x_i}$$

$$f[x_i,x_{i+1},\ldots,x_{i+k+1}]=\frac{f[x_{i+1},\ldots,x_{i+k+1}]-f[x_i,\ldots,x_{i+k}]}{x_{i+k+1}-x_i}$$

(see any elementary numerical analysis text). The reader who experiments a bit will see that when n is small, divided differences are indeed quite simple for hand computations. Another advantage associated with (5) is that it enables one to obtain the Lagrange polynomial $p_{n+1}(x)$ of degree $n+1$, interpolating the same data (2) as $p_n(x)$ plus one additional piece of data $p_{n+1}(x_{n+1})=f(x_{n+1})$ by simply adding one more term,

$$f[x_0,x_1,\ldots,x_{n+1}](x-x_0)(x-x_1)\cdots(x-x_n),$$

to $p_n(x)$.

We now work out a simple example using Method I.

EXAMPLE

Let $x_0=1$, $x_1=3$, and $x_2=5$. Then

$$l_0(x)=\frac{(x-3)(x-5)}{(1-3)(1-5)}=\frac{x^2-8x+15}{8}$$

$$l_1(x)=\frac{(x-1)(x-5)}{(3-1)(3-5)}=\frac{-x^2+6x-5}{4}$$

$$l_2(x)=\frac{(x-1)(x-3)}{(5-1)(5-3)}=\frac{x^2-4x+3}{8}.$$

The polynomial

$$p(x)=f(1)l_0(x)+f(3)l_1(x)+f(5)l_2(x)$$

is the Lagrange polynomial of degree 2 which interpolates f at 1, 3, and 5. Suppose, for example, that $f(x)$ is the integral

$$f(x)=\int_0^1\frac{1}{\sqrt{x^2+g(x,y)}}\,dy$$

of some given function $g(x,y) \geqslant 0$ for all x, y and suppose we know

$$f(1) = 1.5708$$

$$f(3) = 1.5719$$

$$f(5) = 1.5738$$

to four-decimal-place accuracy. It follows that $p(3.5) = 1.5723$. If this number is used to approximate f at 3.5, it is natural to want to know what error is accrued. This leads us to the error estimate for approximation by Lagrange polynomials.

EXERCISES

1. Find the Lagrange polynomial of degree 16, solving the interpolation problem $p(x_i) = 1$, $0 \leqslant i \leqslant 16$.
2. Find the polynomial of degree 4, interpolating $\sin 2\pi x$ at $0 < \pi/8 < \pi/4 < 3\pi/8 < \pi/2$.
3. Prove that $\Sigma_{i=0}^{n} l_i(x) \equiv 1$. Find $\Sigma_{i=0}^{n} x_i^k l_i(x)$ for each k, $0 \leqslant k \leqslant n$.

2.5 ERROR ESTIMATES FOR LAGRANGE INTERPOLATES

To estimate the *error* $\|f - p_n\|$, where $p_n(x)$ is the Lagrange polynomial of degree n interpolating $f(x)$ at $n+1$ distinct knots $x_0, x_1, \ldots, x_n$, we must restrict the smoothness of f. In particular, let $C^k[a,b]$ denote the space of all real-valued functions defined on $[a,b]$ which have a continuous kth derivative on $[a,b]$. Initially we insist that $f \in C^{n+1}[a,b]$. In the next section we alleviate this restriction. To clarify the definition of $C^k[a,b]$, let $[a,b] = [-1,1]$. Note that $g(x) = |x|$ belongs to $C[-1,1]$ but does not belong to $C^1[-1,1]$, since the derivative of g does not exist at 0. Similarly $f(x) = x|x|$ belongs to $C^1[-1,1]$ but does not belong to $C^2[-1,1]$. That is,

$$f(x) = x|x| = \left\{ \begin{array}{ccc} x^2 & \text{if} & x \geqslant 0 \\ -x^2 & \text{if} & x \leqslant 0 \end{array} \right\}$$

while

$$f'(x) = \left\{ \begin{array}{ccc} 2x & \text{if} & x \geqslant 0 \\ -2x & \text{if} & x \leqslant 0 \end{array} \right\} = 2|x|.$$

Clearly $f''(0)$ does not exist. Restricting the smoothness of f is typical in obtaining error estimates both for interpolating polynomials and for interpolating splines. This is not at all surprising considering even such simple functions as $g(x) = |x|$. This function has a smooth graph and seems likely to be decently approximable by an interpolating polynomials $p(x)$.

However, it is clear that neither $p'(x)$ nor any other polynomial can provide a really good approximate to $g'(x)$ because of its jump at 0. In fact, a simple graph of $g'(x)$ will convince one that $\|g-q\| \geqslant 1$ for all polynomials q.

So much for the gloomy side of error estimates. Now let us assume $f \in C^{n+1}[a,b]$ and attempt to estimate $\|f-p\|$. We first recall Rolle's theorem from elementary calculus:

ROLLE'S THEOREM

Let $f(x)$ be continuous on $[a,b]$, differentiable on (a,b), and suppose $f(a)=f(b)=0$. Then there exists a point ζ, $a<\zeta<b$, such that $f'(\zeta)=0$.

That is, if a sufficiently smooth function has zeros at two distinct points a and b, there is at least one point ζ between a and b at which $f'(\zeta)=0$. This theorem suffices to prove

AN ERROR ESTIMATE FOR LAGRANGE INTERPOLATES

Let $f \in C^{n+1}[a,b]$ and let $p(x)$ be the Lagrange polynomial of degree n interpolating $f(x)$ at $a=x_0<x_1<\cdots<x_n=b$. Then for each $x \in [a,b]$ there exists a ζ in (a,b) such that

$$\boxed{f(x)-p(x)=\frac{f^{(n+1)}(\zeta)}{(n+1)!}(x-x_0)(x-x_1)\cdots(x-x_n).}$$

Proof. Let $\bar{x} \neq x_0, x_1, \ldots, x_n$ be a point in $[a,b]$. Define the function $g(t)$ by

$$g(t)=[p(t)-f(t)]-\frac{w(t)}{w(\bar{x})}[p(\bar{x})-f(\bar{x})].$$

where $w(x)=(x-x_0)(x-x_1)\cdots(x-x_n)$. Then $g(t)$ has $(n+2)$ zeros at $x_0, x_1, \ldots, x_n, \bar{x}$. It follows by applying Rolle's theorem repeatedly to $g(t)$ that $g^{(n+1)}(\zeta)=0$ for some $\zeta \in [a,b]$. But

$$g^{(n+1)}(\zeta)=-f^{(n+1)}(\zeta)-\frac{(n+1)!}{w(\bar{x})}[p(\bar{x})-f(\bar{x})],$$

since p is a polynomial of degree n and w is a polynomial of degree $n+1$. Therefore

$$f(\bar{x})-p(\bar{x})=\frac{f^{(n+1)}(\zeta)}{(n+1)!}w(\bar{x}).$$

as was to be proved. ■

If we let $h=\max(x_{i+1}-x_i)$, $0\leqslant i\leqslant n$ and $x\in[x_0,x_n]$, it is evident (Exercises 4 and 7) that

$$|w(x)|=|(x-x_0)(x-x_1)\cdots(x-x_n)|\leqslant\frac{n!h^{n+1}}{4}.$$

Thus

$$\boxed{\|f-p\|\leqslant\frac{\|f^{(n+1)}\|h^{n+1}}{4(n+1)}}$$

is a very coarse estimate of our error. Notice that if $\|f^{(n+1)}\|$ grows too fast as n becomes large, reducing the size of h may prove to be a matter of diminishing returns even though the knots are spaced optimally (i.e., not spaced evenly), to obtain the best fit to f by a polynomial of degree n. Finding this particular $p(x)$ can prove difficult. One means of overcoming this difficulty is to use piecewise fits, which we discuss presently. However, let us first finish with Lagrange polynomials.

By repeatedly invoking Rolle's theorem one can also estimate

$$\|f^{(k)}-p^{(k)}\|\qquad\text{for}\quad 0\leqslant k\leqslant n-1.$$

To illustrate this in the case $k=1$, we observe that $f'(x)-p'(x)$ has at least n zeros $y_1,y_2,\ldots,y_n$ in $[a,b]$, where $x_{i-1}\leqslant y_i\leqslant x_i$. Choosing $\bar{x}$ in $[a,b]$ so that $\bar{x}\neq y_1,y_2,\ldots,y_n$ we can repeat the argument leading to $\|f-p\|$ to find

$$\|f'-p'\|\leqslant\frac{\|f^{(n+1)}\|}{n!}\|(x-y_1)(x-y_2)\cdots(x-y_n)\|,$$

where this time we choose $w(x)=(x-y_1)(x-y_2)\cdots(x-y_n)$. Note that if $y_{i-1}\leqslant x\leqslant y_i$, then $|x-y_{i+k}|\leqslant(k+1)h$ for all $k=0,1,\ldots,n-i$. It follows that

$$\boxed{\|f'-p'\|\leqslant\tfrac{1}{2}\|f^{(n+1)}\|h^n.}$$

In general one can prove (also see Exercises)

THEOREM EXTENDED ERROR ESTIMATES FOR LAGRANGE INTERPOLATES

Let $f\in C^{n+1}[a,b]$ and let $p(x)$ be the Lagrange polynomial interpolating $f(x)$ at $a=x_0<x_1<\cdots<x_n=b$. Then for each $j=1,2,\ldots,n$, we have

$$\boxed{\|f^{(j)}-p^{(j)}\|\leqslant\frac{\|f^{(n+1)}\|n!}{(j-1)!(n+1-j)!}h^{n+1-j}.}$$

In simple illustration of these error estimates we graph the error curve $e(x)=\cos x - p(x)$, where $p(x)$ is the Lagrange interpolate to $\cos x$ of degree 4 interpolating $\cos x$ at 0, $\pi/4$, $\pi/2$, $3\pi/4$, π and $5\pi/4$ (Figure 2.6).

It is well to remark that the Lagrange polynomials are far from the only polynomials solving the interpolation problem $p(x_i)=f(x_i)$, $0 \leqslant i \leqslant n$. There are infinitely many polynomials capable of solving this problem. However, the Lagrange polynomial is the unique polynomial of degree n solving the problem. Another ghost that should be exorcised is the suspicion that by introducing more and more evenly spaced knots $x_0, x_1, \ldots, x_n$ one can force the corresponding sequence $(p_n(x))$ of Lagrange interpolates to $f(x)$ to converge pointwise to $f(x)$. This notion was first dispelled by Méray in his papers of 1884 and 1896, and later by Runge in 1901. In his paper Runge examines this problem for the function $f(x)=1/(1+5x^2)$. To illustrate this divergence Figure 2.7 shows the Lagrange polynomials $p(x)$ of degrees 5 and 15, respectively, with evenly spaced knots $-1=x_0<x_1<\cdots<x_n=1$ interpolating $1/(1+100x^2)$ at these knots and the error curves $f(x)-p(x)$ when $n=5$ and $n=15$. Note the divergence at -1 and 1, which suggests that convergence might be achieved by respacing the knots so that there is a concentration of knots around end points.

The preceding paragraph suggests that if pointwise convergence of the sequence of Lagrange polynomials $(p_n(x): n=1,2,\ldots)$ cannot be

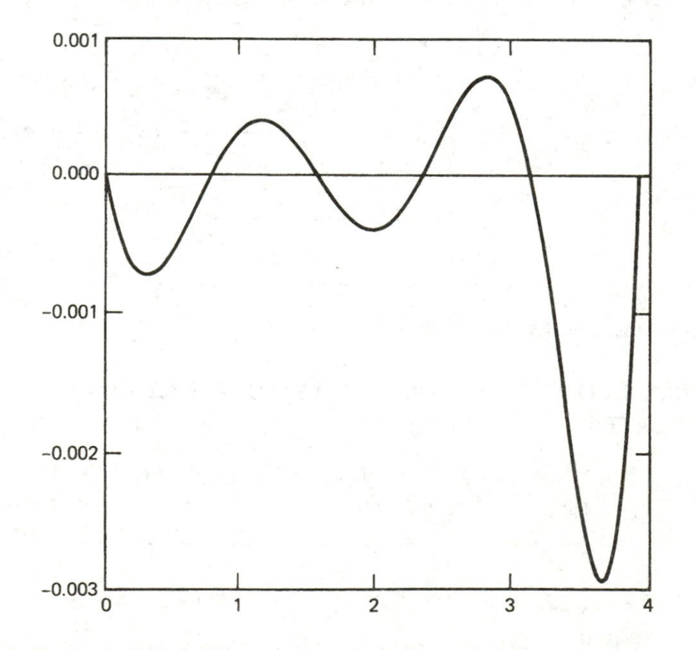

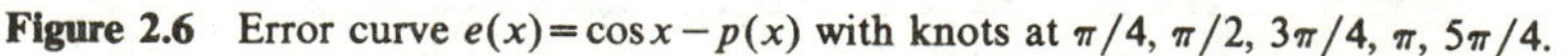

Figure 2.6 Error curve $e(x)=\cos x - p(x)$ with knots at $\pi/4$, $\pi/2$, $3\pi/4$, π, $5\pi/4$.

achieved with evenly spaced knots $a = x_0^n < x_1^n < \cdots < x_n^n = b$, where $\max_{1 \leqslant i \leqslant n}(x_i^n - x_{i-1}^n) \to 0$ as $n \to \infty$, perhaps such pointwise convergence to a given continuous function f can be achieved by a judicious spacing of the knots when $n \to \infty$. The answer to this question is yes. However, as might be suspected from the Runge-Méray syndrome of ideas, the placing of the knots must be tailored to the function f. In fact, it can be shown that given a family $\pi_n : a = x_0^n < x_1^n < \cdots < x_n^n = b$ of subdivisions of $[a,b]$, there exists a continuous function f such that $\lim_n \sup \| p_n(x) \| = \infty$ where $p_n(x)$ is the Lagrange interpolate to f with respect to π. The intersested reader is referred to the book of Rivlin (1969).

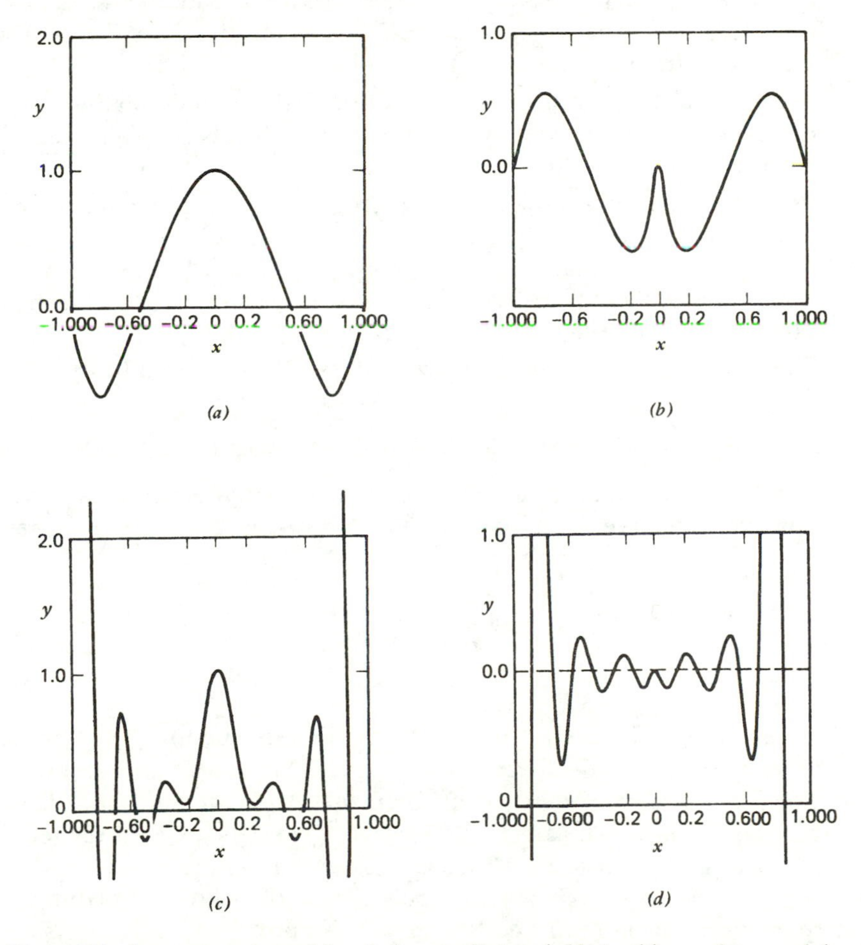

Figure 2.7 Lagrange polynomials of degrees 5(a) and 15(c) with evenly spaced knots interpolating $1/(1+100x^2)$; error curves for degrees 5(b) and 15(d).

EXERCISES

1. Compute the Lagrange polynomial $p(x)$ of degree n interpolating $f(x)=\cos k\pi x$ at evenly spaced knots on the interval $[0,1]$. Estimate $\|f-p\|$. Discuss what happens as both n and k grow large. If $k=500$, how large must n be to guarantee $\|f-p\|<0.01$?
2. Prove the error estimate for $\|f^{(j)}-p^{(j)}\|$.
3. Prove that $\|f''-p''\| \leqslant Kh^{n-1}$ where $f\in C^{n+1}[a,b]$, $h=(b-a)/n$, and K is a positive constant. Find K.
4. Prove that $|(x-x_0)(x-x_1)\cdots(x-x_n)| \leqslant n!(h^{n+1}/4)$.
5. Compute and graph the Lagrange interpolate $p(x)$ of degree 13 to $f(x)=1/1+100x^2$ with (a) evenly spaced knots, (b) knots at $\pm 1/k$, $k=1,2,3,\ldots,7$, and (c) knots at $\pm(1-1/k)$, $k=1,2,3,\ldots,7$. In each case graph $f(x)-p(x)$.
6. A drawback to Lagrange interpolation is that the interpolating polynomial can be excessively oscillatory between knots. Explain what is meant by this, giving illustrations.
7. Prove that the uniform norm of $w(x)=\Pi_{1=1}^{n}(x-x_i)$ is minimized on $[-1,1]$ when $x_i=\cos[2i-1)\pi/2n]$, the n roots of the nth degree Tchebycheff polynomial $T_n(x)=a_0+a_1x+\cdots+a_nx^n$. (Hint: use the trigonometric identity $\cos n\theta=\Pi_{k=0}^{n}a_k\cos^k\theta$, where $a_n=2^{n-1}$, so that $T(\cos\theta)=\cos n\theta$. Then let $w(x)=2^{1-n}T_n$.
8. Find $x_0,x_1,\ldots,x_n$ so as to minimize $w(x)=\Pi_{i=1}^{n}(x-x_i)$ on $[a,b]$.

2.6 BEST APPROXIMATION AND EXTENDED ERROR ESTIMATES

In the last section we proved that if $f\in C^{n+1}[a,b]$ and $p(x)$ is the Lagrange polynomial interpolating f at a set of knots $a=x_0<x_1<\cdots<x_n=b$, then

$$\|f-p\| \leqslant \frac{\|f^{(n+1)}\|}{4(n+1)}h^{n+1},$$

where $h=\max_{0\leqslant i\leqslant n-1}(x_{i+1}-x_i)$.

It is natural to ask whether one can obtain an estimate for $\|f-p\|$ when $f\in C^k[a,b]$, $0\leqslant k\leqslant n$. Such estimates are available to us and depend on the notion of *best approximation*. Preliminary to our discussion of these matters it is useful to estimate $\Sigma_{i=0}^{n}\|l_i\|$, where $\{l_i(x):0\leqslant i\leqslant n\}$ is the set of canonical basis Lagrange polynomials of degree n with knots at $x_0<x_1<\cdots<x_n$. That is, each $l_i(x)$ is a polynomial of degree n solving the interpolation problem $l_i(x_j)=\delta_{ij}$, $0\leqslant i,j\leqslant n$. We prove

THEOREM A USEFUL ESTIMATE

$$\sum_{i=0}^{n} \|l_i\| \leqslant 2^n \left(\frac{h}{\bar{h}}\right)^n, \tag{1}$$

where

$$h = \max_{0 \leqslant 1 \leqslant n-1} (x_{i+1} - x_i)$$

$$\bar{h} = \min_{0 \leqslant i \leqslant n-1} (x_{i+1} - x_i).$$

Proof

$$|l_i(x)| = \frac{|(x-x_0)(x-x_1)\cdots(x-x_{i-1})(x-x_{i+1})\cdots(x-x_n)|}{|(x_i-x_0)(x_i-x_1)\cdots(x_i-x_{i-1})(x_i-x_{i+1})\cdots(x_i-x_n)|}$$

$$\leqslant \frac{|(x-x_0)(x-x_1)\cdots(x-x_{i-1})(x-x_{i+1})\cdots(x-x_n)|}{i!(n-i)!\bar{h}^n}.$$

But $|(x-x_0)\cdots(x-x_{i-1})(x-x_{i+1})\cdots(x-x_n)| \leqslant (j+1)!(n-j)!h^n$ if $x \in [x_j, x_{j+1}]$. Since $(j+1)!(n-j)! \leqslant n!$, it follows that

$$|l_i(x)| \leqslant \frac{n!}{i!(n-i)!}\left(\frac{h}{\bar{h}}\right)^n = \binom{n}{i}\left(\frac{h}{\bar{h}}\right)^n,$$

where $\binom{n}{i}$ is a binomial coefficient. Thus for each $x \in [a,b]$,

$$\sum_{i=0}^{n} |l_i(x)| \leqslant \sum_{i=0}^{n} \binom{n}{i}\left(\frac{h}{\bar{h}}\right)^n = 2^n\left(\frac{h}{\bar{h}}\right)^n,$$

since $(1+1)^n = \sum_{i=0}^{n}\binom{n}{i}$. This completes the proof of the theorem ■

Now suppose f is a bounded function on $[a,b]$ and let P_n denote the set of all polynomials of degree n. We know that a Lagrange polynomial P with $(n+1)$ knots in the interval $[a,b]$ approximates f. By moving the knots about, we try to make the error $\|f-p\|$ smaller. If we do not insist that the polynomial of degree n interpolate f at the knots, we might be able to make $\|f-p\|$ even smaller. This leads us to the idea of *the best approximation $p^*(x)$ to a function $f(x) \in C[a,b]$ by a polynomial of degree n*. In particular

DEFINITION. BEST APPROXIMATION BY POLYNOMIALS OF FIXED DEGREE

Let $f \in C[a,b]$ and let P_n denote the set of all polynomials of degree n. A polynomial $p^*(x) \in P_n$ is said to be a *best approximation* to f, provided $\|f-p^*\| \leqslant \|f-p\|$ for all $p \in P_n$.

One can prove that such a best approximation always exists. It can also be shown that p^* is unique, a result due to Tonelli (1908). Furthermore, we can estimate $\|f-p^*\|$ when $f \in C^k[a,b]$, $0 \leqslant k \leqslant n$. These important error estimates due to Jackson have been extended to many more general settings and are known as Jackson type theorems. We shall content ourselves with a simple statement of that theorem which is of interest to us. Its somewhat delicate proof can be found in the books of Rivlin (1969), and of Davis and Rabinowitz (1967).

The estimates of Jackson's theorem depend on the *modulus of continuity* $\omega(f,h)$ of a function f.

DEFINITION. MODULUS OF CONTINUITY

Let f be a function defined on some interval I and let $h>0$ be a real number. The modulus of continuity $\omega(f,h)$ of f with respect to h on I is defined by

$$\omega(f,h) = \sup\left\{|f(x+\bar{h}) - f(x)| : x, x+\bar{h} \in I, |\bar{h}| \leqslant h\right\}. \tag{2}$$

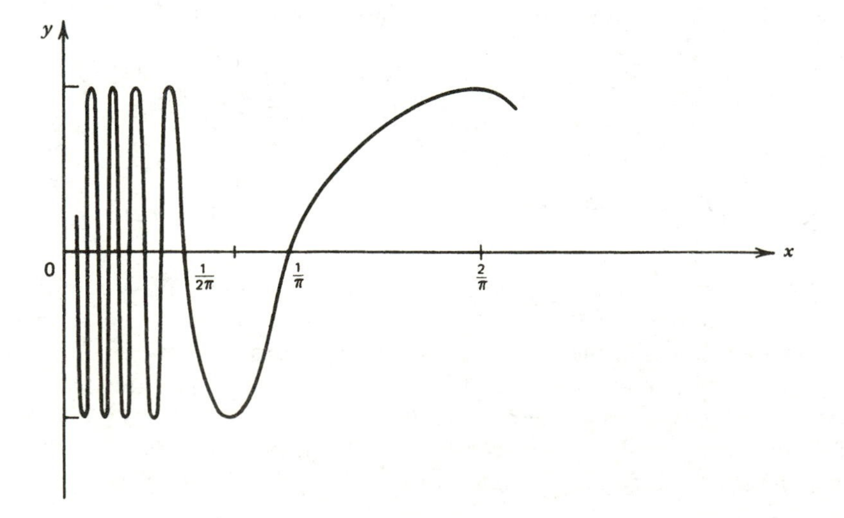

Figure 2.8 $f(x) = \sin 1/x \cdot \omega(f,h) \not\to 0$ as $h \to 0$.

It is clear that if f is continuous on a closed interval, $\lim_{h\to 0}\omega(f,h)=0$. In fact, one can prove that f is uniformly continuous on an interval I if and only if $\omega(f,h)\to 0$ as $h\to 0$. For fixed h, ω is a measure of the oscillatory nature of f. For example, if $f(x)=\sin 1/x$ on $(0,\pi)$, then $\omega(f,h)\to 1$ for all h, no matter how small. Such an f is just too oscillatory to have $\omega(f,h)\to 0$ as $h\to 0$ (see Figure 2.8).We are now in a position to state Jackson's theorem.

JACKSON'S THEOREM FOR POLYNOMIALS

Let $f\in C^k[a,b]$, $0\leqslant k\leqslant n$, and let p^* be the best approximation to f by a polynomial of degree n on the interval $[a,b]$. Then

$$\|f-p^*\|\leqslant g_k(f,n), \tag{3}$$

where

$$g_k(f,n)=\begin{cases} 6\omega\left(f,\dfrac{b-a}{2n}\right), & \text{when } k=0\\ \dfrac{3(b-a)}{n}\|f'\|, & \text{when } k=1\\ \dfrac{6^k(k-1)^{k-1}}{(k-1)!n^k}k(b-a)^k\|f^{(k)}\|, & \text{when } k>1 \text{ and } n>k-1\geqslant 1.\end{cases} \tag{4}$$

What has all this to do with Lagrange polynomials? Letting $a=x_0<x_1<\cdots<x_n=b$ be a partition of $[a,b]$, it is clear that if p^* is a best approximation to f on $[a,b]$ by a polynomial of degree n, and if $p(x)$ is the Lagrange polynomial interpolating f at $x_0,x_1,\ldots,x_n$, then $\|f-p\|\leqslant\|f-p^*\|+\|p^*-p\|$. But since p^* is a polynomial of degree n, $p^*(x)=\Sigma_{i=0}^n p^*(x_i)l_i(x)$. Since $p(x)=\Sigma_{i=0}^n f(x_i)l_i(x)$, we have, invoking (1)

$$|p^*(x)-p(x)|=|\sum_{i=0}^{n}(p^*(x_i)-f(x_i))l_i(x)|$$

$$\leqslant\|p^*-f\|\sum_{i=0}^{n}|l_i(x)|$$

$$\leqslant\|p^*-f\|2^n\left(\frac{h}{\bar{h}}\right)^n.$$

It follows from (3) that if $f \in C^k[a,b]$, then $\|p^* - p\| \leqslant g_k(f,n)2^n(h/\bar{h})^n$. In particular, since $\|f-p\| \leqslant \|f-p^*\| + \|p^*-p\|$, we have proved.

EXTENDED ERROR ESTIMATES FOR LAGRANGE INTERPOLATES

Let $f \in C^k[a,b]$, $0 \leqslant k \leqslant n$ and let $p(x)$ be the Lagrange polynomial interpolating f at $a = x_0 < x_1 < \cdots < x_n$. Then

$$\boxed{\|f-p\| \leqslant \left[1 + \left(\frac{2h}{\bar{h}}\right)^n\right] g_k(f,n),} \tag{5}$$

where $g_k(f,n)$ is given by (4),

$$h = \max_{0 \leqslant i \leqslant n-1} (x_{i+1} - x_i)$$

and

$$\bar{h} = \min_{0 \leqslant i \leqslant n-1} (x_{i+1} - x_i).$$

This estimate is extremely useful in approximating solutions to differential and integral equations by variational methods, as a later chapter reveals.

EXERCISE

1. Let $f(x) = 1/(1+100x^2)$. Then $f \in C^k[-1,1]$ for all $k \geqslant 0$. Let $h = \bar{h}$ and let p be the Lagrange interpolate of degree n to f with knots at $x_i = -1 + ih$, $0 \leqslant i \leqslant n$, and $h = 2/n$. Check whether $[1+2^n]g_k(f,n) \nrightarrow 0$ as $h \to 0$, no matter what $k \geqslant 0$ is chosen.

2.7 PIECEWISE LAGRANGE INTERPOLATION

We see from our error estimates for Lagrange interpolates that even though h approaches zero as n goes to infinity, it simply may not be the case that $\|f-p\|$ goes to zero as h goes to zero. If $f \in C^{n+1}[a,b]$, $\|f-p\| \leqslant \|f^{(n+1)}\| h^{n+1}/4(n+1)$; whereas, if $f \in C^k[a,b]$, $\|f-p\| \leqslant b_k[1+(2h/\bar{h})^n] \|f^{(k)}\| \tilde{h}^k$ where $\tilde{h} = (b-a)/n$, k is fixed, and b_k is a constant dependent on k. Even with even mesh spacing ($h = \bar{h} = \tilde{h}$), the quantity $(1+2^n)$ can outgrow $\tilde{h}^k$-very rapidly when $k < n$. Of course error estimates are pessimistic, and you might choose to live dangerously and ignore the entire theoretic error prediction. This is not a bad thing when you know the function you are approximating is very mundane with no rapid fluctua-

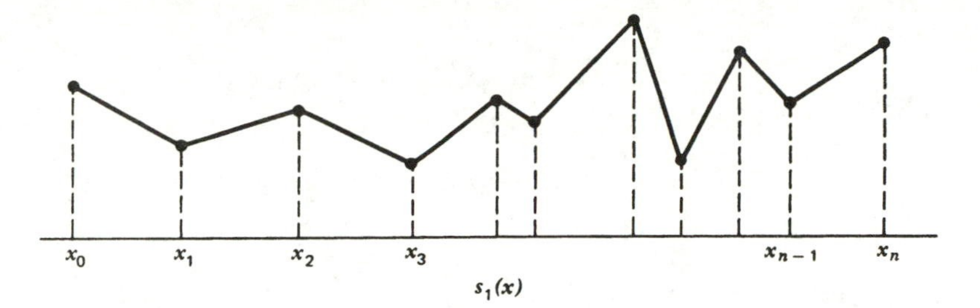

Figure 2.9 A piecewise linear Lagrange polynomial.

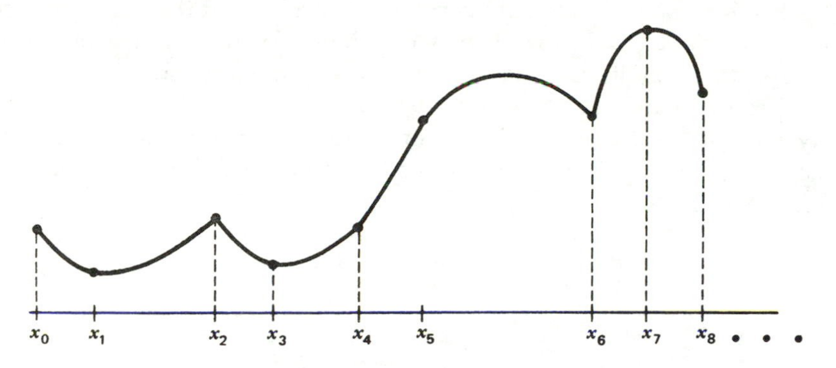

Figure 2.10 A piecewise quadratic Lagrange polynomial.

tions, none too steep a graph, and similar characteristics. However, if you want to exercise a bit more care, one rather crude but very effective and popular means of alleviating the difficulty is to piece together Lagrange polynomials of a fixed degree m and force them to interpolate the data. The resulting function $s(x)$ is known as a *piecewise Lagrange polynomial of degree m*. Specifically,

$$s(x)=\begin{cases} a_0+a_1x+\cdots+a_mx^m, & x_0\leqslant x\leqslant x_m \\ a_{m+1}+a_{m+2}x+\cdots+a_{2m+1}x^m, & x_m\leqslant x\leqslant x_{2m} \\ a_{2m+2}+a_{2m+3}x+\cdots+a_{3m+2}x^m, & x_{2m}\leqslant x\leqslant x_{3m} \\ \vdots \end{cases}$$

where the a_i's are constants to be determined by $s(x_i)=f(x_i)$, $i=0,1,\ldots,$ and n must be some multiple of m. For example, if $m=1$, then $s_1(x)=s(x)$ is a piecewise polygonal line (a piecewise Lagrange polynomial of degree 1—see Figure 2.9). On the other hand, if $m=2$, then $s_2(x)=s(x)$ is a piecewise quadratic (piecewise Lagrange of degree 2—Figure 2.10), which means that

$$s_2(x) = \begin{cases} a_0 + a_1x + a_2x^2 & \text{on } [x_0, x_2] \\ a_3 + a_4x + a_5x^2 & \text{on } [x_2, x_4] \\ a_6 + a_7x + a_8x^2 & \text{on } [x_4, x_6] \\ \text{etc.} \end{cases}$$

Such a fit requires n to be even.

One can go higher obtaining piecewise cubics, quartics, quintics, and so on. In each of these cases $s(x)$ may fail to be differentiable at some or all of its knots x_i (note all the sharp corners in the graphs at the knots). We do something about this in the next two chapters. Meanwhile, for problems amenable to a nondifferentiable approximating function, the error estimate $\|s-f\|$ can be quite good even when $m=1$.

By way of illustration of this error analysis, suppose f is twice continuously differentiable and our mesh spacing is even. Then, invoking our Lagrange estimates,

$$\|f-s_1\| \leqslant \frac{\|f''\|}{4}h^2; \tag{1}$$

whereas if f is three times continuously differentiable,

$$\|f-s_2\| \leqslant \frac{\|f'''\|}{6}h^3. \tag{2}$$

The first estimate abundantly illustrates why such crude fits as polygonal lines can be excellent. After all, h^2 convergence is quite respectable. Of course, if f cannot be differentiated twice, no such rapid convergence is guaranteed. For example, if f is only once continuously differentiable ($f \in C^1[a,b]$), then

$$\|f-s_1\| \leqslant \beta\|f'\|\tilde{h}, \qquad \beta \text{ constant,}$$

and if f is only continuous,

$$\|f-s_1\| \leqslant \gamma\omega(f,h), \qquad \gamma \text{ constant.}$$

In any of these cases we are guaranteed convergence of successive piecewise polygonal fits and piecewise quadratic fits to f as h goes to zero. If f is in C^2 but not in C^3, there may be no point to a piecewise quadratic fit, since the additional computational effort will not guarantee any better convergence than the piecewise linear fit. However, in all cases, you must be prepared to live with sharp corners on the graph.

A very popular application of these functions is *numerical quadrature formulas* (numerical integration formulas), which replaces the definite integral

$$Ef = \int_a^b f(x)\,dx$$

by the integral

$$E_n f = \int_a^b s_n(x)\,dx,$$

where $s_n(x)$ is a piecewise Lagrange polynomial of degree n interpolating $f(x)$ at the knots $a = x_0 < x_1 < \cdots < x_N = b$ of a partition of the interval $[a,b]$. When $n=1$ we obtain the well-known *trapezoidal rule*

$$\boxed{E_1 f = \frac{h}{2}[f(x_0) + 2f(x_1) + 2f(x_2) + \cdots + 2f(x_{n-1}) + f(x_N)],}$$

and when $n=3$ we obtain *Simpson's rule*

$$\boxed{E_3 f = \frac{h}{3}[f(x_0) + 4f(x_1) + 2f(x_2) + 4f(x_3) + \cdots + 4f(x_{2M-1}) + f(x_{2M})].}$$

Letting n vary, we range over a large class of quadrature formulas known as *Newton-Côtes formulas*. Moreover, our error analysis for $\|f - s_n\|$ enables us to estimate the error

$$Ef - E_n f = \int_a^b [f(x) - s_n(x)]\,dx.$$

In the case of the trapezoidal rule, for example, a preliminary estimate when $f \in C^2[a,b]$ is

$$|Ef - E_1 f| \leqslant \frac{\|f''\|}{4}(b-a)h^2,$$

which follows immediately from (2). These formulas can be improved somewhat using Peano kernals (see Davis (1963)).

The *computation* of piecewise Lagrange polynomials $f(x)$ of degree d with $n+1$ knots $x_0 < x_1 < \cdots < x_n$, $n = pd$, for some positive integer p, is a fairly simple matter. One popular means of computing involves the *car-*

dinal basis functions ϕ_i, $0 \leqslant i \leqslant n$, which solve the interpolation problem

$$\phi_i(x_j) = \delta_{ij}, \qquad 0 \leqslant i,j \leqslant n.$$

Each of these functions is itself a piecewise Lagrange polynomial of degree d between knots, and it is clear that

$$f(x) = f(x_0)\phi_0(x) + f(x_1)\phi_1(x) + \cdots + f(x_n)\phi_n(x).$$

For example, when $d=1$, $\phi_i(x)$ is the "hat function"

$$\phi_i(x) = \begin{cases} 0, & x \leqslant x_{i-1} \\ (x - x_{i-1})(x_i - x_{i-1})^{-1}, & x_{i-1} \leqslant x \leqslant x_i \\ (x_{i+1} - x)(x_{i+1} - x_i)^{-1}, & x_i \leqslant x \leqslant x_{i+1} \\ 0, & x \geqslant x_{i+1} \end{cases}$$

whose graph appears in Figure 2.11. At the two end knots $i = 0, n$

$$\phi_n(x) = \begin{cases} (x - x_{n-1})(x_n - x_{n-1})^{-1}, & x \geqslant x_{n-1} \\ 0, & x_0 \leqslant x \leqslant x_{n-1} \end{cases}$$

and

$$\phi_0(x) = \begin{cases} 0, & x \geqslant x_1 \\ (x_1 - x)(x_1 - x_0)^{-1}, & x_0 \leqslant x \leqslant x_1 \end{cases}$$

(see Figure 2.12). The higher degree, $d > 1$, cardinal piecewise Lagrange basis functions are defined in an analogous interval by interval manner. Graphs of piecewise quadratic basis functions appear in Figure 2.13; the algebraic expression for such functions and higher degree ϕ_i's is left as an exercise for the reader. Note the sharp corners on each of the graphs, hence the nondifferentiability of the ϕ_i's. Note too that each of the sets $L_d(\pi)$ of piecewise Lagrange polynomials of degree d between successive knots of π is a *linear space*. In particular the sum of any two functions from $L_d(\pi)$ and a real number times a function in $L_d(\pi)$ are again in $L_d(\pi)$, since such functions decompose to polynomials of degree d on each successive pair of subintervals (t_i, t_{i+1}) and polynomials of degree d are a linear space. In addition,

$$\dim L_d(\pi) = Nd + 1 = n + 1$$

is the dimension of $L_d(\pi)$, and $\mathfrak{B} = \{\phi_0, \phi_i, \ldots, \phi_n\}$ is a basis for $L_d(\pi)$, where $\phi_i(x)$ is the unique function in $L_d(\pi)$ solving

$$\phi_i(x_j) = \delta_{ij}, \qquad 0 \leqslant i,j \leqslant n.$$

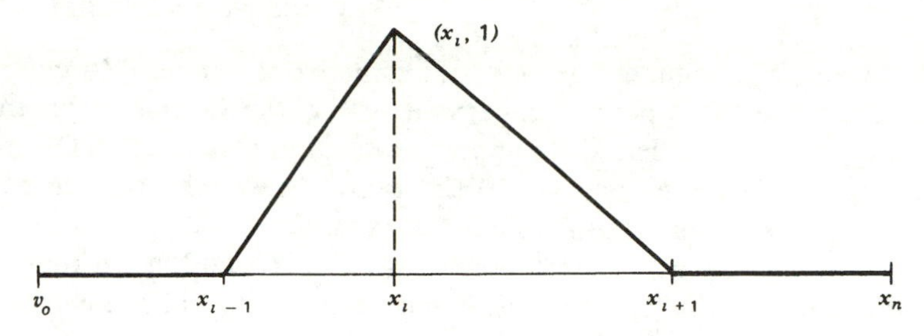

Figure 2.11 Graph of $\phi_i(x)$, $1 \leqslant i \leqslant n-1$.

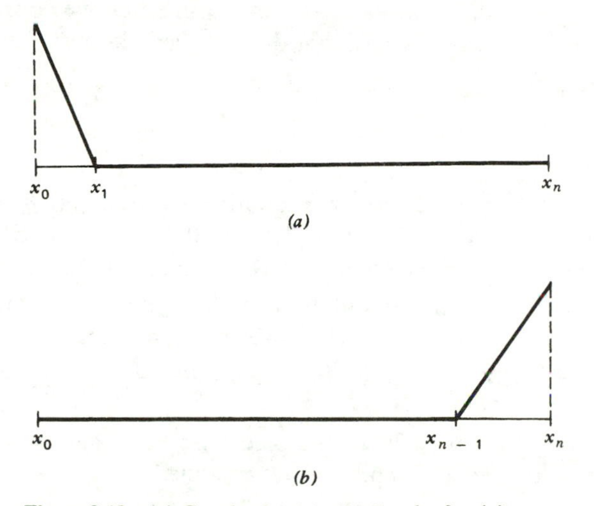

Figure 2.12 (*a*) Graph of $\phi_0(x)$. (*b*) Graph of $\phi_n(x)$.

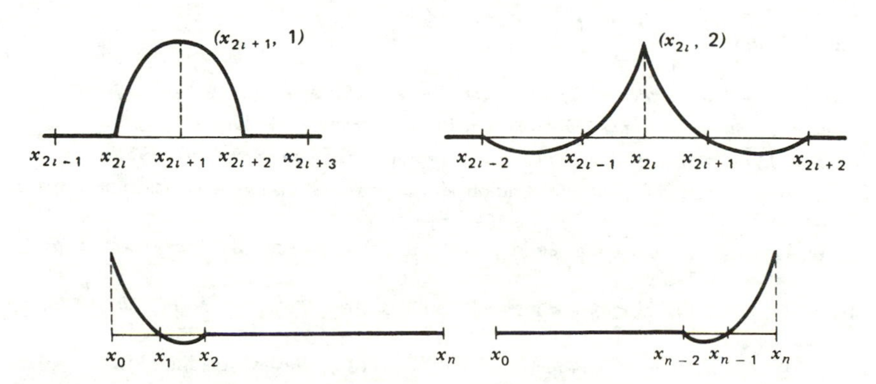

Figure 2.13 Basis piecewise quadratic Lagrange polynomials.

Finally we must comment on our use of the word "knots." *Usually*, the points of any piecewise polynomial $s(x)$ at which the various polynomials pieces are joined are called, quite appropriately, "knots" of $s(x)$. Owing to our more generous definition of *knots* (Section 2.2), we refer to such points as *joints* or *natural knots* of $s(x)$. Thus for example, $x_2, x_4, x_6, \ldots$, are joints of $s_2(x)$, and $x_1, x_2, \ldots, x_{n-1}$ are those of $s_1(x)$. It is readily seen that even this definition can be too catholic, since any finite collection of real numbers can be regarded as joints or natural knots of *any* polynomial, as can any infinite set or real numbers (viz., $0,1,2,3,\ldots$, and $1,\frac{1}{2},\frac{1}{3},\ldots$). In any event the set $x_0, x_1, \ldots, x_n$ is, in our nomenclature, a collection of knots for s_1, s_2 and, in general, s, since it is at these points that we constrain these functions—in the sense of "tying them down" (viz. "knots")—by insisting $s(x_i) = f(x_i)$, $0 \leqslant i \leqslant n$.

EXERCISES

1. Compute and graph the piecewise quadratic fit $s(x)$ to $f(x) = 1/(1+100x^2)$ with knots at -1, $-\frac{1}{10}$, $-\frac{1}{5}$, $-\frac{1}{3}$, 0, $\frac{1}{3}$, $\frac{1}{5}$, $\frac{1}{10}$, 1. Estimate the error $\|f-s\|$ and compare this with the full Lagrange polynomial fit.
2. Give an example of a continuous function you *cannot* approximate with order h^2 accuracy using piecewise polygonal lines.
3. Find a Newton-Côtes formula for $n=2$ and estimate the error $|Ef - E_2 f|$.
4. Compute the cardinal basis for the piecewise Lagrange polynomials of degree 2 and degree 3. Graph these functions.
5. For fixed $d \geqslant 1$, each of the ϕ_i's is nondifferentiable. Let $f(t) = \sum_{i=0}^{n} a_i \phi_i(t)$. Could $f \in C^1[x_0, x_n]$? When? Explain.

REFERENCES

Cheney, E. W. *Introduction to approximation theory*. McGraw-Hill, New York, 1966.

Davis, P. J. *Interpolation and approximation*. Blaisdell, New York, 1963.

Davis, P. J. and P. Rabinowitz. *Numerical integration*, Blaisdell, New York, 1967.

Deutch, Frank. Simultaneous interpolation and approximation in topological linear spaces. *SIAM J. Appl. Math.*, **14**, No. 5 (Sept. 1966).

Méray, C. Observations sur la légitime de l'interpolation. Ann. Sci. Ec. Norm. Super. **1**, (1884), pp. 165–176.

Méray, C. Nouvelles examples d'interpolation illusoires. *Bull. Sci. Math.*, **20** (1896), pp. 266–270.

Rice, J. R. *The approximation of functions*, Vol. 1, Addison-Wesley, Reading, Mass., 1964.

Rivlin, T. R. *An Introduction to the approximation of functions.* Blaisdell, New York, 1969.

Runge, C. Uber die Darstellung willkürlicher Funktionen und die Interpolation zwischen aquidistantent Ordinaten. *Z. Math. Phys.*, **46** 1901, pp. 224–243.

Timan, A. F. *Theory of approximation of functions of a real variable.* Moscow, 1963; translation Macmillan, New York.

Tonelli, L. I polinomi d'approssimazione di Tchebychev. *Ann. Math. Pur. Appl.*, **15** (1908), pp. 47–119.

3

HERMITIAN INTERPOLATES

3.1 INTRODUCTION

The Lagrange polynomial $p(t)$ of degree 1 solved the interpolation problem

$$\begin{aligned} p(a) &= f(a) \\ p(b) &= f(b), \end{aligned} \tag{1}$$

where a and b were two distinct real numbers. We saw that the piecewise linear Lagrange polynomial s_1 obtained by piecing together such polynomials led to a rather good fit to a given function $f(\|f-s_1\|=o(h^2)$ when $f\in C^2[a,b])$, but a drawback to this approximate was the sharp corners on the graph of s_1 occurring at the knots of s_1. One way of overcoming this lack of smoothness is to glue or piece together Hermite polynomials rather than Lagrange polynomials. Hermite interpolation generalizes Lagrange interpolation by fitting a polynomial to a function f which not only interpolates f at each knot but also interpolates a given number of consecutive derivatives of f at each knot. For example, suppose we replace the constraints (1) by

$$p(a)=f(a), \qquad p'(a)=f'(a),$$
$$p(b)=f(b), \qquad p'(b)=f'(b) \tag{2}$$

and ask the following *question*. Does there exist a polynomial satisfying (2)? The answer to this question is "of course." There are, in fact, infinitely many such polynomials, but *there is only one*

$$p(t)=a_0+a_1t+a_2t^2+a_3t^3$$

of degree 3, and it is called the *cubic Hermite interpolate to f*. The graph of such an interpolate appears in Figure 3.1. Note the matching of derivatives as well as function values at a and b: we are again finding an approximating function $\tilde{f}=p$ subject to *pure interpolatory constraints*.

There are a number of ways of proving the existence of p; one is simply to write it down directly as

$$p(t)=p(a)\phi_1(t)+p(b)\phi_2(t)+p'(a)\psi_1(t)+p'(b)\psi_2(t), \tag{3}$$

where

$$\phi_1(t)=\frac{(t-b)^2[(a-b)+2(a-t)]}{(a-b)^3}$$

$$\phi_2(t)=\frac{(t-a)^2[(b-a)+2(b-t)]}{(a-b)^3}$$

$$\psi_1(t)=\frac{(t-a)(t-b)^2}{(a-b)^2} \tag{4}$$

$$\psi_2(t)=\frac{(t-a)^2(t-b)}{(a-b)^2}$$

and $t_1=a$, $t_2=b$. Note that

$$\phi_i(t_j)=\delta_{ij}$$

$$\phi_i'(t_j)=0, \qquad 1\leqslant i,j\leqslant 2$$

and

$$\psi_i(t_j)=0$$

$$\psi_i'(t_j)=\delta_{ij}, \qquad 1\leqslant i,j\leqslant 2.$$

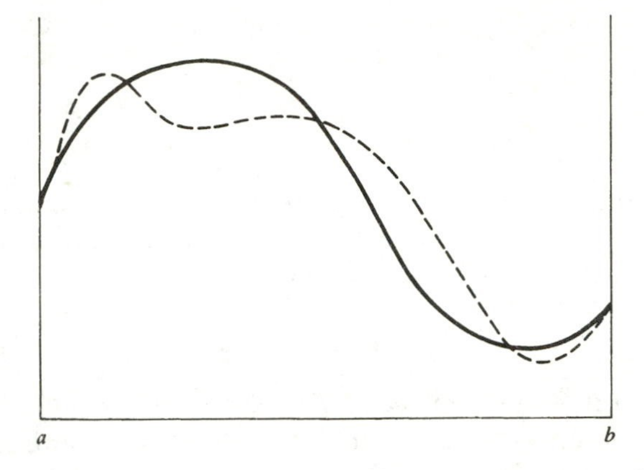

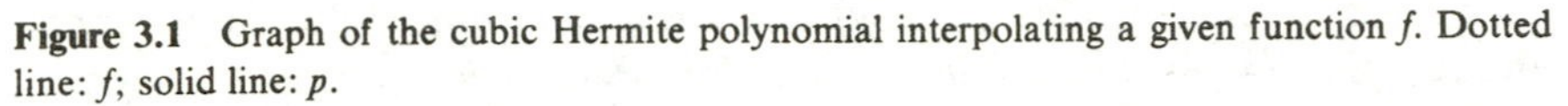
Figure 3.1 Graph of the cubic Hermite polynomial interpolating a given function f. Dotted line: f; solid line: p.

Expression 3 does not, however establish the uniqueness of p. Another approach to this problem, which establishes both the existence and the uniqueness of p, consists of proving that the solution to the boundary value problem $D^4u=0$ subject to boundary conditions (2) is unique, using integration by parts (see Exercises). A third method is given in Section 3.4, where we prove existence and uniqueness of Hermite interpolating polynomials in the most general case. In any event we have the following.

EXISTENCE THEOREM

There exists a unique cubic polynomial $p(t)=a_0+a_1t+a_2t^2+a_3t^3$ solving the interpolation problem $p(t_i)=f(t_i)$, $p'(t_i)=f'(t_i)$, $i=1,2$.

It is now simple to see how to eliminate the sharp corners occurring in the piecewise Lagrange polynomials: simply construct *piecewise cubic Hermite polynomials*. In particular, given $a=t_0<t_1<\cdots<t_n=b$, and i, $1\leqslant i\leqslant n$, we construct the cubic Hermite polynomial $p_i(t)$ solving

$$\begin{cases} p_i(t_{i-1})=f(t_{i-1}), & p_i'(t_{i-1})=f'(t_{i-1}), \\ p_i(t_i)=f(t_i), & p_i'(t_i)=f'(t_i), \end{cases} \tag{5}$$

and let $s(t)$ be the piecewise cubic polynomial

$$s(t)=p_i(t), \qquad t_{i-1}\leqslant t\leqslant t_i.$$

The graph of such an s interpolating f is given in Figure 3.2. Since s is a polynomial on each of the intervals (t_{i-1},t_i), it is infinitely differentiable on each (t_{i-1},t_i). Moreover, since $f(t_i)=p_i(t_i)=p_{i+1}(t_i)=s(t_i)$ and $p_{i+1}'(t_i)$

$= s'(t_i) = p_i'(t_i) = f'(t_i)$ at all interior knots t_i, $1 \leqslant i \leqslant n-1$, it is clear that

$$s \in C^1[a,b].$$

Because we have not matched higher derivatives of s at these points, however, $s''(t_i)$ is unlikely to exist at interior knots. To obtain greater smoothness with piecewise cubics, we must turn to the cubic polynomial splines of the next chapter or to higher degree piecewise Hermite polynomials. Meanwhile, let us finish discussing Hermite and piecewise Hermite polynomials.

We later prove (Section 3.4) that if $f \in C^4[a,b]$, then

$$\|f - s\| \leqslant \frac{\|f^{(4)}\|}{96} h^4.$$

This is to be compared with the estimate

$$\|f - s_3\| \leqslant \frac{\|f^{(4)}\|}{16} h^4,$$

where s_3 is the piecewise cubic Lagrange interpolate to f. Thus s is a slight improvement over s_3 errorwise in addition to being smoother. However, it is slightly more work to compute, as we shall see. You are asking more of s than in the Lagrange case; thus, you must atone for your avarice by doing more computational work.

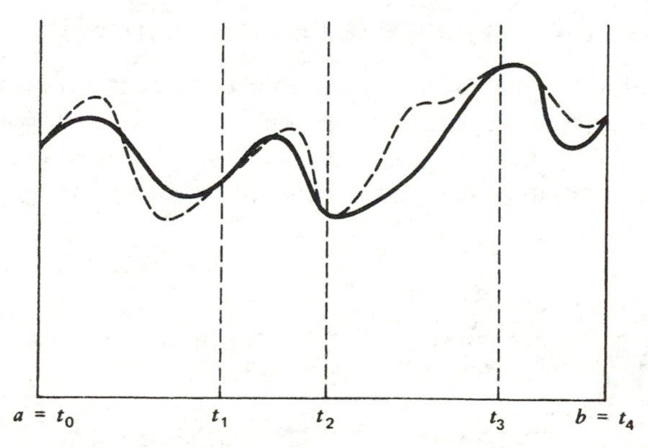

Figure 3.2 A piecewise cubic Hermite polynomial interpolating f. Dotted line: f; solid line: s.

EXERCISES

1. Prove that the differential equation $D^4u=f$, $u(a)=u(b)=u'(a)=u'(b)=0$ has a unique solution by first proving $\int_a^b u(t)D^4u(t)\,dt=0$ if and only if $u\equiv 0$. Then explain how to apply this to prove the existence and uniqueness of the cubic Hermite polynomial $p(t)$.
2. Can one prove the existence and uniqueness of $p(t)$ by showing that the coefficient matrix

$$\begin{bmatrix} \phi_1(a) & \phi_2(a) & \psi_1(a) & \psi_2(a) \\ \phi_1(b) & \phi_2(b) & \psi_1(b) & \psi_2(b) \\ \phi_1'(a) & \phi_2'(a) & \psi_1'(a) & \psi_2'(a) \\ \phi_1'(b) & \phi_2'(b) & \psi_1'(b) & \psi_2'(b) \end{bmatrix}$$

is nonsingular?
3. Quintic Hermite polynomials p also exist. Such polynomials solve the interpretation problem

$$p^{(k)}(a)=f^{(k)}(a)$$

$$p^{(k)}(b)=f^{(k)}(b), \qquad k=0,1,2.$$

Prove the existence and uniqueness of such p's.

3.2 COMPUTATION OF PIECEWISE CUBIC HERMITES

The computation of the piecewise cubic Hermites interpolates $s(t)$ to a given function $f(t)$ with knots at $\pi: a=t_0<t_1<\cdots<t_n=b$ is quite a simple matter with the right choice of *basis*. To this end we simply generalize the functions (4) of the previous section. In particular, for $1\leqslant i\leqslant n-1$, let

$$\phi_{i0}(t)=\begin{cases} \dfrac{(x-x_{i-1})^2}{(x_i-x_{i-1})^3}[2(x_i-x)+(x_i-x_{i-1})], & x_{i-1}\leqslant x\leqslant x_i \\ \dfrac{(x_{i+1}-x)^2}{(x_{i+1}-x_i)^3}[2(x_{i+1}-x)-(x_{i+1}-x_i)], & x_i\leqslant x\leqslant x_{i+1} \\ 0 \quad \text{otherwise} \end{cases} \tag{1}$$

and let

$$\phi_{i1}(t)=\begin{cases}\dfrac{(x-x_{i-1})^2(x-x_i)}{(x_i-x_{i-1})^2}, & x_{i-1}\leqslant x\leqslant x_i\\[2ex] \dfrac{(x-x_{i+1})^2(x-x_i)}{(x_{i+1}-x_i)^2}, & x_i\leqslant x\leqslant x_{i+1}\\[2ex] 0, & \text{otherwise}\end{cases} \tag{2}$$

The graphs of these functions (Figures 3.3, 3.4) indicate they vanish identically outside the interval (t_{i-1}, t_{i+1}). This feature is very important to us in the sequel. The functions ϕ_{00}, ϕ_{n0}, ϕ_{01}, and ϕ_{n1} are similarly defined. In particular

$$\phi_{00}(t)=\begin{cases}\dfrac{(t_1-t)^2}{(t_1-t_0)^2}[2(t_1-t)-(t_1-t_0)], & t_0\leqslant t\leqslant t_1\\[2ex] 0, \quad \text{when} \quad t_n\geqslant t\geqslant t_1\end{cases}$$

$$\phi_{n0}(t)=\begin{cases}\dfrac{(t-t_{n-1})^2}{(t_n-t_{n-1})^2}[2(t_{n-1}-t)+(t_n-t_{n-1})], & t_{n-1}\leqslant t\leqslant t_n\\[2ex] 0, \quad \text{when} \quad t_0\leqslant t\leqslant t_{n-1}\end{cases}$$

$$\phi_{01}(t)=\begin{cases}\dfrac{(t-t_1)^2(t-t_0)}{(t_1-t_0)^3}, & t_0\leqslant t\leqslant t_1\\[2ex] 0, \quad \text{when} \quad t_1\leqslant t\leqslant t_n\end{cases}.$$

$$\phi_{n1}(t)=\begin{cases}\dfrac{(t-t_{n-1})^2(t-t_n)}{(t_n-t_{n-1})^3}, & t_{n-1}\leqslant t\leqslant t_n\\[2ex] 0, \quad t_0\leqslant t\leqslant t_{n-1}\end{cases}$$

These four functions are graphed in Figure 3.5. All the $\phi_{ij}(t)$'s are uniquely determined by the constraints

$$\begin{cases}\phi_{i0}(t)=\delta_{ij}\\ \phi_{i0}'(t)=0,\ 0\leqslant i,j\leqslant n,\end{cases} \tag{3}$$

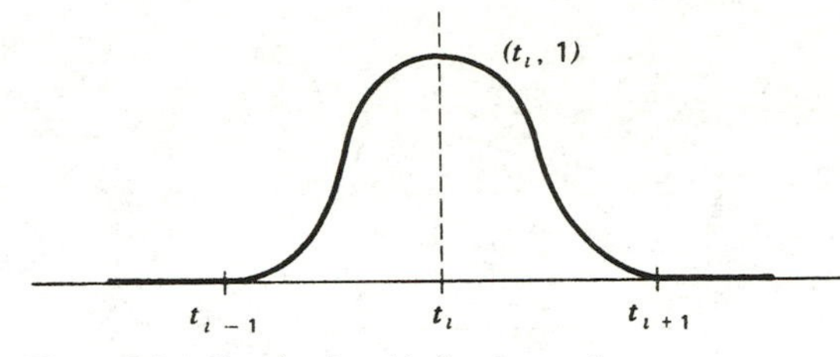

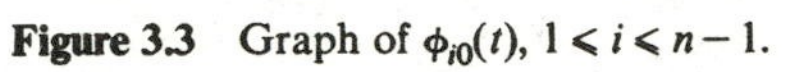
Figure 3.3 Graph of $\phi_{i0}(t)$, $1 < i < n-1$.

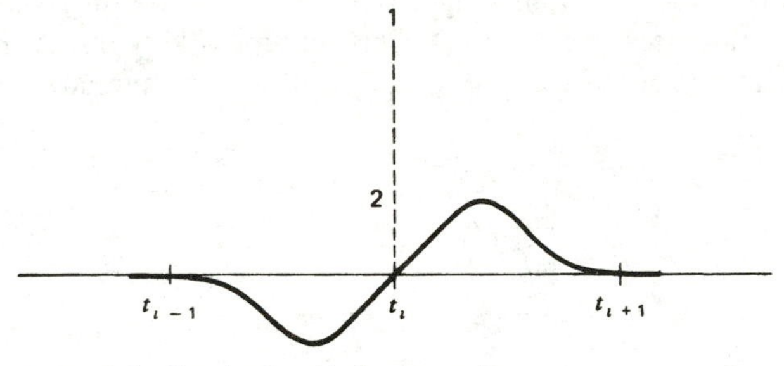

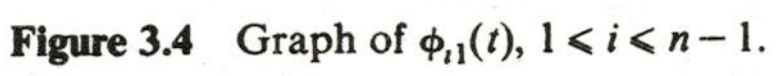
Figure 3.4 Graph of $\phi_{i1}(t)$, $1 < i < n-1$.

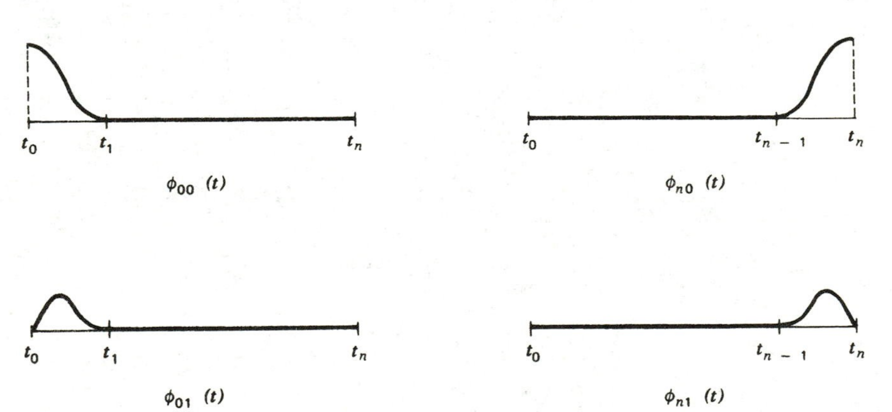

Figure 3.5 Graphs of ϕ_{00}, ϕ_{n0}, ϕ_{01} and ϕ_{n1}.

whereas

$$\begin{cases} \phi_{i1}(t_j)=0,\ 0\leqslant i,j\leqslant n. \\ \phi_{i1}'(t_j)=\delta_{ij}. \end{cases} \tag{4}$$

Thus

$$s(t)=f(t_0)\phi_{00}(t)+f(t_1)\phi_{10}(t)+\cdots+f(t_n)\phi_{n0}(t)$$
$$+f'(t_0)\phi_{01}(t)+f'(t_1)\phi_{11}(t)+\cdots+f'(t_n)\phi_{n1}(t),$$

or more succinctly,

$$\boxed{s(t)=\sum_{i=0}^{n} f(t_i)\phi_{i0}(t)+\sum_{i=0}^{n} f'(t_i)\phi_{i1}(t).} \tag{5}$$

Formula 5 comprises a simple numerical algorithm for computing $s(t)$. Other algorithms are also available (see Exercises).

Let $H_3(\pi)$ denote the linear space of all piecewise cubic Hermite polynomials (it is a linear space—see Exercises). Then $\mathcal{B}=\{\phi_{00},\phi_{01},\ldots,\phi_{1n}\}$ is a *basis* for $H_3(\pi)$; it is known as the *cardinal basis* because of properties (3) and (4). Moreover, since $\mathcal{B}$ is linearly independent (see Exercises), $H_3(\pi)$ has dimension $2n+2$. If $n=3N$, we can also construct $L_3(\pi)$, the linear space of piecewise cubic Lagrange polynnmials with the same knots. Clearly $L_3(\pi)$ has dimension $n+1$, whereas the dimension of $H_3(\pi)$ is $2n+2$. Thus interpolates to f from $H_3(\pi)$ are exactly twice the work of those from $L_3(\pi)$ to compute. This is the price to be paid for the added smoothness. Since both fits give $o(h^4)$ convergence when $f\in C^4[a,b]$ and s is not very much harder to compute, we see that s has a decided advantage over its piecewise Lagrange counterpart. If, however, the values of $f'(t)$ are not given or cannot be estimated with good accuracy, $s(t)$ cannot be computed at all. Ample possibilities exist, however, for overcoming these difficulties if one needs $C^1[a,b]$ smoothness. For example, try computing

$$\hat{f}_n(t)=\sum_{i=0}^{n} x_i\phi_{i0}(t)+\sum_{i=0}^{n} y_i\phi_{i1}(t), \tag{6}$$

where $(x_0,x_1,\ldots,x_n,y_0,y_1,\ldots,y_n)$ are chosen to force $\hat{f}_n(t)$ to satisfy the pure interpolatory constraints

$$\hat{f}_n(\bar{t}_i)=f(\bar{t}_i),$$

where $t_0 < \bar{t}_0 < \bar{t}_1 < t_1 < \bar{t}_2 < \bar{t}_3 < t_2 < \bar{t}_4 < \ldots < t_{n-1} < \bar{t}_{2n-2} < \bar{t}_{2n-1} < t_n$. The reader is left to experiment with existence, uniqueness, and error estimates for such interpolates.

EXERCISES

1. Let $\pi : a = t_0 < t_1 < \cdots < t_n = b$. Prove that $L_3(\pi)$ and $H_3(\pi)$ are linear spaces. With the Tchebycheff norm these two spaces become normed linear spaces. Are they complete? Explain.
2. Construct two other basis for $H_3(\pi)$ and compare them computationally with (6). Is the basis presented in the text preferable for computing?
3. Prove that $\mathfrak{B}$ is linearly independent.
4. (a) Construct the cardinal basis for the piecewise quintic Hermite polynomials $H_5(\pi)$ (see Exercise 3, Section 3.1).
   (b) What is the dimension of $H_5(\pi)$? Prove it.
5. Prove a set of necessary and sufficient conditions for the existence and uniqueness of $\hat{f}_n(t)$. What other possibilities are there when $f'(t_i)$ is not known?

3.3 A SIMPLE APPLICATION

The charitable applied reader is undoubtedly thinking by this time that these "Hermite things" are all very nice; but at the same time he is wondering just what they are good for unless one happens to be fitting curves. As we indicated in Section 1.1, we want to use such functions to approximate solutions of differential equations. As a "preview of coming attractions," let us look for a moment at the differential equation

$$\{ Lu(t) = u''(t) - \sigma(t)u(t) = f(t), \qquad 0 \leqslant t \leqslant 1, \qquad u(0) = u(1) \tag{1}$$

of a mechanical oscillator satisfying the given boundary conditions at $t = 0$ where $\sigma(t) \geqslant 0$ and $f(t)$ are given. We want to approximate the solution $u(t)$ to this problem. As a first try, let $\pi : a = t_0 < t_1 < t_2 = b$ and look for an approximation

$$\tilde{u}(t) = c_{00}\phi_{00}(t) + c_{10}\phi_{10}(t) + c_{20}\phi_{20}(t) + c_{01}\phi_{01}(t)$$
$$+ c_{11}\phi_{11}(t) + c_{21}\phi_{21}(t),$$

where the ϕ_{ij}'s are the *basis piecewise cubic Hermite polynomials* of the last section having knots at π. We want to choose the six unknown c_{ij}'s so that

$u(t)$ is a good approximation to $u(t)$ in that $\|u-\tilde{u}\|$ is small. Since we do not know u, we cannot compute the piecewise cubic Hermite interpolate s to u. Therefore let us try a slight modification of that approach. Since we know $f(t)=Lu(t)$, we can force $L\tilde{u}(t)$ to satisfy the pure interpolatory constraints

$$L\tilde{u}\left(\bar{t}_i\right)=f\left(\bar{t}_i\right), \qquad i \leqslant i \leqslant 6,$$

where $\bar{t}_1<\bar{t}_2<\cdots<\bar{t}_6$ are some six points of $[0,1]$ (six unknowns, six constraints—counts such as these are always necessary).

Some of you will, of course, note that the system just expressed may not even have a solution; indeed, it might be pure rubbish, depending on one's choice of $\bar{t}_i$'s. This, of course, is perfectly correct. If, for example, $\bar{t}_i=t_1$ for any one of the i's, we are in serious trouble because

$$L\phi_{ij}(t_1)=\phi_{ij}''(t_1)+\sigma(t_1)\phi_{ij}(t_1),$$

and $\phi_{ij}''(t_1)$ *does not exist*. But the more pragmatic observer will undoubtedly counter, "Who cares—just choose different $\bar{t}_i$'s." And this is what we do. In fact, we choose only four $\bar{t}_i$'s and force them to satisfy the inequality $t_0<\bar{t}_1<\bar{t}_2<t_1<\bar{t}_3<\bar{t}_4<t_2$. While we are at it, we also throw out ϕ_{01} and ϕ_{21} by setting $c_{00}=c_{20}=0$. Our rationale is that since $u(0)=u(1)=0$, we choose only those ϕ_{ij} satisfying the same boundary values. Thus because $\phi_{01}(0)=\phi_{21}(1)=1$, we discard them. Now we are looking for

$$\begin{aligned}\bar{u}(t)&=c_{00}\phi_{00}(t)+c_{10}\phi_{10}(t)+c_{11}\phi_{11}(t)+c_{20}\phi_{20}(t)\\&=c_1\phi_1(t)+c_2\phi_2(t)+c_3\phi_3(t)+c_4\phi_4(t)\end{aligned} \tag{2}$$

satisfying

$$L\bar{u}\left(\bar{t}_i\right)=f\left(\bar{t}_i\right), \qquad i=1,2,3,4. \tag{3}$$

The second equation in the right-hand side of (2) results from renumbering our remaining ϕ_{ij}'s and c_{ij}'s (i.e., $\phi_{00}=\phi_1$, $\phi_{10}=\phi_2$, etc.). We now have a linear system of four equations

$$Ac=\bar{f} \tag{4}$$

in four unknowns $c=(c_1,c_2,c_3,c_4)^T$, where $\bar{f}=(f(\bar{t}_1),f(\bar{t}_2),f(\bar{t}_3),f(\bar{t}_4))^T$, and

$A=(a_{ij})$, $a_{ij}=L\phi_j(\bar{t}_i)$, is the 4×4 matrix

$$A=\begin{bmatrix} L\phi_1(\bar{t}_1) & L\phi_2(\bar{t}_1) & L\phi_3(\bar{t}_1) & L\phi_4(\bar{t}_1) \\ L\phi_1(\bar{t}_2) & L\phi_2(\bar{t}_2) & L\phi_3(\bar{t}_2) & L\phi_4(\bar{t}_2) \\ L\phi_1(\bar{t}_3) & L\phi_2(\bar{t}_3) & L\phi_3(\bar{t}_3) & L\phi_4(\bar{t}_3) \\ L\phi_1(\bar{t}_4) & L\phi_2(\bar{t}_4) & L\phi_3(\bar{t}_4) & L\phi_4(\bar{t}_4) \end{bmatrix}.$$

The system (4) has a unique solution if and only if the coefficient matrix A is nonsingular, and we can prove quite easily that it is (see Chapter 8).

Thus we obtain in a rather simple manner a function $\bar{u}(t)$ given by (2) which we maintain can be a rather good approximate to $u(t)$. The question is, of course, how good? And how do we know in general that A is nonsingular? For example, we may want to introduce a different L, or more knots, or different ϕ_i's and try the same sort of scheme. As we shall see, for many choices of ϕ_i's and L this method can provide excellent approximations. The method has a name. It is called the *method of collocation*. Note that it is the operator analog of the method of pure interpolatory constraints of Chapter 1. That is, we force the image of $\bar{u}$ under L, $L\bar{u}(t)$, to interpolate given data instead of forcing $\bar{u}(t)$ to interpolate given data. The theory and application of this method are the content of Chapter 8. If we apply the method with four knots as given by (2), (3), and (4) to the equation

$$u''(t)+u(t)=2-6t+t^2-t^3$$

so that $\sigma(t)\equiv 1$ and $f(t)=2(1-3t)+t^2(1-t)$, we find $\bar{u}(t)=u(t)=t^2(1-t)$. Thus the approximate solution is the exact solution in this case (see if you can determine why this must be true). The exercises give you some more complicated problems to think about.

Note also that when $f(t)$ is continuous on $[0,1]$, $u(t)$ is twice continuously differentiable on $[0,1]$. (This follows from the basic theory of differential equations.) However, while $\bar{u}'(t)$ exists and is continuous, $\bar{u}''(t)$ does not exist at t_1. Thus we are approximating a twice continuously differentiable function $u(t)$ by a function that can only be differentiated once and is not properly in the domain of L. We frequently indulge in the sequel in such unseemly mathematical conduct. Moreover, we shall see that such breaches of decorum can be extremely profitable and can also be made mathematically respectable.

EXERCISES

1. Describe the method of collocation using a basis of eight cubic Hermites $\phi_{ij}(t)$ vanishing at $t=0$ and $t=1$ and eight distinct values of $\bar{t}_i$ to approximate the solution of the differential equation $u''-e^t u = \sin \pi t$, $0 \leqslant t \leqslant 1$, $u(0)=u(1)=0$. Program this method, experimenting with different values of $\bar{t}_i$ and print out your answers.
2. Can you apply the method of collocation to the boundary value problem given in Exercise 1 using a sum of piecewise linear Lagrange polynomials? Piecewise quadratic (cubic) polynomials? Explain in complete detail.

3.4 HERMITE INTERPOLATION

Hermite interpolation generalizes Lagrange interpolation by fitting a polynomial to a function f that not only interpolates f at each knot but also interpolates a given number of consecutive derivatives of f at each knot. In particular, given real numbers $x_1 < x_2 < \cdots < x_k$ and integers $m_1, m_2, \ldots, m_k$ all greater than zero, we can find a unique polynomial $p(x)$ of degree $m_1 + m_2 + \cdots + m_k - 1$ solving the interpolation problem

$$p^{(j)}(x_i) = f^{(j)}(x_i) \tag{1}$$

for all $j = 0, 1, \ldots, m_i - 1$ and $i = 1, 2, \ldots, k$. We say $p(x) - f(x)$ has a zero of order m_i at x_i, and we call $p(x)$ the *Hermite interpolate to* f at the points x_i, $i = 1, 2, \ldots, k$. We are asking more of p than in the Lagrange case; thus we must do more computational work to obtain it. Are there any advantages over simple Lagrange interpolation? The answer is yes, provided f is sufficiently smooth. In particular, Hermite interpolating polynomials give better fits to f when f is very smooth than do their Lagrange counterparts. Depending on the problem at hand the added computational work may not be warranted. In any event, they are a useful theoretical tool; so we prove

THEOREM EXISTENCE OF HERMITE INTERPOLATING POLYNOMIALS

Given real numbers $x_1 < x_2 < \cdots < x_k$ and positive integers $m_1, m_2, \ldots, m_k$, there exists a unique polynomial $p(x)$ of degree $m_1 + m_2 + \cdots + m_k - 1 = N$ or less which at each knot x_i, $i = 1, 2, \ldots, k$, solves the interpolation problem

$$p^{(j)}(x_i) = f^{(j)}(x_i),$$

where $j = 0, 1, \ldots, m_i - 1$ and where f is a function having $m_i - 1$ consecutive derivatives at x_i, $i = 1, 2, \ldots, k$.

Proof Let $N = m_1 + m_2 + \cdots + m_k - 1$. We seek a polynomial $p(x) = a_N x^N + a_{N-1} x^{N-1} + \cdots + a_0$ such that at each x_i, $i = 1, 2, \ldots, k$, $p^{(j)}(x_i) = f^{(j)}(x_i)$, where $j = 0, 1, 2, \ldots, m_i - 1$. We thus have $N+1$ unknowns a_N, $a_{N-1}, \ldots, a_0$ to solve for. Holding i fixed, the conditions $p^{(j)}(x_i) = f^{(j)}(x_i)$, where $j = 0, 1, \ldots, m_i - 1$ yield m_i linear equations in our N unknowns. For example $p''(x_2) = f''(x_2)$ gives us

$$N(N-1)x_2^{N-2}a_N + (N-1)(N-2)x_2^{N-3}a_{N-1} + \cdots + a_2 = f''(x_2).$$

Since we have m_i equations at each knot x_i, we accrue a total of $m_1 + m_2 + \cdots + m_k$ linear equations in our $N+1$ unknowns $a_N, a_{N-1}, \ldots, a_0$. But $m_1 + m_2 + \cdots + m_k = N+1$. Thus we have exactly as many equations as we have unknowns. If a unique solution to our problem is to exist, we must prove that the coefficient matrix of our linear system is nonsingular. Let us for simplicity write our equations in matrix notation $Ay = b$, where $y = (a_N, a_{N-1}, \ldots, a_0)^T$ and $b = (f(x_1), f'(x_1), \ldots, f^{(m_1-1)}(x_1), f(x_2), \ldots, f^{(m_k-1)}(x_k))^T$. It suffices to show that the system $Ay = 0$ has only the trivial solution $a_N = a_{N-1} = \cdots = a_0 = 0$; for then A is nonsingular. Thus without loss of generality we assume $f(x) \equiv 0$. But then we are seeking a polynomial $p(x)$ such that at each knot x_i, $p^{(j)}(x_i) = 0$ when $j = 0, 1, \ldots, m_i - 1$. However, it is well known that any polynomial of degree N having a root at $x = r$ and having l successive derivatives vanish at $x = r$ either has a repeated root of order $l+1$ at $x = r$ or is identically equal to zero. In particular $p(x) = \alpha(x-r)^{l+1}q(x)$, where $q(x)$ is a polynomial of degree $N-(l+1)$ or less and α is a constant, possibly equal to zero. Thus each x_i is a repeated root of order m_i so that $p(x)$ factors as

$$p(x) = \alpha(x-x_1)^{m_1}(x-x_2)^{m_2} \cdots (x-x_k)^{m_k},$$

where α is a constant. If $\alpha \neq 0$, then $\alpha(x-x_1)^{m_1}(x-x_2)^{m_2} \cdots (x-x_k)^{m_k}$ is a polynomial of degree $m_1 + m_2 + \cdots + m_k = N+1$, whereas $p(x)$ is a polynomial of degreee N. Thus we must have $\alpha = 0$; so that $p(x) \equiv 0$ when $f(x) \equiv 0$. In particular $a_N = a_{N-1} = \cdots = a_0 = 0$, and the system $Ay = 0$ has only the trivial solution. Thus A is nonsingular, as was to be proved. ■

It is worth noting that one can write down a specific formula for the Hermite polynomial with knots at $x_1 < x_2 < \cdots < x_k$ when $m_1 = m_2 = \cdots = m_k = 2$. In particular, let $l_i(x)$ be the Lagrange polynomial of degree $(k-1)$ solving the interpolation problem $l_i(x_j) = \delta_{ij}$, $i, j = 1, 2, \ldots, k$. Let

$$\phi_i(x) = [1 - 2l_i'(x_i)(x-x_i)]l_i^2(x)$$

and let

$$\psi_i(x) = (x-x_i)l_i^2(x).$$

The reader can check that

$$p(x)=\sum_{i=1}^{k} f(x_i)\phi_i(x)+\sum_{i=1}^{k} f'(x_i)\psi_i(x)$$

satisfies all required conditions; so that the polynomials $\phi_i(x)$, $i=1,2,\ldots,k$ and $\psi_i(x)$, $i=1,2,\ldots,k$, form a *basis* for the set of all such Hermite polynomials with knots at $x_1<x_2<\cdots<x_k$. (The reader should check that the set of ϕ_i's and ψ_i's is indeed linearly independent.)

One can estimate the *error* $\|f^{(j)}-p^{(j)}\|$ for $0\leqslant j\leqslant \text{minimum}\{m_i : 1\leqslant i \leqslant k\}$ in a manner similar to that used for Lagrange interpolation. For example, if $f\in C^{N+1}[a,b]$, we can prove that for each x in $[a,b]$ there exists a ζ in (a,b) such that

$$f(x)-p(x)=\frac{f^{(N+1)}(\zeta)}{(N+1)!}(x-x_1)^{m_1}(x-x_2)^{m_2}\cdots(x-x_k)^{m_k},$$

where $N=\sum_{i=1}^{k} m_i-1$. The key for obtaining this particular estimate is to let

$$E(x)=[f(x)-p(x)]-[f(\bar{x})-p(\bar{x})]\frac{w(x)}{w(\bar{x})},$$

where $\bar{x}$ is some point in (a,b) distinct from $x_1,x_2,\ldots,x_k$, and $w(x)=(x-x_1)^{m_1}(x-x_2)^{m_2}\cdots(x-x_k)^{m_k}$. One then uses Rolle's theorem, repeatedly counting the zeros of $E(x)$ and its derivatives to determine $E^{(N+1)}(\zeta)=0$, where ζ is some point in (a,b). To estimate $f^{(j)}(x)-p^{(j)}(x)$, proceed in an analogous fashion, first counting the zeros of y_i of $f^{(j)}(x)-p^{(j)}(x)$ and letting $w(x)=\Pi_i(y-y_i)$ in the formula for $E(x)$. Since we are most concerned with the case $k=2$ (i.e., two knots), we illustrate the aforementioned error analysis for this specific case, leaving the more general case to the reader (see Exercises).

To this end let $x_0=a$ and $x_1=b$ and suppose $p(x)$ is the quintic (degree 5) Hermite polynomial interpolating $f(x)$ and its first and second derivatives at a and at b. To estimate $f(x)-p(x)$ let

$$E(x)=[f(x)-p(x)]-[f(\bar{x})-p(\bar{x})]\frac{w(x)}{w(\bar{x})},$$

where $w(x)=(x-x_0)^3(x-x_1)^3$ and $\bar{x}$ is some point in (a,b) distinct from a and b. Since $E(x)$ vanishes at a, b, and $\bar{x}$, $E'(x)$ vanishes at $a<z_1<z_2<b$. To see this, invoke Rolle's theorem and note that $w'(x)$ vanishes at a and b, as do $f'(x)-p'(x)$. But then $E''(x)$ vanishes at $a<y_1<y_2<y_3<b$ (for the

same reason). Thus Rolle's theorem implies $E^{(6)}(\zeta)=0$ at some point ζ in (a,b). But

$$E^{(6)}(\zeta)=f^{(6)}(\zeta)-\frac{6![f(\bar{x})-p(\bar{x})]}{w(\bar{x})},$$

since $p^{(6)}(x)=0$; thus letting $x=\bar{x}$, we have

$$f(x)-p(x)=\frac{f^{(6)}(\zeta)}{6!}(x-a)^3(x-b)^3.$$

Therefore, since the maximum of $|(x-a)(x-b)|=(b-a)/2$,

$$\|f-p\|\leqslant\frac{\|f^{(6)}\|}{6!2^6}h^6,$$

where $h=b-a$. To estimate $f'(x)-p'(x)$, note first that $f'(x)-p'(x)$ vanishes at at least three points a, b, and z_1, $a<z_1<b$, of (a,b). This time let

$$E(x)=[f'(x)-p'(x)]-[f'(\bar{x})-p'(\bar{x})]\frac{w(x)}{w(\bar{x})},$$

where $w(x)=(x-a)^2(x-z_1)(x-b)^2$ and $\bar{x}$ is some point of (a,b) distinct from z_1, a, and b. Then $E(x)$ vanishes at $\bar{x}$, a, z_1, and b, which implies (again by Rolle's theorem) that $E'(x)$ vanishes at five points $a<y_1<y_2<y_3<b$. But then $E^{(5)}(\zeta)=0$, where ζ is some point (a,b). Since

$$E^{(5)}(\zeta)=f^{(6)}(\zeta)-[f'(\bar{x})-p'(\bar{x})]\frac{5!}{w(\bar{x})}=0,$$

we have, letting $x=\bar{x}$,

$$f'(x)-p'(x)=\frac{f^{(6)}(\zeta)}{5!}(x-a)^2(x-z_1)(x-b)^2.$$

Thus

$$\|f'-p'\|\leqslant\frac{\|f^{(6)}\|}{5!2^4}h^5.$$

Finally, to estimate $f''(x)-p''(x)$ observe that $f''(x)-p''(x)$ vanishes at at least four points $a<x_1<x_2<b$ of $[a,b]$. Choose $\bar{x}$ distinct from a,x_1,x_2,b

and let

$$E(x)=[f''(x)-p''(x)]-[f''(\bar{x})-p''(\bar{x})]\frac{w(x)}{w(\bar{x})},$$

where $w(x)=(x-a)(x-x_2)(x-x_3)(x-b)$. Arguing as before, we see that $E(x)$ vanishes at five points a, x_2, x_3, b, and $\bar{x}$; thus $E^{(4)}(\eta)=0$ at some η in (a,b). It follows that

$$\|f''-p''\|\frac{\|f^{(6)}\|}{4!2^4}h^4.$$

Now simply carry on in the same monotonous way, ending with an estimate of $\|f^{(5)}-p^{(5)}\|$. Clearly an inductive proof, if available, would be in better taste. See if you can produce one (see Exercise 3). In general, one can prove the following theorem.

ERROR ESTIMATE FOR HERMITE INTERPOLATION ON TWO KNOTS

Let $f\in C^{2m}[a,b]$ and let $p(x)$ be the Hermite polynomial of degree $2m-1$ interpolating f and its first through $(m-1)$st derivatives at a and b. Then

$$\|f^{(j)}-p^{(j)}\|\leqslant\frac{\|f^{(2m)}\|}{(2m-j)12^{2m-[j]}}h^{2m-j},\qquad 0\leqslant j\leqslant 2m-1,$$

where $[j]=j$ if j is even and $[j]=j+1$ if j is odd.

The previous analysis depends on $f\in C^{2m}[a,b]$. If f does not have this much smoothness, say $f\in C^k[a,b]$ $0\leqslant k\leqslant 2m-1$, error estimates are still available using existing spline analysis (see Chapter 4) or the Jackson theory of best approximation, as we did in the Lagrange case. The Peano kernal theorem is also applicable (see Davis (1963)). In general, one can prove that if $f\in C^k[a,b]$, there exists constants γ_j independent of m, a, and b such that

$$\|f^{(j)}-p^{(j)}\|\leqslant\gamma_j h^{k-j},\qquad 0\leqslant j\leqslant 2k-1.$$

EXERCISES

1. Under what conditions would Hermite interpolation be preferable to Lagrange interpolation, and conversely?
2. Find the Hermite polynomial $p(x)$ solving $p(x_i)=1$ and $p'(x_i)=0$, $0\leqslant i\leqslant 8$. Find that solving $p(x_i)=0$ and $p'(x_i)=1$, where $0\leqslant i\leqslant 8$.

3. Try to prove formula (8) using mathematical induction. You first must prove a generalized Rolle's theorem.

3.5 PIECEWISE HERMITE INTERPOLATION

The full Hermite polynomials of the preceding section, although interesting, are rather cumbersome for computational purposes. Moreover, they suffer the same instabilities as the full Lagrange interpolates to a function $f(x)$ (i.e., the Runge-Méray phenomenon) when the successive derivatives of f become very large and the knots are not very carefully placed. For this reason, as we have already seen in the cubic case, it is quite common to take recourse to piecewise Hermite polynomials, of which the piecewise cubic Hermite interpolates of Section 3.1 were a simple but practical example.

In the general case we start with a partition $\pi: a=t_0<t_1<\cdots<t_n=b$ and a positive integer m. On each subinterval $[t_{i-1},t_i]$, $1\leqslant i\leqslant n$, we construct the Hermite polynomial $p_i(x)$ of degree $2m-1$, solving the interpolation problem

$$\begin{aligned} p^{(j)}(t_{i-1}) &= f^{(j)}(t_{i-1}) \\ p^{(j)}(t_i) &= f^{(j)}(t_i), \qquad 0\leqslant j\leqslant m-1. \end{aligned} \tag{1}$$

The piecewise Hermite interpolate of degree $2m-1$ to f is given by

$$s(t)=p_i(t), \qquad t_{i-1}\leqslant t\leqslant t_i, \qquad 1\leqslant i\leqslant n. \tag{2}$$

It is clear from the constraints of (1) that $s\in C^{m-1}[a,b]$, since derivatives of p_{i-1} and p_i up through the $(m-1)$st match at t_i, $1<i<n$. In particular when $m=0$ we have simply the piecewise linear Lagrange polynomials that belong to $C[a,b]$, when $m=2$ we have the piecewise cubic Hermites that belong to $C^1[a,b]$, and when $m=3$ we obtain the piecewise quintic Hermites that belong to $C^2[a,b]$.

Moreover, the error estimates of Section 3.4 obtain. Thus when $f\in C^{2m}[a,b]$

$$\|f^{(j)}-s^{(j)}\|\leqslant\frac{\|f^{(2m)}\|}{(2m-j)!\,2^{2m-[j]}}h^{2m-j}, \qquad 0\leqslant j\leqslant 2m-1,$$

where for fixed f, $\|f^{(2m)}\|$ is a constant independent of the partition π, hence of h. It follows that $\|f^{(j)}-s^{(j)}\|\to 0$, $0\leqslant j\leqslant 2m-2$ as $h\to 0$ and the Runge-Méray phenomenon is avoided. In particular s and its successive derivatives up to order $2m-2$ converge uniformly to f in the uniform norm on $[a,b]$ with a specific rate of convergence.

If $f \in C^k[a,b]$, where k is some integer between 0 and $2m-1$, we have mentioned that one can use the theory of best approximation (Jackson's theorems) to obtain estimates for $\|f-s\|$. In general, we would expect

$$\|f-s\| \leqslant M_k h^k, \qquad k>1$$

and

$$\|f-s\| \leqslant M_o \omega(f,h), \qquad k=0,$$

where the numbers M_k are positive constants dependent on k and m but independent of h. We have already obtained such formulas for piecewise Lagrange polynomials. These methods indicate possible means of obtaining such formulas in the Hermite and piecewise Hermite cases. The reader is invited to exercise his ingenuity.

The *computation* of piecewise Hermite polynomials is important and is taken up in the next section. One method is by way of the cardinal basis functions. The theoretic underpinnings for this method are linear space notions, which we discuss now, leaving computational details for the following section. The basic idea is as follows: If we fix a partition π and choose an m, the set of piecewise Hermite polynomials we have been discussing forms a linear space which we call $H_2^m(\pi)$. That is,

DEFINITION $H_2^m(\pi) = H_{2m-1}(\pi)$

Given m and π, the set of all $(m-1)$-times continuously differentiable piecewise Hermite polynomials of degree $2m-1$ on successive intervals $[x_i, x_{i+1}]$ with knots from π is denoted $H_2^m(\pi)$ or $H_{2m-1}(\pi)$.

The m in the notation $H_2^m(\pi)$ indicates interpolation up through the $(m-1)$st derivative. The 2 indicates fitting on successive *pairs* of knots. Note that when either m or π changes, $H_2^m(\pi)$ changes. However, for fixed m and π it is clear that $H_2^m(\pi)$ forms a *linear space*, since the *sum* of any two functions from $H_2^m(\pi)$ is again in $H_2^m(\pi)$, as is any *scalar multiple* of a member of $H_2^m(\pi)$.

By constructing the cardinal basis for $H_2^m(\pi)$ we easily see $H_2^m(\pi)$ has dimension $m(n+1)$. A proof of this is left as an exercise. What we have said so far can be succinctly summarized in the parlance of Chapter 1. Let $\phi_1, \phi_2, \ldots, \phi_N$, and let $N = m(n+1)$ be a basis for $H_2^m(\pi)$. Then there is a unique

$$\tilde{f}(x) = c_1\phi_1(x) + c_2\phi_2(x) + \cdots + c_N\phi_N(x) \tag{3}$$

satisfying the *pure interpolatory constraints*

$$\tilde{f}^{(j)}(x_i) = f^{(j)}(x_i), \qquad 0 \leqslant i \leqslant n;\ 0 \leqslant j \leqslant m-1.$$

In particular $\tilde{f}(x)=s(x)$, and the error $\|f^{(j)}-\tilde{f}^{(j)}\|$ is given by (3).

EXERCISES

1. Prove that $H_2^m(\pi)$ has dimension $m(n+1)$.
2. Find a function $f(x)$ whose piecewise cubic Hermite interpolate $s(x)$ belongs to $C^2[x_0,x_N]$.
3. Let $\mathfrak{B}=\{\phi_1,\phi_2,\ldots,\phi_N\}$ be a basis for $H_2^m(\pi)$. Is $\mathfrak{B}$ a *Haar system* (see Section 1.6) with respect to $C[x_0,x_N]$? Prove your answer.

3.6 COMPUTATION OF PIECEWISE HERMITE POLYNOMIALS

In this section we delineate two methods for computing the piecewise Hermite polynomials of the previous section. Each approach derives from a different basis for $H_2^m(\pi)$. In particular, given a partition $\pi: a=x_0<x_1<\cdots<x_n=b$ of $[a,b]$ and a fixed positive integer m, we seek the piecewise polynomial $s(x)$ of degree $2m-1$ on successive intervals $[x_{i-1},x_i]$, $1\leqslant i\leqslant n$, solving the interpolation problem

$$D^k s(x_i)=D^k f(x_i), \qquad 0\leqslant k\leqslant m-1;\ 0\leqslant i\leqslant n, \tag{1}$$

where $f(x)$ is given.

METHOD I (THE DIRECT METHOD)

Let

$$s(x)=\begin{cases} a_{10}+a_{11}x+\cdots+a_{1,2m-1}x^{2m-1}=p_1(x), & x_0\leqslant x\leqslant x_1\\ a_{20}+a_{21}x+\cdots+a_{2,2m-1}x^{2m-1}=p_2(x), & x_1\leqslant x\leqslant x_2\\ \vdots & \\ a_{n0}+a_{n1}x+\cdots+a_{n,2m-1}x^{2m-1}=p_n(x), & x_{n-1}\leqslant x\leqslant x_n \end{cases} \tag{2}$$

and force

$$D^k s(x_i)=D^k f(x_i), \qquad 0\leqslant k\leqslant m-1;\ 0\leqslant i\leqslant n,$$

to solve for the a_{ij}'s. We thus must solve *n systems of 2m equations in 2m unknowns*, where each of the n coefficient matrices is full. To take a simple example, suppose $m=2$, so that $2m-1=3$. Then

$$p_i(x)=a_{i3}x^3+a_{i2}x^2+a_{i1}x+a_{i0}.$$

To solve for a_{i3}, a_{i2}, a_{i1}, a_{i0}, force

$$\begin{cases} D^k p_i(x_{i-1})=D^k f(x_{i-1})\\ D^k p_i(x_i)=D^k f(x_i), \qquad k=0,1. \end{cases} \tag{3}$$

The coefficient matrix for this system is

$$\begin{bmatrix} x_{i-1}^3 & x_{i-1}^2 & x_{i-1} & 1 \\ 3x_{i-1}^2 & 2x_{i-1} & 1 & 0 \\ x_i^3 & x_i^2 & x_i & 1 \\ 3x_i^2 & 2x_i & 1 & 0 \end{bmatrix}.$$

In this case we would have to solve n 4×4 linear systems. For large n, this could prove bothersome. Method II is preferable.

METHOD II (THE CARDINAL BASIS METHOD)

Method II generalizes the method of Section 3.1. In particular we must first find the cardinal basis functions $\phi_{ik}(x)$, $0 \leqslant i \leqslant n$ and $0 \leqslant k \leqslant m-1$, solving the interpolation problem

$$D^l \phi_{ik}(x_j) = \delta_{kl}\delta_{ij}, \tag{4}$$

$0 \leqslant k, 1 \leqslant m-1$, $0 \leqslant i,j \leqslant n$, and $D^l = d^l/dx^l$, where each of the ϕ_{ik}'s is a piecewise Hermite polynomial of degree $2m-1$ between the knots. Once having found the ϕ_{ik}'s, we can immediately write down $s(x)$ using

$$s(x) = \sum_{k=0}^{m-1} \sum_{i=0}^{n} f^{(k)}(x_i)\phi_{ik}(x). \tag{5}$$

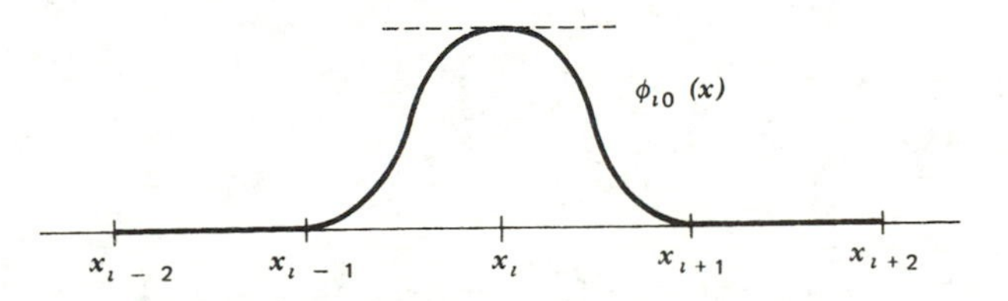

Figure 3.6 Graph of $\phi_{i0}(x)$.

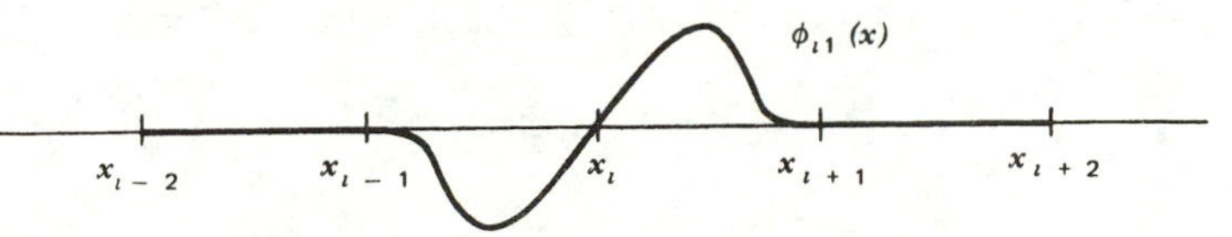

Figure 3.7 Graph of $\phi_{i1}(x)$.

The method is best illustrated by our previous simple example, the piecewise cubic Hermites. Some details were omitted earlier; we include them now in all their laborious entirety for those faced with hand computation. It is of course simplest to put the system (4) on the computer and let the machine do the work.

Since we are looking for piecewise cubics, $m=2$ in system (4) and $k=0$ or $k=1$. Thus for each i, ϕ_{ik} is either ϕ_{i0} or ϕ_{i1}. We begin by choosing *interior knots* x_i (i.e., $1 \leqslant i < n$) and actually computing $\phi_{i0}(x)$, and $\phi_{i1}(x)$. Note that these two functions interpolate the following data:

$$\begin{cases} \phi_{i0}(x_j) = \delta_{ij} \\ \phi'_{i0}(x_j) = 0 \end{cases}$$

and

$$\begin{cases} \phi_{i1}(x_j) = 0 \\ \phi'_{i1}(x_j) = \delta_{ij} \qquad 0 \leqslant j \leqslant n. \end{cases} \tag{6}$$

In particular $\phi_{i0}(x)$ is 1 when $x = x_i$, but it is zero at all other knots. Moreover, $\phi_{i0}(x)$ has zero slope at each knot and is a cubic polynomial between each successive pair of knots. Also, since $s(x)$ interpolates zero data on all intervals outside $[x_{i-1}, x_{i+1}]$, $\phi_{i0}(x) \equiv 0$ *for all* $x \geqslant x_{i+1}$ *and* $x \leqslant x_{i-1}$. (This fact is very important later in the book.) Knowing the general shape of a cubic, ϕ_{i0} is easily graphed (see Figure 3.6). Analogous reasoning applies to $\phi_{i1}(x)$. In particular $\phi_{i1}(x) = 0$ at all knots x_j whereas $\phi'_{i1}(x)$ is 1 when $x = x_i$ and is zero at all other knots. Again, since $\phi_{i1}(x)$ satisfies zero data outside the interval $[x_{i-1}, x_{i+1}]$, $\phi_{i1}(x)$ *must vanish identically outside* $[x_{i-1}, x_{i+1}]$. A graph of $\phi_{i1}(x)$ appears in Figure 3.7. The graphs of the ϕ_{ij}'s make it easy to see that (5) is certainly correct. We now compute ϕ_{i0} and ϕ_{i1}, beginning with $\phi_{i0}(x)$, which has the form

$$\phi_{i0}(x) = \begin{cases} a_{i3}x^3 + a_{i2}x^2 + a_{i1}x + a_{i0}, & x_{i-1} \leqslant x \leqslant x_i \\ a_{i+1,3}x^3 + \cdots + a_{i+1,0}, & x_i \leqslant x \leqslant x_{i+1} \\ 0, & \text{elsewhere.} \end{cases}$$

But since ϕ_{i0} has a double zero at x_{i+1}, we know that

$$\phi_{i0}(x) = (x - x_{i+1})^2(\alpha x + \beta), \qquad x_i \leqslant x \leqslant x_{i+1}.$$

Forcing $\phi_{i0}(x_i) = 1$ and $\phi'_{i0}(x_i) = 0$ in the foregoing equation, we find

$$\phi_{i0}(x) = \frac{(x_{i+1} - x)^2}{(x_{i+1} - x_i)^3}[2(x_{i+1} - x) + (x_{i+1} - x_i)], \qquad x_i \leqslant x \leqslant x_{i+1}.$$

An analogous argument for $x_{i-1} \leq x \leq x_i$ serves to prove

$$\phi_{i0}(x) = \begin{cases} \dfrac{(x-x_{i-1})^2}{(x_i-x_{i-1})^3}[2(x_i-x)+(x_i-x_{i-1})], & x_{i-1} \leq x \leq x_i \\ \dfrac{(x_{i+1}-x)^2}{(x_{i+1}-x_i)^3}[2(x_{i+1}-x)-(x_{i+1}-x_i)], & x_i \leq x \leq x_{i+1} \\ 0, \quad \text{elsewhere.} \end{cases} \tag{7}$$

Now note that $\phi_{i1}(x)$ has a double zero at x_{i-1} and a single zero at x_i. Thus $\phi_{i1}(x) = \alpha(x-x_{i-1})^2(x-x_i)$ on the interval $[x_{i-1}, x_i]$. Forcing $\phi'_{i1}(x_i) = 1$ we find $\alpha = 1/(x_i - x_{i-1})^2$. Proceeding analogously on $[x_i, x_{i+1}]$, we find

$$\phi_{i1}(x) = \begin{cases} \dfrac{(x-x_{i-1})^2(x-x_i)}{(x_i-x_{i-1})^2}, & x_{i-1} \leq x \leq x_i \\ \dfrac{(x-x_{i+1})^2(x-x_i)}{(x_{i+1}-x_i)^2}, & x_i \leq x \leq x_{i+1} \\ 0, & \text{elsewhere.} \end{cases} \tag{8}$$

We now can write down $s(x)$. Namely,

$$s(x) = \sum_{i=0} [f(x_i)\phi_{i0}(x) + f'(x_i)\phi_{i1}(x)], \tag{9}$$

where ϕ_{i0} and ϕ_{i1} are given in (7) and (8) with appropriate deletions at $i=0$ and $i=n$. It is now easy to see how the method extends for arbitrary m. Some computational work is involved in each case to obtain the basis functions ϕ_{ik}, but once they are obtained, they can be stored, and no further inversion of linear systems is involved. Exercise 1 requires the computation and graphing of ϕ_{i0}, ϕ_{i1}, and ϕ_{i2} when $m=3$. Working this laborious problem should aid in your understanding of the method.

EXERCISES

1. Let $m=2$ and let $s(x)$ be the piecewise quintic Hermite polynomial with four knots $x_1 < x_2 < x_3 < x_4$ solving

$$D^k s(x_i) = D^k f(x_i), \qquad 1 \leq i \leq 4;\ 0 \leq k \leq 2,$$

where $f(x) = x^8 - 1$.

(a) Compute $s(x)$ using Method I.

(b) Compute the basis quintic Hermite polynomials $\phi_{i0}(x)$, $\phi_{i1}(x)$, and $\phi_{i2}(x)$.
(c) Compute $s(x)$ using Method II.
(d) Estimate $\|f-s\|$.
(f) Let $p(x)$ be the Lagrange polynomial of degree 3 solving $p(x_i) = f(x_i)$, $1 \leqslant i \leqslant 4$. Estimate $\|f-p\|$. Which is smaller, $\|f-s\|$ or $\|f-p\|$? What happens if you introduce more knots?

2. What is the dimension of $H_2^3(\pi)$? Prove your answer.
3. Each of the methods of this section amounts to computing the piecewise Hermite polynomials with respect to a particular basis $\mathfrak{B}$. Find the basis associated with each method.
4. Suppose you would like a piecewise quintic Hermite approximate to a given function f but do not know either the first or the second derivatives of f. How could you determine such an approximate uniquely and estimate the error? Explain.

3.7 THE HERMITE–BIRKOFF INTERPOLATION PROBLEM

The Lagrange and Hermite interpolates to a function $f(x)$ have the property that they fit a unique polynomial $p(x)$ degree $n-1$ to m_i consecutive pieces of data

$$D^j f(x_i) = D^j p(x_i), \qquad 0 \leqslant j \leqslant m_i - 1$$

on $f(x)$ at each of the knots $x_1, x_2, \ldots, x_k$, where $n = \sum_{i=1}^{k} m_i$ and each m_i is an integer whose value is 1 or greater. The possibility of generalization immediately arises. Namely, can we find a unique polynomial $p(x)$ of degree $n-1$ fitting any n pieces of data on $f(x)$ at a set of knots $x_1, x_2, \ldots, x_k$. In particular can we find a unique $p(x)$ of degree $n-1$ such that

$$D^k f(x_i) = D^k p(x_i), \qquad k = i_1, i_2, \ldots, i_{n_i}$$

for each $i = 1, 2, \ldots, k$; $\{i_1, i_2, \ldots, i_{n_i}\}$ are integers with $0 \leqslant i_1 < i_2 < i_3 < \cdots < i_{n_i}$; and the cardinality of the set $\{i_j : i = 1, 2, \ldots, k \text{ and } j = n_1, n_2, \ldots, n_i\}$ is n. This question is known as *the Hermite–Birkoff interpolation problem* (HB problem). In general the answer is *NO*! For example, there does not exist a unique polynomial $p(x)$ of degree 2 solving the interpolation problem $p(-1) = 0$, $p(1) = 0$, and $p'(0) = 1$. If $p(x)$ exists, then since $n = 3$, $n-1 = 2$ and $p(x) = ax^2 + bx + c$. Thus $p'(x) = 2ax + b$. Since $p'(0) = 0$, $b = 0$ and $p(x) = ax^2 + c$. The two conditions $p(1) = p(-1) = 0$ give us the single linear equation.

$$a + c = 0$$

which has infinitely many solutions. Thus there are infinitely many polynomials of degree 2 solving this interpolation problem. Can we find conditions under which the Hermite-Birkoff problem has a unique solution? Certainly! One possible condition is that the problem be m-poised as defined below. We first give

DEFINITION 3.1.

A matrix $E=(E_{ij})$, $E_{ij}=0$ or 1, $1 \leqslant i \leqslant r$ and $1 \leqslant j \leqslant s$ is called an *incidence matrix* if each row of E and the last column of E contain some nonzero element. Let $e=\{(i,j): E_{ij}=1\}$. Couched in the language of incidence matrices, the Hermite-Birkhoff problem is: given distinct real numbers $x_1 < x_2 < \cdots < x_k$, does there exist a unique polynomial $p(x)$ of degree $n-1$ for which $p^{(j)}(x_i)=f^{(j)}(x_i)$ for $(i,j) \in e$ and $n=\sum_{i,j} E_{ij}$?

For Lagrange interpolation we have the incidence matrix

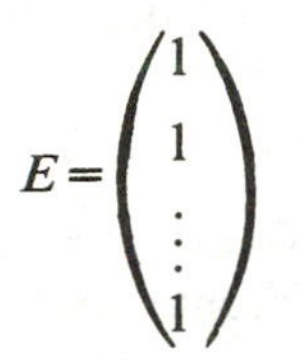

$$E=\begin{pmatrix} 1 \\ 1 \\ \vdots \\ 1 \end{pmatrix}$$

with one row for every point. For Hermite interpolation on three points $x_1 < x_2 < x_3$ with $m_1=2$, $m_2=4$, and $m_3=5$ we have

$$E=\begin{pmatrix} 1 & 1 & 0 & 0 & 0 \\ 1 & 1 & 1 & 1 & 0 \\ 1 & 1 & 1 & 1 & 1 \end{pmatrix}.$$

DEFINITION 3.2.

The HB problem is *normal* provided it can be solved uniquely by a polynomial of degree $n-1$.

It is clear that a necessary condition for a normal HB problem is that $s < n$. That is, if $s \geqslant n$ we can arrange conditions so that a unique polynomial of degree $n-1$ satisfying the HB problem does not exist. It is also clear that the HB problem is normal if and only if the only polynomial $p(x)$ of degree $n-1$ solving the interpolation problem

$$D^k p(x_i)=0, \qquad E_{ik}=1$$

is $p \equiv 0$. This leads to:

DEFINITION 3.3.

The HB problem is *m-poised* (m is a positive integer) if and only if the only polynomial p of degree $m-1$ solving the interpolation problem $D^k p(x_i)=0$ when $E_{ik}=1$ is $p\equiv 0$.

Thus the HB problem is normal if and only if it is n-poised. The same problem is m-poised if and only if $m \leqslant n$ and if the problem is m-poised it is also m'-poised for each $1 \leqslant m' \leqslant m$. One can prove:

HERMITE–BIRKHOFF POLYNOMIAL EXISTENCE THEOREM.

The HB problem is m-poised if and only if the $n\times m$ matrix

$$\frac{x_i^{k-j}}{(k-j)!}$$

has rank m, where $k=0,1,\ldots,m-1$ indicates the column and each (i,j) in e determines a row.

The interested reader is referred to the papers of Schoenberg (1966) Polya (1931) and Ferguson (1969) for results on existence and uniqueness and to the (1969) paper of Jerome and Varga for error estimates. In general, error estimates depend on Rolle's theorem, thus on the spacing of the knots $x_{i_1}, x_{i_2}, \ldots, x_{i_k}$ at which the polynomial $p(x)$ interpolates $f(x)$.

REFERENCES

Cheney, E. W. *Approximation theory*. McGraw-Hill, New York, 1966.

Ferguson, David. The question on uniqueness for G. D. Birkhoff interpolation problems. *J. of Approx. Theory* **2**, pp. 1–28 (1969).

Jerome, J. W. and R. S. Varga. Generalizations of spline functions and applications to nonlinear boundary value and eigenvalue problems. In *Theory and application of spline functions*, Academic Press, New York, 1969.

Polya, G. Bemerkungen zur Interpolation und zur Naherungstheorie der Balken biegung. *Z. Angew. Math. u. Mech.*, **11** (1931), pp. 445–449.

Schoenberg, I. J. On Hermite-Birkhoff interpolation, M.R.C. Technical Report 659, May, 1966.

Timan, A. F. *Theory of approximation of functions of a real variable*. Moscow, 1963; translation, Macmillan, New York.

4

POLYNOMIAL SPLINES AND GENERALIZATIONS

4.1 INTRODUCTION

We now turn to the study of approximating functions

$$\tilde{f}(t) = c_1\phi_1(t) + c_2\phi_2(t) + \cdots + c_N\phi_N(t)$$

that satisfy sets of *mixed* (i.e., both interpolatory and smoothness) *constraints*. The functions $\phi_1, \phi_2, \ldots, \phi_N$ will be linearly independent on the interval $[a,b]$ and will span an N-dimensional subspace X_N of $C[a,b]$. Usually we are able to suppress the smoothness constraints by building the desired smoothness into the basis $\mathfrak{B} = \{\phi_1, \phi_2, \ldots, \phi_N\}$, leaving ourselves with only a set of pure interpolatory constraints to fulfill. An example of this imbedding of smoothness into the basis is provided by the B splines of the next and latter sections. At other times we simply start with the space X_N and ask what kind of interpolation problems can be uniquely solved by members of this linear space. This is our approach in developing the polynomial splines. It then behooves the reader to discover the underlying structure of the space X_N and to realize that he is uniquely solving a mixed constraint problem, not a pure interpolation problem. All these techniques are best illustrated by an example, and the cubic splines are just such an example.

4.2 CUBIC SPLINES

A cubic spline function is a piecewise cubic polynomial that is twice continuously differentiable. Such functions arise in a natural way in trying to improve on the piecewise cubic Lagrange and piecewise cubic Hermite interpolates $s_3(t) \in C[a,b]$ and $\tilde{s}(t) \in C^1[a,b]$, respectively, to a given function $f(t)$. We have seen that these functions exist and that

$$\|f - s_3\| \leqslant \frac{\|f^{(4)}\|}{16} h^4$$

while

$$\|f - \tilde{s}\| \leqslant \frac{\|f^{(4)}\|}{96} h^4.$$

It is only natural to ask if one can do anything better with piecewise cubics. In particular, does there exist a twice continuously differentiable piecewise cubic polynomial $s(t)$ interpolating values of $f(t)$ at the knots t_i, $0 \leqslant i \leqslant n$, of a partition π of $[a,b]$ that is an equally good approximate to $f(t)$? The answer is "yes." There are several such $s(t)$'s, and they all belong to the collection $S_3(\pi)$ of *cubic spline functions* defined by

DEFINITION CUBIC POLYNOMIAL SPLINES

The space $S_3(\pi)$ is the set of all functions $s(t) \in C^2[a,b]$ that reduce to cubic polynomials on each subinterval (t_i, t_{i+1}), $0 \leqslant i \leqslant n-1$, of $[a,b]$.

The space $S_3(\pi)$ is a linear space (why?), and since it contains the set of all cubic polynomials, there are infinitely many functions in $S_3(\pi)$. However, we shall prove that there exists a *unique* function $s(t)$ in $S_3(\pi)$ satisfying the pure interpolatory constraints

$$\begin{aligned} &s'(t_0) = f'(t_0) \\ &s(t_i) = f(t_i), \qquad 0 \leqslant i \leqslant n \\ &s'(t_n) = f'(t_n). \end{aligned} \tag{1}$$

This being the case, $s(t)$ is referred to as *the cubic spline interpolate to* $f(t)$.

We have already replaced a mixed constraint problem with a pure interpolatory constraint problem by seeking $s \in C^2[a,b]$ satisfying (1). That is, we have insisted that the smoothness be built into $s(t)$ a priori. An equivalent approach to the cubic spline (see the book of Ahlberg, Nilson and Walsh (1967)) is to seek a piecewise cubic polynomial $s(t)$ satisfying (1) plus the additional $3(n-1)$ smoothness constraints

$$s^{(j)}(t_1)=f^{(j)}(t_i), \qquad 1 \leqslant i \leqslant n-1; j=0,1,2.$$

To establish the existence of $s(t)$ for evenly spaced knots $x_i = x_0 + i(b-a)/n$, we *construct* it. In particular, we introduce four additional knots $t_{-2} < t_{-1} < t_0$ and $t_{n+2} > t_{n+1} > t_n$, and the functions $B_i(t)$ defined by

$$B_i(t) = \frac{1}{h^3}\begin{cases} (t-t_{i-2})^3, & \text{if } t \in [t_{i-2}, t_{i-1}] \\ h^3 + 3h^2(t-t_{i-1}) + 3h(t-t_{i-1})^2 - 3(t-t_{i-1})^3, & \text{if } t \in [t_{i-1}, t_i] \\ h^3 + 3h^2(t_{i+1}-t) + 3h(t_{i+1}-t)^2 - 3(t_{i+1}-t)^3, & \text{if } t \in [t_i, t_{i+1}] \\ (t_{i+2}-t)^3, & \text{if } t \in [t_{i+1}, t_{i+2}] \\ 0, & \text{otherwise,} \end{cases} \tag{2}$$

which is graphed in Figure 4.1. The reader will note by simply substituting into (2) that each of the functions $B_i(t)$ is twice continuously differentiable on the entire real line, that

$$B_i(t_j) = \begin{cases} 4 & \text{if } \ j=i \\ 1 & \text{if } \ j=i-1 \quad \text{or} \quad i+1 \\ 0 & \text{if } \ j=i+1 \quad \text{or} \quad j=i-1, \end{cases}$$

and that $B_i(t) \equiv 0$ for $t \geqslant t_{i+2}$ and $t \leqslant t_{i-2}$. It has been proved by I. J. Schoenberg (1966) that the B splines are the unique nonzero splines of smallest compact support with knots at $t_{-2} < t_{-1} < \cdots < t_n < t_{n+1} < t_{n+2}$. That is, any (cubic) spline $s(t)$ with these knots that vanishes identically outside every interval (t_{j-1}, t_{j+2}) must be identically equal to zero. Moreover, since each $B_i(t)$ is also a piecewise cubic with knots at π, each

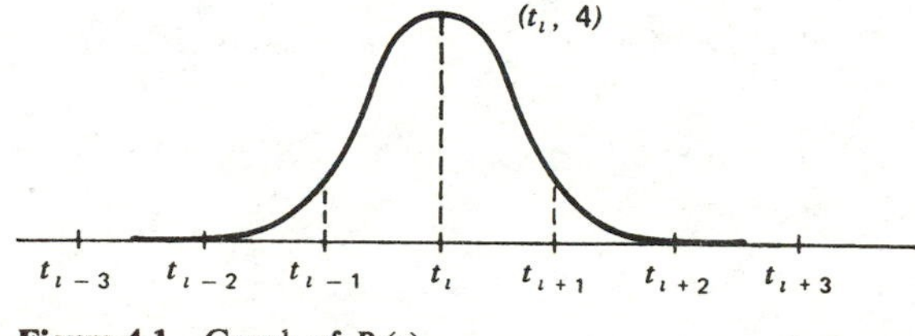

Figure 4.1 Graph of $B_i(t)$.

$B_i(t) \in S_3(\pi)$. To compute $s(t)$ we use Table 4.1 giving values of $B_i(t)$ and its derivatives at the knots. Since $B_i(t)$ and its derivatives vanish at all other knots, these values are omitted from our table.

TABLE 4.1

	t_{j-2}	t_{j-1}	t_j	t_{j+1}	t_{j+2}
$B_j(t)$	0	1	4	1	0
$B_j'(t)$	0	$\frac{3}{h}$	0	$-\frac{3}{h}$	0
$B_j''(t)$	0	$\frac{6}{h^2}$	$-\frac{12}{h^2}$	$\frac{6}{h^2}$	0

Let $\mathfrak{B} = \{B_{-1}, B_0, \ldots, B_{n+1}\}$ and let $B_3(\pi) = \text{span}\ \mathfrak{B}$. The functions $\mathfrak{B}$ are linearly independent on $[a,b]$ (see Exercises); thus $B_3(\pi)$ is $(n+3)$-dimensional. We prove the following theorem.

THEOREM 4.1

There exists a unique function $s(t)$ in $B_3(\pi)$ solving the interpolation problem (1).

Proof. Let $s(t)$ belong to $B_3(\pi)$. Then

$$s(t) = x_{-1}B_{-1}(t) + x_0B_0(t) + \cdots + x_{n+1}B_{n+1}(t). \tag{3}$$

Forcing $s(t)$ to satisfy the constraints (1) in the order given, we have

$$s'(t_0) = x_{-1}B'_{-1}(t_0) + x_0B'_0(t_0) + \cdots + x_{n+1}B'_{n+1}(t_0) = f'(t_0) \tag{4a}$$

$$s(t_i) = x_{-1}B_{-1}(t_i) + x_0B_0(t_i) + \cdots + x_{n+1}B_{n+1}(t_i) = f(t_i) \tag{4b}$$

$$s'(t_n) = x_{-1}B'_{-1}(t_n) + x_0B'_0(t_n) + \cdots + x_{n+1}B'_{n+1}(t_n) = f'(t_n), \qquad 0 \leq i \leq n. \tag{4c}$$

This comprises a set of $n+3$ linear equations $Ax = b$, where $x = (x_{-1}, x_0, \ldots, x_{n+1})^T$, $b = (f'(t_0), f(t_0), f(t_1), \ldots, f(t_n), f'(t_n))^T$, and the coefficient matrix A (from Table 4.1) is given by

$$A=\begin{bmatrix} -\frac{3}{h} & 0 & \frac{3}{h} & 0 & 0 & 0 & \cdots & & \\ 1 & 4 & 1 & 0 & 0 & 0 & \cdots & & \\ 0 & 1 & 4 & 1 & 0 & 0 & \cdots & & \\ 0 & 0 & 1 & 4 & 1 & 0 & \cdots & & \\ \vdots & & & & & & & & \\ 0 & 0 & 0 & \cdots & 0 & 1 & 4 & 1 \\ 0 & 0 & 0 & \cdots & 0 & -\frac{3}{h} & 0 & \frac{3}{h} \end{bmatrix}.$$

Since the matrix A is diagonally dominant with strict diagonal dominance on all but the first and last rows, it follows from Gershgorin's theorem (see Varga's book (1962)) that A is nonsingular. Hence the system (4) has a unique solution. ■

Note that this theorem not only gives us the existence of a unique spline in $S_3(\pi)$ solving (1), it also gives us a *numerical algorithm* for computing $s(t)$. In particular, $s(t)$ is given by (3), whose coefficients x_i, $0 \leqslant i \leqslant n$ are simply the solution of the linear system $Ax=b$. The *banded* structure of the matrix A is a particularly desirable feature of this algorithm and derives from our choice of basis.

Since every function from $B_3(\pi)$ is automatically in $S_3(\pi)$ (why?), the theorem just stated guarantees the existence of at least one cubic spline solving (1). It does not guarantee uniqueness. It would if we knew that $B_3(\pi)=S_3(\pi)$, however, and that is what we prove next.

THEOREM 4.2 $B_3(\pi)=S_3(\pi)$.

Proof. It is clear from the definition of $B_3(\pi)$ that $B_3(\pi) \subset S_3(\pi)$. To prove the converse containment, let $f(t) \in S_3(\pi)$. Then $f'(t_0)$, $f'(t_n)$, and $f(t_i)$, $0 \leqslant i \leqslant n$, are all defined. Let $s(t) \in B_3(\pi)$ be the unique spline from that set solving the interpolation problem (1). Then $f(t)-s(t)=g(t)$ vanishes at each t_i, $0 \leqslant i \leqslant n$, as do $g'(t_0)$ and $g'(t_n)$. Since both f and s belong to $C^2[a,b]$, $g(t) \in C^2[a,b]$. It follows from Rolle's theorem that $g'(t)$ has at least n zeros at y_i, $t_i<y_i<t_{i+1}$ in addition to the zeros at t_0 and t_n. Thus $g'(t)$ has at least $n+2$ zeros, which means that $g''(t)$ has at least $n+1$ zeros z_i, $t_0<z_0<y_0$, $y_0<z_1<y_1$, $y_1<z_2<y_2,\ldots,y_{n-1}<z_n<t_n$.

But $g''(t)$ is a piecewise linear Lagrange polynomial with knots at π. It follows from this smooth, piecewise linearity of $g''(t)$ together with the distribution of the zeros z_i, $0 \leqslant i \leqslant n$, of $g''(t)$ that $g''(t) \equiv 0$ on $[t_0, t_n]$. But then $g(t) = \alpha t + \beta$, and since $g(t_0) = g(t_n) = 0$, $g(t) \equiv 0$. Thus $s(t) = f(t) \in B_3(\pi)$ and $S_3(\pi) \subset B_3(\pi)$. ■

As a by-product of these two theorems, we have proved Corollaries 1 and 2.

COROLLARY 1

$\dim S_3(\pi) = n+3$ and $\mathfrak{B} = \{B_{-1}, B_0, \ldots, B_{n+1}\}$ is a basis for $S_3(\pi)$.

COROLLARY 2

There exists a unique cubic spline $s(t)$ solving the interpolation problem (1). The function s is called *the cubic spline interpolate to f.*

The unique cubic interpolate to a given function $f(t)$ is not the only cubic polynomial interpolating $f(t)$ at the knots t_i, $0 \leqslant i \leqslant n$. There are, in fact, infinitely many such splines (see Exercises). For example, one can prove in a manner completely analogous to the proof of Theorem 4.1 that there exists a unique spline $\bar{s}(t)$ given by (3) solving the interpolation problem

$$\begin{aligned} \bar{s}''(t_0) &= f''(t_0) \\ \bar{s}(t_i) &= f(t_i), \qquad 0 \leqslant i \leqslant n \\ \bar{s}''(t_n) &= f''(t_n). \end{aligned} \tag{5}$$

The spline $\bar{s}(t)$ is called the *natural cubic spline interpolate to $f(t)$*. This name is chosen because such splines are mathematical models of draftmen's splines. The matrix $\bar{A}$ used in proving the existence of $\bar{s}(t)$ and in computing $\bar{s}(t)$ differs from the matrix A only in the first and last rows, which are modified to interpolate second instead of first derivatives. These entries are easily provided by Table 4.1. In general, given any $n+3$ distinct real numbers $\bar{t}_1 < \bar{t}_2 < \cdots < \bar{t}_{n+3}$ and any $n+3$ positive integers n_i, one can ask whether there exists a cubic spline $\hat{s}(t)$ solving the interpolation problem

$$\hat{s}^{(n_i)}(\bar{t}_i) = f^{(n_i)}(\bar{t}_i), \qquad 1 \leqslant i \leqslant n+3.$$

Often, of course, the answer is no, and often the answer is a unique but totally uninteresting yes. One important case, however, where the answer to the problem is yes involves the question of how to search for the cubic spline interpolate to $f(t)$, knowing neither $f'(t_0)$ nor $f'(t_n)$. One possibility is to construct the two cubic Lagrange interpolates $\lambda_0(t)$ and $\lambda_1(t)$ to $f(t)$ at the extreme end knots $t_0<t_1<t_2<t_3$ and $t_{n-3}<t_{n-2}<t_{n-1}<t_n$. Then compute $s_L(t)$ given by (3) solving

$$s_L'(t_0)=\lambda_0'(t_0) \tag{6a}$$

$$s_L(t_i)=f(t_i), \qquad 0\leqslant i\leqslant n \tag{6b}$$

$$s_L'(t_n)=\lambda_n'(t_n). \tag{6c}$$

Existence and uniqueness of $s_L(t)$ solving (6) is guaranteed by our previous theory, since we have merely changed the right-hand side of the system $Ax=b$ of Theorem 4.1. What will require reaccessing is the error estimate $\|f-s\|$. We now need $\|f-s_L\|$. Common sense dictates that if $\|f-s\|$ is $o(h^4)$ (and we shall prove when this is the case) and if $\max|f(t)-\lambda_0(t)| = o(h^4)=\max|f(t)-\lambda_n(t)|$ on the intervals $[t_0,t_3]$ and $[t_{n-3},t_n]$, respectively, then since $|f(t)-s_L(t)|\leqslant|f(t)-\lambda_0(t)|+|\lambda_0(t)-s(t)|$ with similar inequalities holding at $[t_{n-3},t_n]$, we should be able to prove that $\|f-s_L\|=o(h^4)$ if $\|f-s\|=o(h^4)$. Swartz and Varga (1972) have written a paper on just this topic. They call the splines s_L cubic *Lagrangian splines*.

One can prove (see Hall (1968)) that when $f\in C^4[a,b]$

$$\|f-s\|\leqslant\frac{5}{384}\|f^{(4)}\|h^4.$$

Thus it turns out that the cubic spline interpolate with evenly spaced knots to a given function $f(t)$ more than competes with the cubic Lagrange and cubic Hermite interpolates $s_3(t)$ and $\tilde{s}(t)$, respectively, as an approximate to $f(t)$ (see Exercise 5). It has other highly desirable variation diminishing properties as well, which we discuss. We must also show how we obtained the B splines, $B_i(t)$, to permit the reader to generalize our constructive methods to higher order splines.

EXAMPLE

As a simple example we compute the cubic spline interpolate $s(t)$ to $f(t)=5t+1$ on the interval $[0,1]$ with evenly spaced knots $t=i/5$, $0\leqslant i\leqslant 5$. Letting $t_{-2}=-2/5$,

$t_{-1} = -1/5$, $t_6 = 6/5$, and $t_7 = 7/5$, we have

$$s(t) = x_{-1}B_{-1}(t) + x_0B_0(t) + \cdots + x_{n+1}B_{n+1}(t). \tag{7}$$

Since $h = 1/5$, $1/h = 5$. Moreover

$$s'(0) = f'(0) = 5$$

$$s\left(\frac{i}{5}\right) = f\left(\frac{i}{5}\right) = i + 1, \qquad 0 \leqslant i \leqslant n$$

$$s'(1) = f'(1) = 5,$$

Thus the system $Ax = b$ is simply

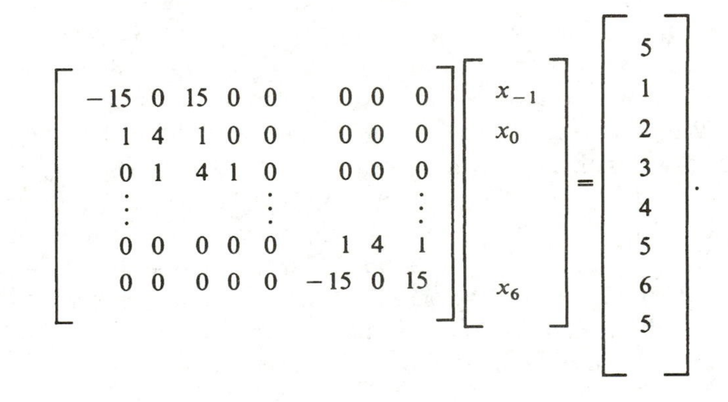

$$\begin{bmatrix} -15 & 0 & 15 & 0 & 0 & & 0 & 0 & 0 \\ 1 & 4 & 1 & 0 & 0 & & 0 & 0 & 0 \\ 0 & 1 & 4 & 1 & 0 & & 0 & 0 & 0 \\ \vdots & & & & \vdots & & & & \vdots \\ 0 & 0 & 0 & 0 & 0 & & 1 & 4 & 1 \\ 0 & 0 & 0 & 0 & 0 & & -15 & 0 & 15 \end{bmatrix} \begin{bmatrix} x_{-1} \\ x_0 \\ \\ \\ \\ x_6 \end{bmatrix} = \begin{bmatrix} 5 \\ 1 \\ 2 \\ 3 \\ 4 \\ 5 \\ 6 \\ 5 \end{bmatrix}.$$

If we simplify our answer, we find $s(t) = 5t + 1$. Why must this be the case?

Figure 4.2 presents the graphs of $\cos 2\pi x$ and of the cubic spline interpolate to $\cos 2\pi x$ on $[0, \pi]$ with 5 and 10 evenly spaced knots. Clearly the introduction of more knots seems to improve the accuracy of the approximation. A more interesting example is presented in Figure 4.3. Here we compare the cubic spline interpolates to $\ln x$ on the interval $[0.01, 1]$ with 5 and 10 knots. It is evident that there is more to the accuracy problem than having the knots close together. The difficulty here, of course, arises because the spline not only interpolates $\ln x$ but also its derivative $1/x$ at the extreme left-hand knot. Thus the spline on $[0.01, 1]$ must take off with a steep initial slope. We need an in-depth analysis of error in terms of mesh spacing and smoothness of the function f being approximated, and this is the topic of Section 4.5.

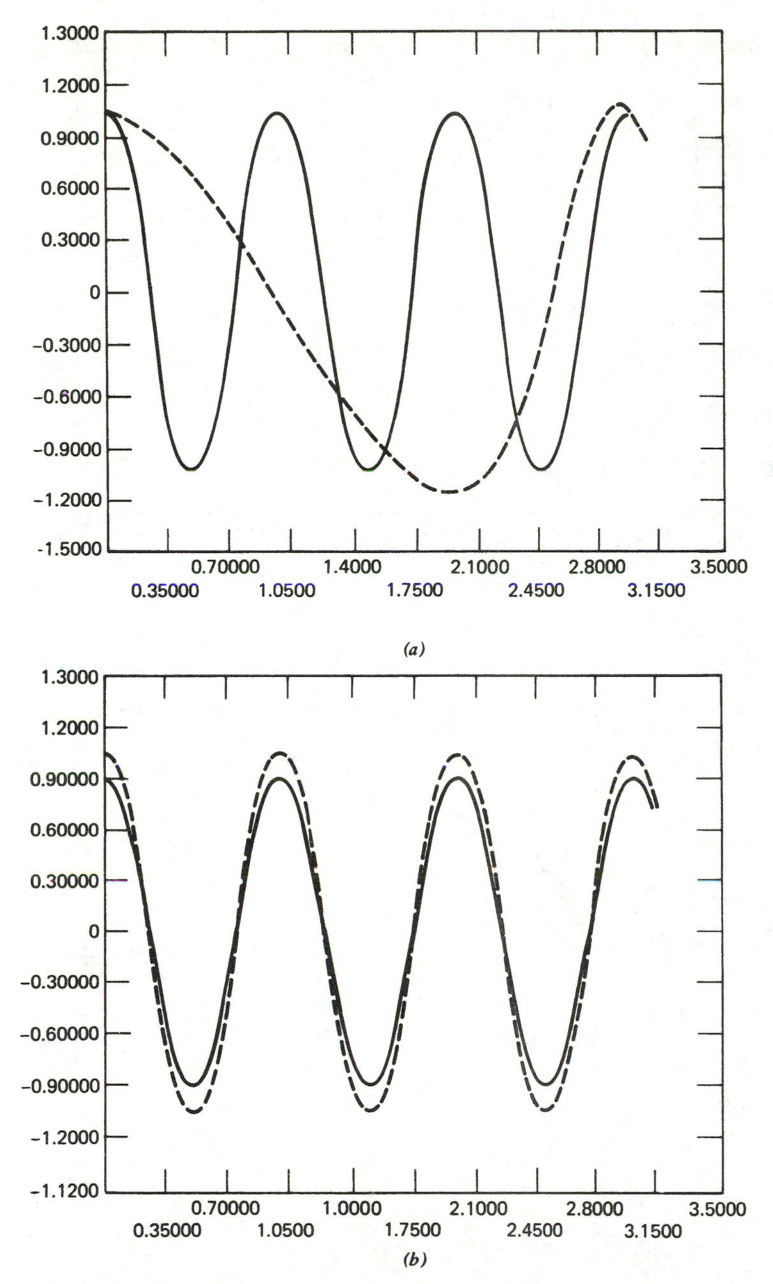

Figure 4.2 Graphs of $\cos 2\pi x$ (solid curves) and the cubic spline interpolate (dotted curves) to $\cos 2\pi x$ with (*a*) 5 and (*b*) 10 evenly spaced knots on $[0, \pi]$.

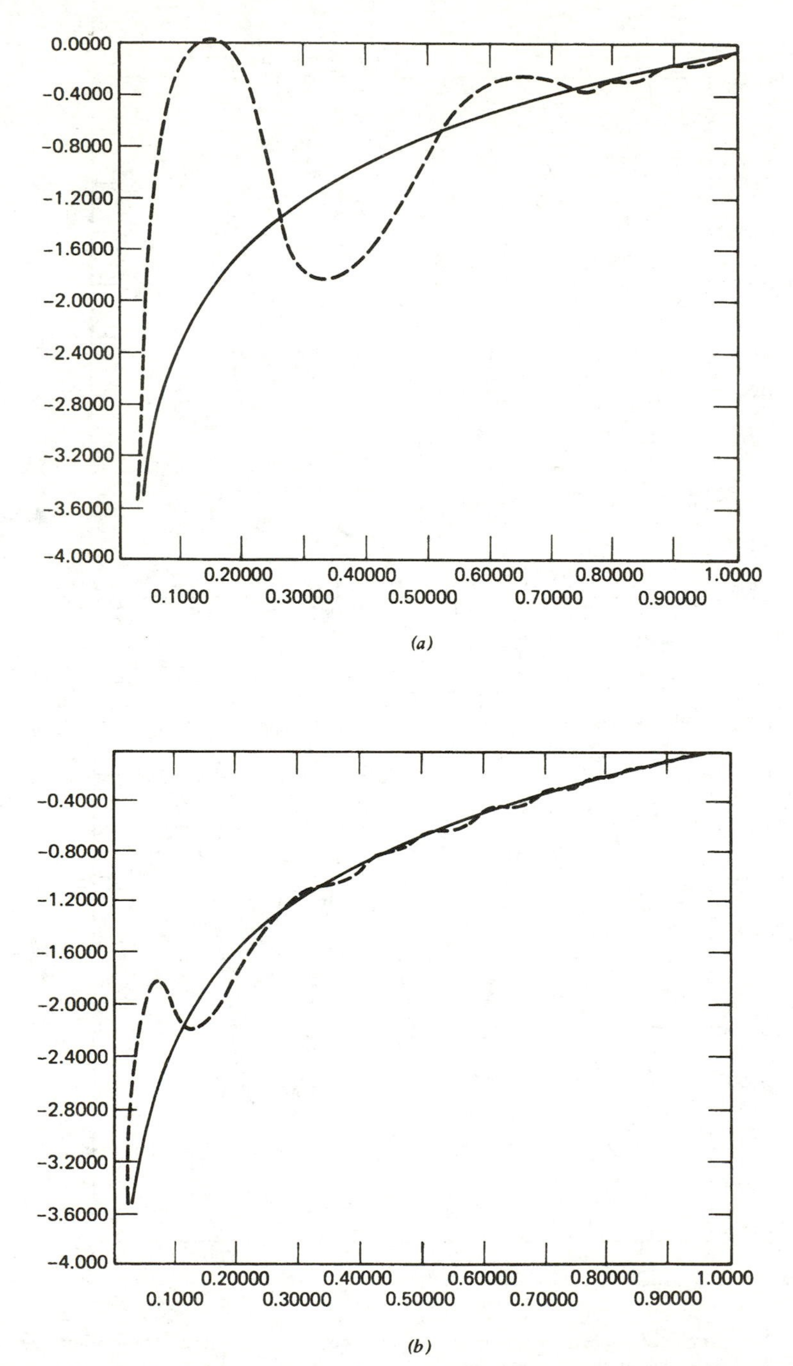

Figure 4.3 Graph of the natural log of x and the cubic spline interpolate to $\ln x$ using (*a*) 5 and (*b*) 10 knots on $[0.01, 1]$.

EXERCISES

1. Prove that the set $\mathfrak{B} = \{B_{-1}, B_0, \ldots, B_{n+1}\}$ is linearly independent on $[a,b]$. Is it independent on all of R? Why?
2. Let $s(t)$ be a cubic spline with knots at the points of a partition $\pi: a = x_0 < x_1 < \cdots < x_n = b$ of $[a,b]$, and suppose $s(t) \equiv 0$ outside the interval $[t_i, t_{i+1}]$. Prove that $s(t) \equiv 0$.
3. Try to construct a set of cubic B splines for unevenly spaced knots.
4. Program your computer to compute and graph the cubic spline interpolates to a given function $f(t)$ at n knots, $5 \leqslant n \leqslant 20$, and to graph $f(t) - s(t) = e(t)$. Then graph f, s, and e when $f(t) = \cos n\pi t$ for $n = 5$, 10, 15, and 20.
5. Compare the *size* of the linear systems and the *computational simplicity* of the *matrices* of the systems arising from computing the piecewise cubic Lagrange, the piecewise cubic Hermite, and the cubic spline interpolates to an arbitrary function $f(t)$. Assume knots $a = t_0 < t_1 < \cdots < t_n$, where $n = 3N$.
6. For what values of k, $k = 1,2,3$, or 4 does the cubic spline interpolate $s(t)$ with k knots $t_0 < t_1 < \cdots t_k$ to a function $f(t)$ exist? Is it unique? Explain.
7. Try to estimate $\|f - s\|$, $\|f' - s'\|$, $\|f'' - s''\|$ when $f \in C^4[a,b]$. When $f \in C^3[a,b]$.
8. Prove that there exist infinitely many cubic splines $s(t) \in S_3(\pi)$ solving the interpolation problem $s(t_i) = f(t_i)$, $0 \leqslant i \leqslant n$.
9. Does there exist a unique cubic spline solving the interpolation problems $s'(t_i) = f'(t_i)$, $0 \leqslant i \leqslant n$, and $s''(t_0) = f''(t_0) = f''(t_n) = s''(t_n)$? Prove your answer.

4.3 DERIVATION OF THE B SPLINES

Our algorithm for computing cubic splines interpolates with evenly spaced knots depended on having the B splines $B_i(t)$, $-1 \leqslant i \leqslant n+1$ given by formula (2) of the previous section, as did our existence and uniqueness analysis for such splines. The reader is undoubtedly wondering where we got these functions. Since any generalization of the arguments of Section 4.2 to polynomial and other splines of degree m other than 3 would depend on our having the analogs of the cubic B splines, we must show the derivation of these functions. To this end, recall that the kth forward difference $f(x_0)$ of a given function $f(x)$ at x_0 is defined recursively by

$$\begin{aligned} \Delta f(x_0) &= f(x_1) - f(x_0) \\ \Delta^{k+1} f(x_0) &= \Delta^k f(x_1) - \Delta^k f(x_0). \end{aligned} \tag{1}$$

In particular

$$
\begin{aligned}
\Delta^2 f(x_0) &= f(x_2) - 2f(x_1) + f(x_0) \\
\Delta^3 f(x_0) &= f(x_3) - 3f(x_2) + 3f(x_1) - f(x_0) \\
\Delta^4 f(x_0) &= f(x_4) - 4f(x_3) + 6f(x_2) - 4f(x_1) + f(x_0),
\end{aligned}
\tag{2}
$$

and so forth. The coefficient of $f(x_k)$ in $\Delta^n f(x_0)$ is simply the binomial coefficient $(-1)^k\binom{n}{k}$. It is well known that with *evenly spaced knots* $x_i = x_0 + ih$, Δ^n annihilates all polynomials of degree $n-1$ (see Exercises). That is

THEOREM A

The nth forward difference $\Delta^n p(x)$ with evenly spaced knots of any polynomial of degree $n-1$ is identically equal to zero.

This theorem is *not* true when the knots are not evenly spaced as is illustrated by $\Delta^2 p(x_0)$ when $p(x) = x$, $x_0 = 0$, $x_1 = 1$, and $x_2 = 3$.

To derive the cubic B splines we compute

$$K(t) = \Delta^4 F_t(x_0), \tag{3}$$

where for each fixed t we have

$$F_t(x) = (x-t)_+^3$$

The function $(x-t)_+^3$ is defined by

$$(x-t)_+^3 = \begin{cases} (x-t)^3 & \text{when} \quad t \leqslant x \\ 0 & \text{when} \quad t > x \end{cases} \tag{4}$$

(see Figure 4.4). It is clear that $F_t(x)$ is twice continuously differentiable on all of R but is not three times differentiable. In fact $F_t'''(x)$ is the jump function given by

$$F_t'''(x) = \frac{d^3 F_t(x)}{dx^3} = \begin{cases} 6 & \text{when} \quad x > t \\ 0 & \text{when} \quad x < t \\ \text{undefined at} & x = t. \end{cases}$$

Since $F_t(x)$ is piecewise cubic, it is easy to see that each of the functions $F_t(x)$, $t \in R$, is a cubic spline and, as such, belongs to $S_3(\pi)$.

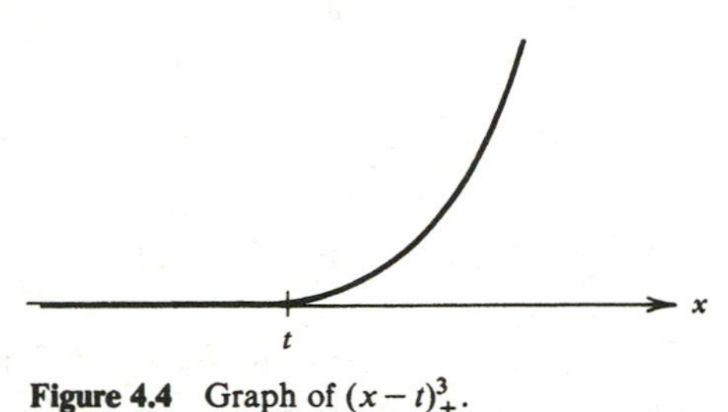

Figure 4.4 Graph of $(x-t)^3_+$.

What has this function $K(t)$ to do with B splines? It has, of course, everything to do with them, since we can prove $h^3K(t)=B_2(t)$. To see this we must combine a little algebra with Theorem A. In particular, by the definition given in (2)

$$\begin{aligned} K(t) &= \Delta^4 F_t(x_0) \\ &= F_t(x_4) - 4F_t(x_3) + 6F_t(x_2) - 4F_t(x_1) + F_t(x_0) \\ &= (x_4-t)^3_+ - 4(x_3-t)^3_+ + 6(x_2-t)^3_+ \\ &\quad - 4(x_1-t)^3_+ + (x_0-t)^3_+. \end{aligned}$$

It is clear from the definition (3) of $(x-t)^3_+$ that $K(t)\equiv 0$ for all $t \geqslant x_4$. Moreover, for fixed t and $x<t$, $F_t(x)=(x-t)^3$ is a polynomial of degree 3. Thus any fourth forward difference $\Delta^4 F_t(x_0)$ with evenly spaced knots vanishes identically when $x_0 \geqslant t$. That is,

$$K(t)\equiv 0 \qquad \text{when} \quad t \geqslant x_4 \qquad \text{and} \qquad t \leqslant x_0.$$

Note that the cubic B spline $B_2(t)$ also has this property. Moreover, being a sum of cubic splines $(x_i-t)^3_+$, $K(t)$ must be a cubic spline. In particular,

$$K(t)=\begin{cases} (x_4-t)^3, & x_3<t\leqslant x_4 \\ (x_4-t)^3-4(x_3-t)^3, & x_2<t\leqslant x_3 \\ (x_4-t)^3-4(x_3-t)^3+6(x_2-t)^3, & x_1<t\leqslant x_2 \\ (x_4-t)^3-4(x_3-t)^3+6(x_2-t)^3-4(x_1-t)^3, & x_0\leqslant t\leqslant x_1 \\ 0, & \text{for all other } t. \end{cases}$$

We now do a bit of algebra, recalling that $x_i = x_0 + ih$ and $x_{i+j} = x_i + jh$. For example,

$$
\begin{aligned}
(x_4 - t)^3 - 4(x_3 - t)^3 &= [(x_3 - t) + h]^3 - 4(x_3 - t)^3 \\
&= (x_3 - t)^3 + 3(x_3 - t)^2 h + 3(x_3 - t)h^2 + h^3 - 4(x_3 - t)^3 \\
&= h^3 + 3h^2(x_3 - t) + 3h(x_3 - t)^2 - 3(x_3 - t)^3.
\end{aligned}
$$

The observant reader will note that this is simply the third line of the formula (2) of the previous section for $h^3 B_2(t)$. Continuing with a bit more algebra, it is easily proved that

$$\frac{1}{h^3} K(t) = \frac{1}{h^3} \Delta^4 F_t(x_0) = B_2(t).$$

It is now readily apparent how to obtain the other cubic B splines. In particular, it must be the case that

$$\boxed{B_i(t) = \frac{1}{h^3} \Delta^4 F_t(x_{i-2})} \tag{5}$$

is the cubic B spline $B_i(t)$ with evenly spaced knots.

To generalize this formula, we let

$$F_t(x) = (x - t)_+^m$$

and

$$\boxed{K(t) = \Delta^{m+1} F_t(x) = \sum_{i=0}^{m+1} \binom{m+1}{i} (-1)^i (x_i - t)_+^m,} \tag{6}$$

where $m = 1, 2, 4, 5, \ldots$. In every case the vanishing of $(x_i - t)_+^m$ for $t \geqslant x_i$ coupled with Theorem A proves that $K(t) \equiv 0$ when $t \leqslant x_0$ and when $t \geqslant x_{m+1}$. Moreover, $K(t)$, being a sum of $(m-1)$-times continuously differentiable functions is itself $(m-1)$-times continuously differentiable. That is, each $K(t)$ belongs to $S_m(\pi)$.

DEFINITION THE POLYNOMIAL *m* SPLINES $S_m(\pi)$

Let π be a partition $t_0 < t_1 < \cdots < t_n (n \geqslant m+1)$ of $[a,b]$. The space $S_m(\pi)$ of all functions from $C^{m-1}[a,b]$ that reduce to polynomials of degree m on each subinterval (t_i, t_{i+1}) of $[a,b]$ is called the set of *m^{th}-degree polynomial splines with knots at π.*

We compute and graph the functions $K(t)$ for a few values of m, leaving further computations to the reader. Those familiar with the Peano kernal theorem will recognize the functions $K(t)$ as the Peano kernals associated with the linear functionals $\text{Lf} = \Delta^{m+1} f(x_0)$.

EXAMPLE 1

Let $m=1$. Then $F_t(x) = (x-t)_+$ and

$$K(t) = \Delta^2 F_t(x_0) = (x_2 - t)_+ - 2(x_1 - t)_+ + (x_0 - t)_+$$

$$= \begin{cases} (x_2 - t), & x_1 < t \leqslant x_2 \\ (x_2 - t) - 2(x_1 - t), & x_0 \leqslant t \leqslant x_1 \\ 0, & \text{otherwise} \end{cases}$$

$$= \begin{cases} (x_2 - t), & x_1 < t \leqslant x_2 \\ h - (x_1 - t), & x_0 \leqslant t \leqslant x_1 \\ 0, & \text{otherwise.} \end{cases}$$

Thus

$$\phi_i^1(t) = \frac{1}{h} \Delta^2 F_t(x_{i-1}) = \frac{1}{h} K(t)$$

is the simple cardinal basis piecewise Lagrange polynomial solving

$$\phi_i^1(t_j) = \delta_{ij}, \qquad 0 \leqslant j \leqslant n,$$

whose graph is the *hat function* depicted in Figure 4.5.

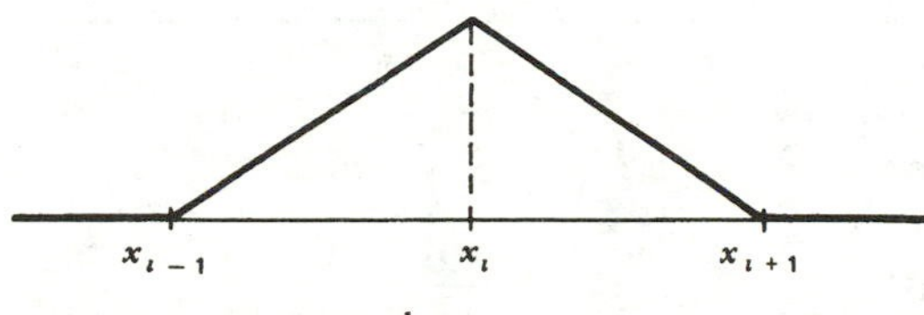

Figure 4.5 Graph of $\phi_i^1(t)$.

EXAMPLE 2

Let $m=2$. Then $F_t(x)=(x-t)_+^2$ and

$$\begin{aligned}K(t)&=\Delta^3F_t(x_0)\\&=(x_3-t)_+^2-3(x_2-t)_+^2+3(x_1-t)_+^2-(x_0-t)_+^2.\end{aligned}$$

Leaving the algebra to the reader, the graph of

$$\phi_i^2(t)=\frac{1}{h^2}\Delta^3F_t(x_i)$$

is the piecewise quadratic belonging to $C^1[a,b]$ (Figure 4.6).

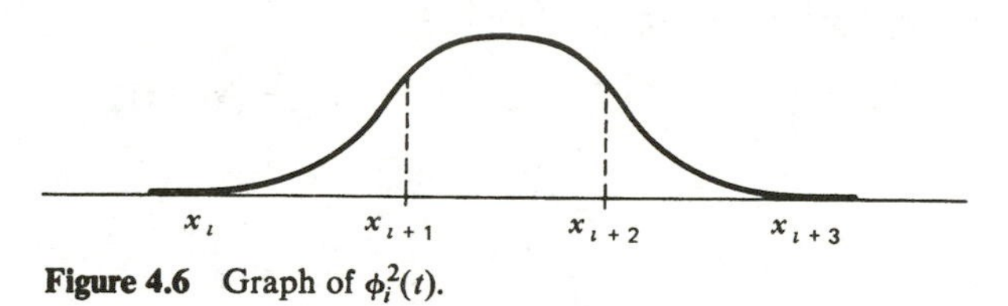

Figure 4.6 Graph of $\phi_i^2(t)$.

EXAMPLE 3

Let $m=3$. In this case $B_i(t)=\phi_i^5(t)=(1/h^5)\Delta^6F_t(x_{i-3})$ is *the quintic B spline with evenly spaced knots.*

It follows that

$$\begin{aligned}B_i(t)=\frac{1}{h^5}\big[&(t-x_{i-3})_+^5-6(t-x_{i-2})_+^5+15(t-x_{i-1})_+^5\\&-20(t-x_i)_+^5+15(t-x_{i+1})_+^5-6(t-x_{i+2})_+^5\\&+(t-x_{i+3})_+^5\big].\end{aligned}$$

Table 4.2 presents values of the $B_i(t)$'s. Since $B_i(t)=0$ outside the interval (t_{i-3},t_{i+3}), where $x_j=t_j$, $0\leqslant j\leqslant n$, there is no need to tabulate B_i for other values of t.

TABLE 4.2 VALUES OF QUINTIC *B* SPLINES AT KNOTS

t	t_{i-3}	t_{i-2}	t_{i-1}	t_i	t_{i+1}	t_{i+2}	t_{i+3}
$B_i(t)$	0	1	26	66	26	1	0
$B_i'(t)$	0	$\frac{5}{h}$	$\frac{50}{h}$	0	$-\frac{50}{h}$	$-\frac{5}{h}$	0
$B_i''(t)$	0	$\frac{20}{h^2}$	$\frac{40}{h^2}$	$-\frac{120}{h^2}$	$\frac{40}{h^2}$	$\frac{20}{h^2}$	0

THE QUINTIC SPLINE INTERPOLATE

$$s(t)=c_{-2}B_{-2}(t)+c_{-1}B_{-1}(t)+\cdots+c_{n+2}B_{n+2}(t)$$

to a given function $f(t)$, where the B_i's are the quintic B splines with evenly spaced knots $t_{-4}<t_{-3}<\cdots<t_{n+5}$ is the unique solution (see Exercise 2) to the interpolation problem $D^js(t_0)=D^jf(t_0), j=1,2$; $s(t_i)=f(t_i)$, $0\leqslant i\leqslant n$; and $D^js(t_n)=D^jf(t_n), j=1,2$.

Each collection $\mathfrak{B}_{mk}=\{\phi_0^m,\phi_1^m,\ldots,\phi_k^m\}$ of B splines spans a linear subspace X_k of $C^{m-1}[a,b]$ consisting of polynomial splines of degree m from $S_m(\pi)$. The dimension of X_k is k in each case and one can ask what kind of approximations

$$\tilde{f}=c_0\phi_0^m+c_1\phi_1^m+\cdots+c_k\phi_k^m \tag{7}$$

to a given function $f(t)$ subject to k independent constraints are available. We have some answers to this question when $m=1$ (the piecewise linear Lagrange polynomials) and when $m=3$ (various types of cubic interpolating spline). If the constraints are to be interpolatory, some answers about existence and uniqueness can be obtained, using the methods of the last section. This approach, however, can soon become laborious for large m, even though it will always provide a numerical algorithm for computing $\tilde{f}(t)$ with evenly spaced knots, when it exists. Because of this difficulty we give existence and uniqueness arguments for a wide variety of odd-degree interpolating polynomial splines, as well as more general splines, in the next section. Our arguments use integration by parts procedures rather than Gershgorin arguments and provide existence and uniqueness regardless of uneven mesh spacing. Unfortunately, these arguments do not give us as efficient a numerical algorithm as the B spline algorithm.

EXERCISES

1. Derive the quintic B spline with evenly spaced knots explicitly, checking Table 4.2.
2. Let $f(t)$ be given. Prove that the quintic spline interpolate to $f(t)$ exists and is unique, assuming that $f(t_i)$, $0\leqslant i\leqslant n$, as well as f' and f'' at t_0 and t_n exist.
3. Let $B_i(t)=(1/h^2)\Delta^2F_t(x_i)$, $F_t(x)=(x-t)_+^2$, be a quadratic B spline. Let $f(t)$ be a given function of t defined on $[a,b]$, and suppose $a=x_0<x_1<\cdots<x_n=b$. Does there exist a unique function

$$s(t)=c_0B_0(t)+c_1B_1(t)+\cdots+c_nB_n(t)$$

solving $s(x_i)=f(x_i)$, $0 \leqslant i \leqslant n$? Why? Does a unique solution exist solving $s(t_i)=f(t_i)$, where $t_0 < t_1 < \cdots < t_n$ are $n+1$ arbitrary points of $[a,b]$? Explain.

4.4 SPLINES AND ORDINARY DIFFERENTIAL EQUATIONS

A totally different approach to proving existence and uniqueness of odd-degree interpolating polynomial splines $s(t)$ consists in viewing such splines as solutions to very general ordinary differential boundary value problems. For example, we can view the cubic spline interpolate $s(t)$ to a given function $f(t)$ as a solution to the differential equation

$$D^4 s(t)=0, \qquad \text{a.b.f. on } [a,b] \tag{1}$$

subject to "*boundary constraints*"

$$\begin{aligned} s'(t_0)&=f'(t_0) \\ s(t_i)&=f(t_i), \qquad 0 \leqslant i \leqslant n \\ s'(t_n)&=f'(t_n), \end{aligned} \tag{2}$$

where $\pi: a=t_0 < t_1 < \cdots < t_n = b$ is a partition of $[a,b]$, $f(t)$ is a given function of t, and $D^4 = d^4/dt^4$. The a.b.f. in (1), which stands for "at all but finitely many points," is necessary because the third and fourth derivatives of $s(t)$ may not exist at the interior knots t_i, $1 \leqslant i \leqslant n-1$. The existence of a unique solution that is twice continuously differentiable on $[a,b]$ and is four times continuously differentiable on each of the intervals (t_{i-1}, t_i), $1 \leqslant i \leqslant n-1$, to the *boundary value problem* (1) coupled with (2) with evenly spaced knots was the content of Theorems 4.1 and 4.2 of the previous section. We have not yet proved that this problem has a solution when the knots are *not* evenly spaced, but we provide such a proof in this section. It is not necessary, however, to confine ourselves to cubic splines or even to polynomial splines. In fact, simply by using integration by parts, we can prove the existence and uniqueness of a large variety of odd-degree polynomial and more general interpolating splines. The approach we take was first observed independently by Greville (1964) and by Ahlberg, Nilson, and Walsh (1964).

We begin our ascent to this more catholic viewpoint by generalizing our definition of *polynomial splines*.

DEFINITION A. POLYNOMIAL SPLINES

A polynomial spline of degree d, $d=0,1,2,3,\ldots$, and smoothness $k (k \geqslant 0$ an integer) is any function $s(t)$ belonging to $C^k[a,b]$ that reduces to a polynomial of degree d on each subinterval (t_{i-1}, t_i) of $[a,b]$. The set of all such piecewise polynomials is denoted $S_d(\pi,k)$. The set $S_d(\pi,-1)$ is simply the set of all functions defined on each of the intervals (t_{i-1}, t_i), which reduce to polynomials of degree d on each of these intervals.

Some of these spaces are old friends. The reader will observe, after a moment's reflection, that $S_3(\pi,2) = S_3(\pi)$, the space of *cubic polynomial splines*, and that $S_3(\pi,1) = H_3(\pi)$, the space of *piecewise cubic Hermite polynomials*, while $S_3(\pi,0) = L_3(\pi)$ is the space of *piecewise cubic Lagrange polynomials*. The space $S_0(\pi,-1)$ is simply the space of all *step functions* with knots at π. Moreover, for each $k \geqslant d > 0$, $S_d(\pi,k)$ is never empty, since it always contains the space $\mathcal{P}_d$ of polynomials of degree d on $[a,b]$. From the point of view of one working in differential equations, each function $s(t) \in S_d(\pi,k)$, for fixed d and any integer $k \geqslant -1$, is a "solution" to the differential equation

$$D^{d+1}s(t) = 0, \qquad \text{on } (t_{i-1}, t_i), \qquad 1 \leqslant i \leqslant n. \tag{3}$$

Just what we are calling a "solution" becomes apparent by examining $Ds(t)$ when $s(t)$ is a step function. (The winds of distribution theory will be recognizable to the initiated but is ignored by us, since we do not need these ideas.) We can easily see that Definition A is equivalent to

DEFINITION B POLYNOMIAL SPLINES

$S_d(\pi,k)$ is the set of all k times continuously differentiable functions $s(t)$ satisfying the differential equation

$$D^{d+1}s(t) = 0, \qquad \text{on} \quad (t_{i-1}, t_i), \qquad 1 \leqslant i \leqslant n,$$

which are $d+1$ times continuously differentiable on each of the subintervals (t_{i-1}, t_i), $\quad 1 \leqslant i \leqslant n$.

The reason for the equivalence follows from the theory of ordinary differential equations. Since $s(t)$ is k times continuously differentiable and $D^{d+1}s(t) = 0$ on (t_{i-1}, t_i),

$$s(t) = c_{id}t^d + \cdots + c_{i1}t + c_{i0}$$

on (t_{i-1}, t_i). Thus $s(t)$ is a piecewise polynomial of degree $d+1$ belonging to $C^k[a,b]$.

The analog of viewing cubic interpolating splines from the vantage of boundary value problems presented by (1) coupled with (2) in this more general setting follows. An interpolating polynomial d spline $s(t)$ of smoothness k is a function that is k times continuously differentiable on

$[a,b]$ and satisfies the differential equation (3) coupled with a set of (interpolatory) constraints such as

$$\begin{cases} s(t_i)=f(t_i), & 0\leqslant i\leqslant n \\ J \text{ other constraints,} \end{cases} \tag{4}$$

where J is some positive integer to be determined and $f(t)$ is a given function of t defined on $[a,b]$.

This problem, it turns out, is already too much for us to handle, so we confine ourselves to odd-degree polynomial splines, $d=2m-1$, together with $J=2m-2$ specific constraints besides the $n+1$ interpolatory constraints $f(t_i)=s(t_i)$, $0\leqslant i\leqslant n$. We can prove the existence and uniqueness of a $2m-2$ times continuously differentiable polynomial spline of degree $2m-1$ solving the interpolation problem

$$\begin{cases} D^js(t_0)=D^jf(t_0), & 1\leqslant j\leqslant m-1 \\ s(t_i)=f(t_i), & 0\leqslant i\leqslant n \\ D^js(t_n)=D^jf(t_n), & 1\leqslant j\leqslant m-1, \end{cases} \tag{5}$$

and can indicate some generalizations of this existence and uniqueness theory to conditions more general than (5). We see that when $m=2$, we are simply solving (1) plus (2); thus, in this case, $s(t)$ is *the* cubic spline interpolate to $f(t)$, where the knots need no longer be evenly spaced. We call any solution to (5) from $S_{2m-1}(\pi,2m-2)$ *a polynomial spline interpolate to* $f(t)$ *of degree* $2m-1$. Looking ahead to further generalizations of the notion of a spline function, we introduce the notation

$$\begin{aligned} Lg(t)&=D^mg(t) \\ L^*g(t)&=(-1)^mD^mg(t). \end{aligned} \tag{6}$$

Assuming the existence of a polynomial spline interpolate of degree $2m-1$ to $f(t)$, we can prove two important theorems known as the first and second integral relations.

THEOREM THE FIRST INTEGRAL RELATION

Let $f\in C^m[a,b]$ and let s be a polynomial spline interpolate of degree $2m-1$ to f. Then

$$\boxed{\int_a^b(Lf)^2=\int_a^b(Ls)^2+\int_a^b(Lf-Ls)^2.} \tag{7}$$

THEOREM THE SECOND INTEGRAL RELATION

Let $f \in C^{2m}[a,b]$ and let s be a polynomial spline interpolate of degree $2m-1$ to f. Then

$$\boxed{\int_a^b [Lf-Ls]^2\,dx=\int_a^b (f-s)L^*Lf\,dx.} \tag{8}$$

Proof (First Integral Relation). Since $L=D^m$

$$(Lf)^2=[f^{(m)}]^2=[\{f^{(m)}-s^{(m)}\}+s^{(m)}]^2$$

$$=[f^{(m)}-s^{(m)}]^2+2s^{(m)}[f^{(m)}-s^{(m)}]+[s^{(m)}]^2.$$

If we can prove that $\int_a^b s^{(m)}[f^{(m)}-s^{(m)}]dx=0$, we will be done. The proof of this is no more than an exercise in integration by parts. In particular

$$\int_a^b s^{(m)}[f^{(m)}-s^{(m)}]\,dx$$

$$=s^{(m)}(x)[f^{(m-1)}-s^{(m-1)}](x)|_a^b-\int_a^b s^{(m+1)}[f^{(m-1)}-s^{(m-1)}]\,dx$$

$$=-\int_a^b s^{(m+1)}[f^{(m-1)}-s^{(m-1)}]\,dx,$$

since $f^{(m-1)}(a)=s^{(m-1)}(a)$ and $f^{(m-1)}(b)=s^{(m-1)}(b)$. Continuing to integrate by parts and observing that $f^{(m-p)}(x)=s^{(m-p)}(x)$ when $x=a$, b and $p=1,2,\ldots,m-2$, we find

$$\int_a^b s^{(m)}[f^{(m)}-s^{(m)}]\,dx=(-1)^{2m-2}\int_a^b s^{(2m-2)}[f''-s'']\,dx.$$

At this point a little caution must be exercised. Recall that $s(x)$ is a polynomial of degree $2m-1$ on each subinterval $[x_{i-1},x_i]$, $1\leqslant i\leqslant n$. It follows that $s^{(2m-1)}(x)$ exists and is a constant c_i on each interval (x_{i-1},x_i). Thus $s^{(2m-1)}$ is a step function that is usually not continuous on $[a,b]$ but is at least intrgrable on $[a,b]$. Furthermore $s^{(2m)}(x)\equiv 0$ on each interval (x_i,x_{i+1}). To ensure the correctness of further integrations by parts (when $s^{(2m-1)}$ is not continuous), observe that

$$\int_a^b s^{(2m-2)}[f''-s'']\,dx=\sum_{i=1}^n\int_{x_{i-1}}^{x_i} s^{(2m-2)}[f''-s'']\,dx.$$

But

$$\int_{x_{i-1}}^{x_i} s^{(2m-2)}[f''-s'']\,dx = s^{(2m-2)}(x)[f'(x)-s'(x)]|_{x_{i-1}}^{x_i}$$

$$-\int_{x_{i-1}}^{x_i} s^{(2m-1)}[f'-s']\,dx.$$

But

$$\sum_{i=1}^{n} s^{(2m-2)}(x)[f'(x)-s'(x)]|_{x_{i-1}}^{x_i}$$

is a telescoping series equal to $s^{(2m-2)}(x)[f'(x)-s'(x)]|_a^b$. Since $f'(a)=s'(a)$ and $f'(b)=s'(b)$ this term also vanishes. Furthermore $s^{(2m-1)}(x)=c_i$ on (x_{i-1},x_i). Thus

$$\int_a^b s^{(m)}[f^{(m)}-s^{(m)}]\,dx = (-1)^{2m-1}\sum_{i=1}^{n} c_i \int_{x_{i-1}}^{x_i}[f'(x)-s'(x)]\,dx$$

$$=(-1)^{2m-1}\sum_{i=1}^{n} c_i[f(x)-s(x)]|_{x_{x-1}}^{x_i}=0$$

since $s(x_i)=f(x_i)$ at each knot $x_0,x_1,\ldots,x_n$ and since we can integrate by parts on each subinterval (x_{i-1},x_i). This completes the proof of the theorem. ■

Proof (Second Integral Relation). Again $L=D^m$ and the proof is another exercise in integration by parts. In particular if $g=f-s$, then

$$A=\int_a^b [f^{(m)}-s^{(m)}]^2\,dx=\int_a^b g^{(m)}g^{(m)}\,dx$$

$$=g^{(m)}(x)g^{(m-1)}(x)|_{x=a}^{x=b}-\int_a^b g^{(m+1)}g^{(m-1)}\,dx.$$

Since $g^{(m-1)}(a)=g^{(m-1)}(b)=0$, the first term on the right vanishes. Continuing to integrate by parts and to invoke the boundary conditions

$f^{(k)}(a)=s^{(k)}(a)$ and $f^{(k)}(b)=s^{(k)}(b)$, $0 \leqslant k \leqslant m-1$, we find

$$\int_a^b [f^{(m)}-s^{(m)}]^2\,dx = -\int_a^b g^{(m+1)}g^{(m-1)}\,dx$$

$$= -g^{(m+1)}(x)g^{(m-2)}(x)\Big|_{x=a}^{x=b} + \int_a^b g^{(m+2)}g^{(m-2)}\,dx$$

$$= \int_a^b g^{(m+2)}g^{(m-2)}\,dx$$

$$\vdots$$

$$= \int_a^b g^{(2m-2)}g''\,dx.$$

Note that since $s\in C^{2m-2}[a,b]$ and f can be differentiated $2m$ times almost everywhere, the last integrand is well defined. Since $s^{(2m-1)}$ is a step function (a piecewise constant) with $s^{(2m-1)}(x)=c_i$ on $[x_{i-1},x_i]$, we must write

$$\int_a^b g^{(2m-2)}g''\,dx = \sum_{k=1}^{n}\int_{x_{k-1}}^{x_k} g^{(2m-2)}g''\,dx.$$

Integrating by parts once again, we see that

$$\int_{x_{k-1}}^{x_k} g^{(2m-2)}g''\,dx = g^{(2m-2)}(x)g'(x)\Big|_{x_{k-1}}^{x_k} - \int_{x_{k-1}}^{x_k} g^{(2m-1)}g'\,dx.$$

Noting that

$$\sum_{k=1}^{n} g^{(2m-2)}g'(x)\Big|_{x_{k-1}}^{x_k}$$

is a telescoping series equal to $g^{(2m-2)}(b)g'(b)-g^{(2m-2)}(a)g'(a)$, we find

$$\int_a^b [f^{(m)}-s^{(m)}]^2\,dx = -\sum_{k=1}^{n}\int_{x_{k-1}}^{x_k} g^{(2m-1)}g'\,dx.$$

But

$$-\int_{x_{k-1}}^{x_k} g^{(2m-1)}g'\,dx = -g^{(2m-1)}(x)g(x)\Big|_{x_{k-1}}^{x_k} + \int_{x_{k-1}}^{x_k}[f^{(2m)}-s^{(2m)}][f-s],$$

where $g(x_i)=f(x_i)-s(x_i)=0$. Adding, we have

$$\int_a^b [f^{(m)}-s^{(m)}]^2\,dx = \sum_{k=1}^n \int_{x_{k-1}}^{x_k}[f^{(2m)}-s^{(2m)}][f-s]\,dx$$

$$= \int_a^b [f^{(2m)}-s^{(2m)}][f-s]\,dx. \qquad \blacksquare$$

The first integral relation provides an existence and uniqueness proof for polynomial spline interpolates of degree $2m-1$ to functions $f(t)$. We begin by using the first integral relation to prove

THEOREM A UNIQUENESS THEOREM

$s(t)\equiv 0$ is the only polynomial spline interpolate of degree $2m-1$ to $f(t)\equiv 0$.

Proof. Let $s(t)$ be a spline interpolate of degree $2m-1$ to the zero function. Then $s(t)$ satisfies the first integral relation. In particular

$$\int_a^b [s^{(m)}]^2\,dt \leqslant \int_a^b [f^{(m)}]^2\,dt = 0.$$

It follows that $s^{(m)}(t)=0$ on $[a,b]$. But then $s(t)$ is a polynomial of degree $m-1$

$$s(t)=c_{m-1}t^{m-1}+\cdots+c_1 t+c_0$$

on $[a,b]$. Recalling that $s^{(k)}(a)=0$ when $k=0,1,\ldots,m-1$, we find $s(t)\equiv 0$ on $[a,b]$. $\blacksquare$

It is now easy to prove

THEOREM EXISTENCE AND UNIQUENESS OF SPLINE INTERPOLATES

There exists a unique polynomial spline interpolate of degree $2m-1$ to each function $f(t)$ for which the right-hand side of (5) exists.

If $s\in S_{2m-1}(\pi,2m-2)$ then $s(x)=c_{i,2m-1}x^{2m-1}+\cdots+c_{i,1}x+c_{i0}$ on the interval (x_{i-1},x_i), where $1\leqslant i\leqslant n$. If s is our interpolate to f, we must be able to solve for the $2mn$ unknowns $\{c_{ik}: 1 \leqslant i \leqslant n \text{ and } 0 \leqslant k \leqslant 2m-1\}$. The conditions $D^k s(a) = D^k f(a)$, and $D^k s(b) = D^k f(b)$, $0 \leqslant k \leqslant m-1$, yield $2m$ linear equations in our $2mn$ unknowns. The conditions $s(x_i) = f(x_i)$, $1 \leqslant i \leqslant n-1$ yield an additional $n-1$ linear equation. Finally, the requirement $s \in C^{2m-2}[a,b]$ implying

$$D^k s(x_i+) = D^k s(x_i-),\ 1 \leqslant i \leqslant n-1;\ 0 \leqslant k \leqslant 2m-2$$

yields $(n-1)(2m-1)$ additional linear equations in the unknowns $\{c_{ik}\}$. We thus have

$$2m + (n-1) + (n-1)(2m-1) = 2mn$$

linear equations in $2mn$ unknowns to attempt to solve uniquely. Suppose we let our system of equations be

$$Ac = b,$$

where A is our $2mn$ square coefficient matrix. It is clear from the data giving rise to this system that the matrix A is independent of f and that b is a single column vector whose entries are zeros, values of f at the knots, and derivatives of f up to and including the $(m-1)$st at the end points a and b. Thus since A is independent of f we can take $f \equiv 0$ to see whether A is singular or nonsingular. Since $f \equiv 0$ implies $Ac = 0$ and this has a unique solution $c = 0$, we know that A is nonsingular so that A^{-1} exists. Thus the system $Ac = b$ has a unique solution for each b. This completes the proof of the theorem. ■

The first integral relation also provides a *variation diminishing* property for spline functions plus a "*Pythagorean theorem*" for such functions. The variation diminishing property is simply the observation that of all functions $f(t)$ that interpolate a given set of data

$$\begin{cases} D^j f(t_0) = \gamma_{j0}, & i \leqslant j \leqslant m-1 \\ f(t_i) = \gamma_i, & 0 \leqslant i \leqslant n \\ D^j f(t_n) = \gamma_{jn}, & 1 \leqslant j \leqslant m-1, \end{cases} \tag{9}$$

the spline interpolate $s(t)$ to the same data minimizes the integral of the square of the mth derivative. Specifically

$$\int_a^b (s^{(m)})^2 \leqslant \int_a^b (f^{(m)})^2. \tag{10}$$

Thus in a sense $s(t)$ is the smoothest function from $C^m[a,b]$ interpolating our data (7), in that it "wiggles around" less than any other function interpolating the same data. This property of splines was much heralded from the outset. The reader should look at exactly what it says in the cubic case to get a good insight into what is going on.

The first integral relation itself constitutes a *Pythagorean theorem* for an appropriate pre-Hilbert space (we define Hilbert spaces in Chapter 6). In particular, let $H = K_2^m[a,b]$ denote the set of all real-valued functions defined on $[a,b]$ that possess an absolutely continuous $(m-1)$st derivative and a Lebesgue square integrable mth derivative on $[a,b]$. That is if $f \in H$, we assume that $\int_a^b [f^{(m)}]^2$ exists and that $f^{(m-1)}(x_2) - f^{(m-1)}(x_1) = \int_{x_1}^{x_2} f^{(m)}(x)\,dx$ for all x_1 and x_2 in $[a,b]$. We then define the symbol (f,g) by

$$(f,g) = \int_a^b f^{(m)} g^{(m)}\,dx$$

and let $|f| = \sqrt{(f,f)}$. Then $|f|$ is a norm in every sense except $|f| = 0$ does not imply $f \equiv 0$. For this reason $|f|$ is called as a semi- or pseudonorm. Let $H = K_2^m[a,b]$ together with the seminorm $|f|$ and let $f \in H$. If we agree that two functions f and g from H are perpendicular when $(f,g) = 0$, the spline interpolate s of type I to f is perpendicular to $f - s$ in the sense that

$$(s, f-s) = 0,$$

and the Pythagorean theorem

$$|f|^2 = |f-s|^2 + |s|^2$$

holds.

There is nothing sacred about the smoothness constraint $s \in C^{2m-2}[a,b]$ or the interpolatory constraints (5) satisfied by polynomial spline interpolates of degree $2m-2$. For any k, we can find splines from $S_{2m-1}(\pi,k)$ satisfying quite a variety of interpolatory constraints for which the first and second integral relations still obtain. The existence and uniqueness of these splines also follows from the first integral relation in exactly the same manner as that of the polynomial spline interpolate of degree $2m-1$ to a given function f. We list some of these splines below, remarking that the list is far from complete.

1. *Cubic spline interpolates* (already discussed).
2. *Natural cubic spline interpolates* (already discussed).
3. *Piecewise cubic Hermite polynomials* (already discussed).
4. *Piecewise Hermite polynomials of fixed degree between knots* (already discussed).

There is also no reason to insist on the same smoothness between successive knots of our partition. We could, for example, let $z=(z_0,z_1,\ldots,z_n)$ be any $n+1$-tuple, $1 \leqslant z_i \leqslant m$, of positive integers and look for the following piecewise polynomials of degree $2m-1$ between knots:

5. *General piecewise Hermite polynomials*

$$D^k s(x_i)=D^k f(x_i), \qquad 0 \leqslant k \leqslant z_i-1,\ 1 \leqslant i \leqslant n-1$$

$$D^k s(x_i)=D^k f(x_i), \qquad i=0,\ n;\ 0 \leqslant k \leqslant m-1$$

$$D^k s(x_i-)=D^k s(x_i+), \quad 1 \leqslant i \leqslant n-1;\ 0 \leqslant k \leqslant 2m-1-z_i;\ f \in C^{m-1}[a,b].$$

6. *Periodic polynomials splines*

$$D^k s(x_i)=D^k f(x_i), \qquad 0 \leqslant k \leqslant z_i-1;\ 1 \leqslant i \leqslant n-1$$

$$D_s^k(x_i)=D^k f(x_i), \qquad 0 \leqslant k \leqslant z_i-1;\ i=0,n$$

$$D^k s(a)=D^k s(b), \qquad z_0 \leqslant k \leqslant 2m-1$$

$$D^k s(x_i-)=D^k s(x_i+), \qquad 1 \leqslant i \leqslant n-1;\ 0 \leqslant k \leqslant 2m-1-z_i,$$

where it is known that $D^k f(a)=D^k f(b)$, $0 \leqslant k \leqslant 2m-1$ and $f \in C^{2m-1}[a,b]$.

Existence and uniqueness of these splines has been proved in a variety of cases by Schultz and Varga (1967) using the integration by parts techniques leading to the first integral relation.

All the foregoing analyses can be considered generalized. The most immediate generalization is to think of spline functions $s(t)$ as solutions to the differential equation

$$L^* Ls(t)=0, \qquad \text{a.b.f. on} \quad [a,b] \tag{11}$$

subject to given "boundary values," where L is some mth-order differential operator

$$Lu(t)=a_m(t)u^{(m)}(t)+\cdots+a_1(t)u'(t)+a_0(t)u(t), \tag{12}$$

whose coefficients $a_j(t) \in C^j[a,b]$, and where L^* is the formal adjoint of L

$$L^* u(t)=\sum_{j=0}^{m}(-1)^j D^j\{a_j(t)u(t)\} \tag{13}$$

obtained by integrating $Lu(t)v(t)$ by parts. If $a_m(t)$ does not vanish on the interval $[a,b]$, we know from the theory of differential equations that (11) has $2m$ linearly independent solutions $u_1, u_2, \ldots, u_{2m}$ on the interval $[a,b]$. These $2m$ functions span a linear space $N(L^*L)$ known as the null space of L^*L. The generalization from polynomial splines is simply to piece together functions from $N(L^*L)$ interval by interval and see what kind of smoothness and interpolatory constraints can be solved with such functions. The reason one chooses L^*L is that integration by parts works nicely. Generalization to non-self adjoint operators L may be available. Thus you could also try piecing together things from the null space $N(L)$ interval by interval and see what kind of approximates can be built subject to a given set of interpolatory and smoothness constraints. *All such functions would be spline functions*. However, although the first and second integral relations together with their concomitant existence and uniqueness theorems carry over almost verbatim to piecewise fits of functions from $N(L^*L)$; such is not the case for piecewise fits of functions from $N(L)$. Piecewise fits of functions from $N(L^*L)$ are known as *L splines* or *generalized splines* in the literature. The interested reader is referred to the book of Ahlberg, Nilson, and Walsh (1967) and the papers of Schultz and Varga (1967), of Lucas (1970), and of Jerome and Schumaker (1969). Examples of these more general splines are given below. Moreover, the splines we have mentioned do not begin to categorize the extant species and subspecies of splines. There exist L–g splines, piecewise Euler splines, and many others, some of which are of computational importance. Many may appear esoteric to the very applied individual, and splines of possible numerical promise undoubtedly await discovery.

EXAMPLE 1

In this example we let $[a,b]=[0,\pi]$, have only two knots $t_0=0$ and $t_n=\pi$, and let

$$Lu=(D+i)(D+2i)u=(D^2+3iD-2)u,$$

where $i=\sqrt{-1}$. Substituting $a_0=-2$, $a_1=3i$, and $a_2=1$ into (13), we compute L^*.

$$\begin{aligned}L^*v&=(-1)^0 a_0 v+(-1)D[a_1 v]+D^2[a_2 v]\\&=a_0 v-a_1 Dv+a_2 D^2 v\\&=-2v-3iDv+D^2 v\\&=(D-2i)(D-i)v.\end{aligned}$$

Thus

$$L^*Lu=(D-i)(D-2i)(D+i)(D+2i)u$$
$$=(D^2+1)(D^2+4)u.$$

Any solution to the differential equation $L^*Lu=0$ is of the form

$$u(t)=c_1\sin t+c_2\cos t+c_3\sin 2t+c_4\cos 2t.$$

The unique generalized spline $s(t)$, with just two knots $a<b$ solving

$$s(0)=2,\ s'(0)=3,\ s(\pi)=0, \text{ and } s'(\pi)=1,$$

is given by

$$s(t)=\sin t+\cos t+\sin 2t+\cos 2t.$$

That is, $L^*Ls=0$ on $[0,\pi]$ and $s(t)$ satisfies our boundary values. This function looks suspiciously like a truncated Fourier series. It is not in the classical sense a Fourier series (i.e., it may not be a least squares fit). However, it is a Fourier series in the same sense that least squares fits by polynomials and fits by interpolating polynomials are all polynomials.

EXAMPLE 2

Let $Lu=du/dx-2u=(D-2)u$. This time one can show—again using (13)—that $L^*v=-dv/dx-2v=-(D+2)v$. Thus

$$L^*Lu=L^*\left[\frac{du}{dx}-2u\right]$$
$$=\frac{-d}{dx}\left(\frac{du}{dx}-2u\right)-2\left(\frac{du}{dx}-2u\right)$$
$$=\frac{-d^2u}{dx^2}+2\frac{du}{dx}-2\frac{du}{dx}+4u$$
$$=-\left(\frac{d^2u}{dx^2}-4u\right)$$
$$=-(D^2-4)u$$
$$=-(D+2)(D-2)u.$$

We now seek a solution to

$$L^*Lu=0.$$

As every student of elementary differential equations knows, equation $L^*Lu=0$ has exactly two linear independent solutions e^{2t} and e^{-2t}. Thus any solution to the equation can be written as $d_1e^{2t}+d_2e^{-2t}$. The operator L is of order 1. Letting

$m=1$ so that $2m-1=1$, $f\in C^1[a,b]$, and $\pi: a=x_0<x_1<x_2=b$, the unique spline $s(x)$ of type I to f satisfies $L^*Lu=0$ on $[x_0,x_1]$ and on $[x_1,x_2]$. Thus

$$s(x)=\left\{\begin{array}{lll} c_{11}e^{2x}+c_{12}e^{-2x} & \text{on} & [x_0,x_1] \\ c_{21}e^{2x}+c_{22}e^{-2x} & \text{on} & [x_1,x_2]. \end{array}\right\}$$

Furthermore $s(x)$ also satisfies the boundary values

$$s(x_i)=f(x_i), \qquad i=0,1,2$$

and the continuity conditions $s(x_1+)=s(x_1-)$. We are thus led to the system of linear equations

$$c_{11}e^{2x_0}+c_{12}e^{-2x_0}=f(x_0)$$

$$c_{11}e^{2x_1}+c_{12}e^{-2x_1}=c_{21}e^{2x_1}+c_{22}e^{-2x_1}$$

$$c_{21}e^{2x_1}+c_{22}e^{-2x_1}=f(x_1)$$

$$c_{21}e^{2x_2}+c_{22}e^{-2x_2}=0$$

with the coefficient matrix

$$\begin{pmatrix} e^{2x_0} & e^{-2x_0} & 0 & 0 \\ e^{2x_1} & e^{-2x_1} & -e^{2x_1} & -e^{-2x_1} \\ 0 & 0 & e^{2x_1} & e^{-2x_1} \\ 0 & 0 & e^{2x_2} & e^{-2x_2} \end{pmatrix},$$

which the reader should observe is *banded.*

EXAMPLE 3

Let $\pi: 0=x_0<x_1<x_2=1$ be a partition of $[0,1]$. Let $s(x)$ be the piecewise Hermite polynomial of degree 3 interpolating $f(x)$ and $f'(x)$ at x_0, x_1 and x_2. Then

$$s(x)=a_{i3}x^3+a_{ii}x^2+a_{i1}x+a_{i0} \qquad \text{on} \quad [x_{i-1},x_i], \qquad i=1,2.$$

Thus $D^4s=0$ on each of these subintervals. Furthermore $s\in C^1[0,1]$ since $s'(x_1-)=s'(x_1+)$. But if we let $Lu=D^2u$, then $a_2=1$, $a_1=a_0=0$. It follows that $L^*u=D^2[a_2u]=D^2[u]$. Thus $L^*Lu=D^4u$. Note that $L^*Ls=0$ on each of the intervals $[x_{i-1},x_i]$. Letting $z_0=z_1=z_2=2$, it follows that $s\in Sp(L,\pi,z)$, where $z=(2,2,2)$.

EXERCISES

1. Find the formal adjoint L^* when

$$Lu=t^3u'''-2\sin tu''+u.$$

2. Find a differential operator L for which $N(L^*L)$ is spanned by

$$\{1,t,t^2,t^3,\sin t,\cos t,\sin 3t,\cos 3t\}.$$

3. Is $s(x)=a_0+a_1\sin t+a_2\sin 2t+a_3\sin 3t+a_4\sin 4t$ a spline function? Explain.

4. Let L be the differential operator (12) and let L^* given by (12) be its formal adjoint. Let $s(t)\in N(L^*L)$ solve the interpolatory constraints (5). Prove that the first and second integral relations are true in this case and prove that $s(t)$ is unique.

4.5 ERROR ANALYSIS

The most immediate error analysis for spline approximates s to a given function f defined on an interval $[a,b]$ follows from the first and second integral relations. Throughout our discussion $a=t_0<t_1<\cdots<t_n=b$ is some partition of $[a,b]$ and

$$h=\max_{1\leqslant i\leqslant n}(t_i-t_{i-1})$$

is the mesh of our partition. To see how the analysis works in a simple case, we start with *the* cubic spline interpolate s to f. Ideally, we would like $\|f-s\|\leqslant \beta h^4$ when $f\in C^4[a,b]$ and β is some constant independent of h. If this were the case, spline approximates would compete favorably error-wise with both cubic piecewise Lagrange and cubic piecewise Hermite interpolates to f in addition to having greater smoothness than either of these approximates. It can be proved (Hall, 1968) that when $f\in C^4[a,b]$

$$\|f-s\|\leqslant\frac{5}{384}\|f^{(4)}\|h^4.$$

We do not prove this estimate specifically, but we prove that when $f\in C^4[a,b]$

$$\|f^{(j)}-s^{(j)}\|\leqslant M_j\|f^{(j)}\|_2 h^{3.5-j},\qquad j=0,1,$$

and

$$\|f^{(j)}-s^{(j)}\|_2\leqslant \beta_j\|f^{(4)}\|_2 h^{4-j},\qquad j=0,1,$$

where the M_j's and β_j's are explicit constants independent of h.

We start by locating the zeros of $f(t)-s(t)$. Since $f(t_i)=s(t_i)$, $0\leqslant i\leqslant n$, it follows from Rolle's theorem that $f'-s'$ has n zeros ξ_i, $t_{i-1}<\xi_i<t_i$, in addition to the zeros $\xi_0=t_0$ and $\xi_{n+1}=t_{n+1}$. Since each t in $[a,b]$ belongs to

$[t_i, t_{i+1}]$ for some $0 \leqslant i \leqslant n-1$ and to $[\xi_i, \xi_{i+1}]$ for some $0 \leqslant i \leqslant n$, we have

$$f(t) - s(t) = \int_{t_i}^{t} (f' - s') \tag{1}$$

and

$$f'(t) - s'(t) = \int_{\xi_i}^{t} (f'' - s''). \tag{2}$$

We shall also need Schwarz's inequality

$$\int_a^b fg \leqslant \left[\int_a^b f^2 \right]^{1/2} \left[\int_a^b g^2 \right]^{1/2}$$

on $C[a,b]$. Defining $(f,g) = \int_a^b fg$ and $[\int_a^b f^2]^{1/2} = \|f\|_2$, we can prove

THEOREM SCHWARZ'S INEQUALITY

Let f and g belong to $C[a,b]$. Then

$$\int_a^b fg\,dx \leqslant \left[\int_a^b f^2\,dx \right]^{1/2} \left[\int_a^b g^2\,dx \right]^{1/2}.$$

Proof. Let t be a real number and consider

$$\int_a^b (f + tg)^2\,dx = \int_a^b (f^2 + 2tfg + t^2 g^2)\,dx$$

$$= \int_a^b f^2\,dx + 2t \int_a^b fg\,dx + t^2 \int_a^b g^2\,dx \geqslant 0.$$

Note that this is a quadratic equation $At^2 + Bt + C$. If $f \neq -tg$, then $At^2 + Bt + C > 0$ for all t; thus the only roots are complex, and the discriminant $B^2 - 4AC < 0$. But this is equivalent to

$$\int_a^b fg\,dx < \left[\int_a^b f^2\,dx \right]^{1/2} \left[\int_a^b g^2\,dx \right]^{1/2}.$$

If $\int_a^b (f + tg)^2\,dx = 0$ for some real numbers t, then $f(x) + tg(x) = 0$ for all x in $[a,b]$ and $f = -tg$. But in this case the theorem is also true, and we are done. ∎

Rolle's theorem coupled with Schwarz's inequality suffices to prove

THEOREM ERROR ESTIMATE

Let $s(t)$ be the cubic spline interpolate to $f(t) \in C^2[a,b]$. Then

$$\begin{cases} \|f-s\| \leqslant \sqrt{8}\ \|f''\|_2 h^{1.5} \\ \|f'-s'\| \leqslant \sqrt{8}\ \|f''\|_2 h^{0.5} \end{cases} \tag{3}$$

while

$$\begin{cases} \|f-s\|_2 \leqslant 16\|f''\|_2 h^2, \qquad \beta = \sqrt{1.28(b-a)} \\ \|f'-s'\|_2 \leqslant 4\|f''\|_2 h \\ \|f-s\|_2 \leqslant \|f''\|_2 \end{cases} \tag{4}$$

Proof. Combining (2) with Schwarz's inequality, we have

$$|f'(t)-s'(t)| \leqslant \left[\int_{\xi_i}^{t}(f''-s'')^2\right]^{1/2}\left(\int_{\xi_i}^{t_1}\right)^{1/2}$$

$$\leqslant \left[\int_a^b (f''-s'')^2\right]^{1/2}(t-\xi_i)^{1/2}$$

$$\leqslant \|f''-s''\|_2\sqrt{2h}\ .$$

But $\|f''-s''\|_2 + \|f''\|_2 = \|s''\|_2$ is the first integral relation for cubic spline interpolates $(L = D^2)$; therefore $\|f''-s''\|_2 \leqslant \|f''\|_2$. Thus taking the maximum of the left-hand side of the foregoing inequality over all t in $[a,b]$, we have

$$\|f'-s'\| \leqslant \|f''-s''\|_2\sqrt{2h}\ \leqslant 2\|f''\|_2\sqrt{2h}\ . \tag{5}$$

Since t_i is a zero of $f-s$, we have for each t in $[t_i, t_{i+1}]$

$$|f(t)-s(t)| = |\int_{t_i}^{t}(f'-s')|$$

$$\leqslant \|f'-s'\| \int_{t_i}^{t} 1$$

$$\leqslant \|f'-s'\|(t-t_i)$$

$$\leqslant \|f'-s'\| h.$$

Taking the maximum over all t in $[a,b]$ we find

$$\|f-s\| \leqslant \|f'-s'\|h. \tag{6}$$

Combining this with (5) we find

$$\|f-s\| \leqslant \sqrt{8}\,\|f''\|_2 h^{1.5}.$$

To obtain the L_2 estimate, note that

$$\frac{d}{dx}[f'(x)-s'(x)]^2 = 2[f'(x)-s'(x)][f''(x)-s''(x)]. \tag{7}$$

Let $t \in [\xi_i, \xi_{i+1}]$, and define

$$\|g\|_{i2} = \left[\int_{\xi_i}^{\xi_{i+1}} g^2(x)\,dx\right]^{1/2}.$$

Integrating (7) we find

$$\begin{aligned}[f'(t)-s'(t)]^2 &= \int_{\xi_i}^{t} \frac{d}{dx}[f'(x)-s'(x)]^2\,dx\\ &= 2\int_{\xi_i}^{t} (f'-s')(f''-s'')\\ &\leqslant 2\|f'-s'\|_{i2}\|f''-s''\|_{i2}.\end{aligned}$$

Integrating this inequality from ξ_i to t, we have

$$(\|f'-s'\|_{i2})^2 \leqslant 4\|f'-s'\|_{i2}\|f''-s''\|_{i2}h.$$

Dividing by $\|f'-s'\|_{i2} \neq 0$ and squaring both sides, we have

$$(\|f'-s'\|_{i2})^2 \leqslant 16(\|f''-s''\|_{i2})^2 h^2.$$

Summing these inequalities from $i=0$ to $i=n$, we see

$$\begin{aligned}(\|f'-s'\|_2)^2 &= \sum_{i=0}^{n} (\|f'-s'\|_{i2})^2\\ &\leqslant 16\left(\sum_{i=0}^{n} \|f''-s''\|_{i2}^2\right)h^2\\ &\leqslant 16(\|f''-s''\|_2)^2 h^2.\end{aligned}$$

Thus

$$\|f'-s'\|_2 \leqslant 4\|f''-s''\|_2 h$$
$$\leqslant 4\|f''\|_2 h, \tag{8}$$

invoking the first integral relation. An analogous argument observing

$$[f(t)-s(t)]^2 = \int_{t_i}^{t} \frac{d}{dx}[f(x)-s(x)]^2$$
$$= \int_{t_i}^{t} 2[f(x)-s(x)][f'(x)-s'(x)]$$

suffices to prove

$$\|f-s\|_2 \leqslant 4h\|f'-s'\|_2. \tag{9}$$

Combining this with (8) yields

$$\|f-s\|_2 \leqslant 16h^2\|f''\|_2. \qquad \blacksquare$$

If $f \leqslant C^4[a,b]$ we can considerably improve these estimates by invoking the second integral relation. In particular,

THEOREM

Let $s(t)$ be the cubic spline interpolate to $f \in C^4[a,b]$. Then

$$\|f-s\| \leqslant 16\sqrt{2}\ \|f^{(4)}\|_2 h^{3.5}$$

$$\|f'-s'\| \leqslant 16\sqrt{2}\ \|f^{(4)}\|_2 h^{2.5}.$$

Proof. The second integral relation for cubic splines ($L = D^2 = L^*$) gives us

$$(\|f''-s''\|_2)^2 = \int_a^b (f-s)f^{(4)}.$$

Invoking Schwarz's inequality we have

$$\|f''-s''\|_2^2 \leqslant \left[\int_a^b (f-s)^2\right]^{1/2}\left[\int_a^b [f^{(4)}]^2\right]^{1/2}$$
$$\leqslant \|f-s\|_2\|f^{(4)}\|_2. \tag{10}$$

But from the inequalities (8) and (9)

$$\begin{cases} \|f' - s'\|_2 \leqslant 4\|f'' - s''\|_2 h \\ \|f - s\|_2 \leqslant 4\|f' - s'\|_2 h \end{cases} \tag{11}$$

Thus

$$\|f - s\|_2 \leqslant 16h^2\|f'' - s''\|_2. \tag{12}$$

Combining (10) and (12), we find

$$\|f'' - s''\|_2^2 \leqslant \|f^{(4)}\|_2 16h^2\|f'' - s''\|_2$$

or

$$\|f'' - s''\|_2 \leqslant 16\|f^{(4)}\|_2 h^2.$$

But from (1) and (2) with Schwarz's inequality

$$\|f' - s'\| \leqslant \|f'' - s''\|_2\sqrt{2h}$$

$$\leqslant 16\sqrt{2}\,\|f^{(4)}\|_2 h^{2.5}$$

and

$$\|f - s\| \leqslant \|f' - s'\|h$$

$$\leqslant 16\sqrt{2}\,\|f^{(4)}\|_2 h^{3.5}. \qquad \blacksquare$$

These estimates can be sharpened greatly using the Peano kernal theorem. In particular, one can prove (see Hall (1969) and de Boor (1968)).

THEOREM

If $f \in C^4[a,b]$ then

$$\|f - s\|_\infty \leqslant \frac{5}{384}\|f^{(4)}\|_\infty h^4$$

$$\|f' - s'\|_\infty \leqslant \left[\frac{\sqrt{3}}{216} + \frac{1}{24}\right]\|f^{(4)}\|_\infty h^3$$

$$\|f'' - s''\|_\infty \leqslant \left[\frac{1}{12} + \frac{1}{3}\underline{h}^{-1}h\right]\|f^{(4)}\|_\infty h^2$$

$$\|f^{(3)} - s^{(3)}\|_\infty \leqslant \tfrac{1}{2}\left[1 + \underline{h}^{-2}h^2\right]\|f^{(4)}\|_\infty h, \tag{13}$$

where

$$\|g\|_\infty = \max_{1\leqslant i\leqslant n} \sup_{x_{i-1}<x<x_i} |g(x)|$$

$$\underline{h} = \left[\min_{1\leqslant i\leqslant n}(x_i - x_{i-1})\right]^{-1}$$

Our methods also extend to higher odd degree interpolating polynomial and other splines (see Exercises and Schwarz (1970)).

EXERCISES

1. Let $s(t)\in N(L)$, where

$$Lu(t) = (D^2-1)(D^2-2)u(t)$$

$$Lu(t) = (D^2+1)D^2u(t).$$

Let $a\leqslant t_1<t_2<t_3<t_4=b$. Does there always exist a unique solution to the interpolation problem $s(t_i)=\gamma_i$, $1\leqslant i\leqslant 4$, where the γ_i's are any four real numbers? Explain.

2. Let L and L^* be given by (12) and (13). Let $u_1, u_2, \ldots, u_{2m}$ span $N(L^*L)$ (i.e., $L^*Lu_i(t)=0$ for all t in $[a,b]$ for each $1\leqslant i\leqslant 2m$). Let $s(t)$ be a function satisfying the following generalized boundary value problem:

$$\begin{cases} L^*Ls(t)=0, & \text{on each interval } (t_i, t_{i+1}) \\ s \text{ satisfies boundary conditions (5),} \end{cases}$$

where $f(t)$ is a given function.
   (a) Prove that s satisfies the first integral relation.
   (b) Prove that s always exists and is unique regardless of the smoothness class to which f belongs interior to (a,b).
   (c) Prove that s satisfies the second integral relation.

3. Let $s(t)$ be the unique polynomial spline of degree $2m-1$ between knots satisfying (5). Prove that
   (a) $\|f^{(k)}-s^{(k)}\|\leqslant M_k h^{m-k-1/2}$,
   where $M_k=\sqrt{m-1}\,(m-2)(m-3)\cdots k\|f^{(m)}\|_2$, and $f\in C^m[a,b]$.
   (b) $\|f^{(k)}-s^{(k)}\|\leqslant \beta_k\|D^m(f-s)\|_2 h^{m-k}\leqslant \beta_k\|f^{(m)}\|_2 h^{m-k}$,
   where $M_k=\dfrac{2^{m-k}(m-k-1)!}{k!}$, and $f\in C^m[a,b]$.
   (c) $\|f^{(k)}-s^{(k)}\|\leqslant A_k\|f^{(2m)}\|_2\,\underline{h}^{2m-k-1/2}$,
   where $A_k=\dfrac{2(m-1)m!}{mk!}$ and $f\in C^{2m}[a,b]$.

REFERENCES

Ahlberg, J. H. and E. N. Nilson. Convergence properties of the spline fit. *SIAM J. Appl. Math.*, **11**, (1963), 95–104.

Ahlberg, J. H., E. H. Nilson, and J. L. Walsh. Fundamental properties of generalized splines. *Proc. Nat. Acad. Sci. (U.S.)*, **52** (1964), 1412–1419.

Ahlberg, H. H., E. N. Nilson, and J. L. Walsh. *The theory of splines and their applications*. Academic Press, New York, 1967.

Anselone, P. J. and P. J. Laurent. A general method for the construction of interpolating or smoothing spline-functions. *Numer. Math.*, **12** (1968), 66–82.

Atkinson, K. E. On the order of convergence of natural cubic spline interpolation. *SIAM J. Numer. Anal.*, **5** (1968), 89–101.

Birkhoff, G. and C. de Boor. Piecewise polynomial interpolation and approximation. In *Approximation of functions*, American Elsevier, New York, 1965, pp. 164–190.

Birkhoff, G. and C. de Boor. Error bounds for spline interpolation. *J. Math. Mech.*, **13** (1964), 827–836.

Cheney, E. W. and F. Schurer. A note on the operators arising in spline approximation. *J. Approx. Theory*, **1** (1968), 94–102.

de Boor, Carl. Bicubic spline interpolation. *J. Math. Phys.*, **41** (1962) 212–218.

de Boor, C. On uniform approximation by splines. *J. Approx. Theory*, **1** (1968), 219–235.

de Boor, C. On the convergence of odd degree spline interpolation. *J. Approx. Theory*, **1** (1968), 452–463.

de Boor, C. Subroutine package for calculating with *B* splines. Technical report LA-4728-MS, Los Alamos Scientific Laboratory, August 1971.

Forsythe, G. Solving linear equations can be interesting. *Bull. AMS*, **59** (1953), 299–329.

Greville, T. N. E. Interpolation by generalized spline functions. MRC Technical Report, University of Wisconsin, 1964.

Greville, T. N. E. and H. Vaughn. Smoothest moving average interpolation formulas for equally spaced data. *Trans. Soc. Actuar.*, **6** (1959), 413–476.

Greville T. N. E. and H. Vaughn. Polynomial interpolation in terms of symbolic operators. *Trans. Soc. Actuar.*, **6** (1954), 413–476.

Hall, C. A. On error bounds for spline interpolation. *J. Approx. Theory*, **1** (1968), 209–218.

Hedstrom, G. W. and R. S. Varga. Application of Besov spaces to spline approximation. *J. Approx. Theory*, **4** (1971), 295–327.

Jerome, J. W. and R. S. Varga. Generalizations of spline functions and applications to nonlinear boundary value and eigenvalue problems. In *Theory and application of spline functions*, Greville, ed. Academic Press, New York, 1969, 103–155.

Jerome, J. W. and L. L. Schumaker. On *L–g* splines. *J. Approx. Theory*, **2** (1969b), 29–49.

Lucas, T. R. A theory of generalized splines. *Numer. Math.*, **15** (1970), 359–370.

Lyche, Tom, and L. Schumaker. Computation of smoothing and interpolating natural splines via local bases. Report of Center of Numerical Analysis, University of Texas, April 1971.

Marsden, M. and I. J. Schoenberg. On variation diminishing approximation methods. MRC Report 694, June 1966.

Meir, A. and A. Sharma. Degree of approximation of spline interpolation. *J. Math Mech.* **15**, no. 5 (1966), 759–768.

Meir, A. and A. Sharma. Convergence of a class of interpolatory splines. *J. Approx. Theory*, **1** (1968), 243–250.

Meir, A. and Sharma, A. On uniform approximation by cubic splines. *J. Approx. Theory*, **2** (1969), 270–274.

Noble, B. *Applied linear algebra*, Prentice-Hall, Englewood Cliffs, N. J., 1969.

Nord, S. Approximation properties of the spline fit. *BIT* **7** (1967) 132–144.

Schoenberg, I. J. Contributions to the problem of approximation of equidistant data by analytic functions. *Quart. Appl. Math.*, **4** (1946), 345–369.

Schoenberg, I. J. On Polya frequency functions. IV: The fundamental splines and their limits. MRC Report 567, 1965.

Schoenberg, I. J. On spline functions. MRC Report 625, University of Wisconsin, May 1966.

Schoenberg, I. J. Spline functions and the higher derivatives. MRC Report 761, July 1967.

Schoenberg, I. J. *Spline analysis*, Prentice-Hall, Englewood Cliffs, N. J., 1973.

Schultz, M. H. Error bounds for polynomial spline interpolation. *Math. Comp.*, **24** (1970), 507–515.

Schultz, M. H. and R. S. Varga. *L*-splines. *Numer. Math.*, **10** (1967), 345–369.

Schumaker, L. L. Work on splines (to be published), University of Texas.

Schurer, F. A note on interpolating periodic quintic splines with equally spaced nodes. *J. Approx. Theory*, **1** (1968), 493–500.

Subotin, Yu. N. Piecewise polynomial interpolation. *Math. Zametki*, **1** (1967).

Swartz, B. $0(h^{2n+2-1})$ Bounds on some spline interpolation errors. *Bull. AMS*, **74** (1968), 1072–1078.

Swartz, Blair K. $(h^{k-j}(D^k f, h))$ Bounds on some spline interpolation errors. LA-4477, Los Alamos Scientific Laboratory, 1970.

Swartz, B. K. and R. S. Varga. Error bounds for spline and *L*-spline interpolation. *J. Approx. Theory*, **6** (1972) 6–49.

Varga, R. S. *Matrix iterative analysis*, Prentice-Hall, Englewood Cliffs, N. J., 1962.

Varga, R. S. Error bounds for spline interpolation. In *Approximations with special emphasis on spline functions*, I. J. Schoenberg, ed. Academic Press, New York, 1969, pp. 367–388.

Walsh, J. H., J. H. Ahlberg and E. N. Nilson. Best approximation properties of the spline fit. *J. Math. Mech.*, **11** (1962), 225–234.

5

APPROXIMATING FUNCTIONS OF SEVERAL VARIABLES

5.1 SURFACE FITTING

We now look at the problem of *approximating functions of two variables* or the problem of surface fitting. Specifically, suppose Ω is a closed, bounded set of points in the real plane and $f(p)=f(x,y)$ is a real-valued function defined in Ω. It follows that $f(x,y)$ graphs as a surface in 3-space and that this graph is smooth if f is continuous on Ω (see Figure 5.1).

It is only natural to attempt to generalize the methods of approximating functions of one variable to the problem of approximating functions of two variables. However, this can be done with ease only in very special cases, and the general theory of approximation of functions of several variables is nowhere near as advanced as its one-variable counterpart; thus many simply described and important unsolved problems await the interested researcher.

The natural first step in any generalization from the one-variable case would be to attempt to find the Lagrange and Hermite polynomials in two or more variables. Specifically, let $p_0,p_1,p_2,\ldots,p_n$ be any $n+1$ distinct,

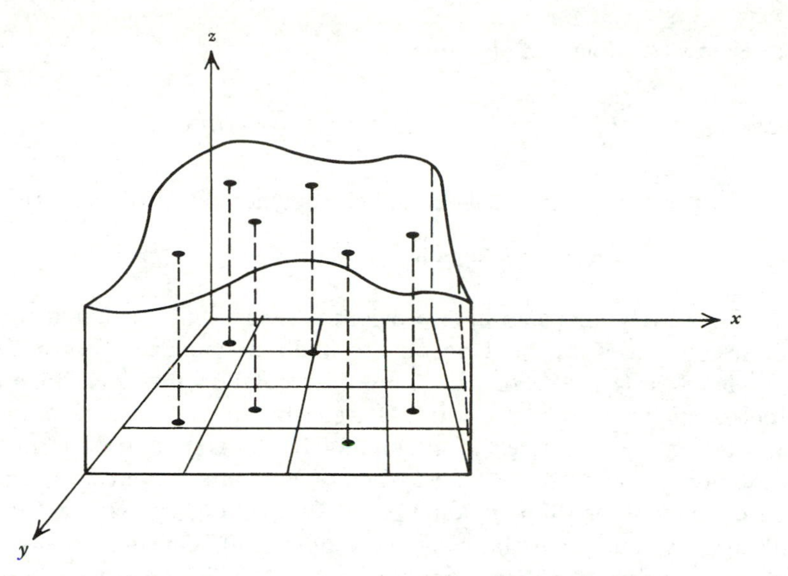

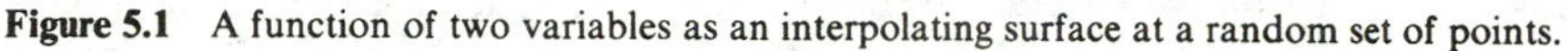
Figure 5.1 A function of two variables as an interpolating surface at a random set of points.

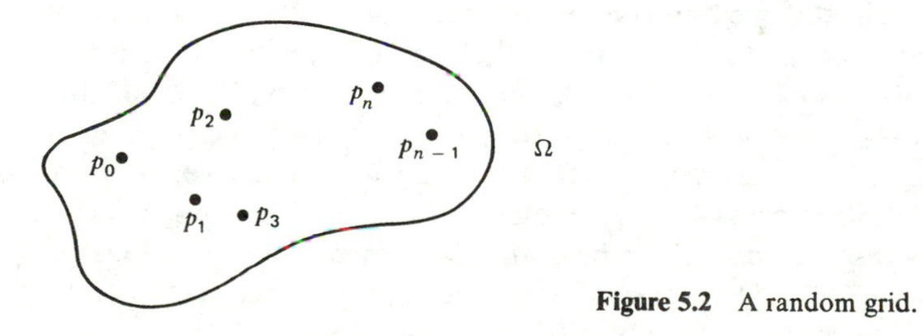

Figure 5.2 A random grid.

randomly placed points of Ω (known as a *random grid*), and let $y_0, y_1, \ldots, y_n$ be any $n+1$ real numbers (see Figure 5.2). The problem, then, is to find a polynomial $P(x,y)$ of degree n in the two variables x and y solving the interpolation problem

$$P(p_i) = f(p_i), \qquad p_i = (x_i, y_i), \qquad 0 \leqslant i \leqslant n,$$

where a polynomial of degree n in x and y is any algebraic function of the form

$$P(x,y) = \sum_{i,j=0}^{n} a_{ij} x^i y^j.$$

This problem always has a solution (see Prenter (1971)) and it is a simple matter to write down an algorithm

$$P(x,y)=\sum_{i=0}^{n} f(p_i)l_i(p)=\sum_{i=0}^{n} f(x_i,y_i)l_i(x,y)$$

for P, where the l_i's are easily computed polynomials of degree n solving

$$l_i(p_j)=\delta_{ij}, \qquad 0\leqslant i\leqslant n.$$

Unfortunately, this algorithm provides existence but not uniqueness. In fact, in most cases it is quite possible to find a polynomial of lower degree or with fewer terms solving the same interpolation problem. Analogous criticism applies to the known full Hermite interpolate on a random grid. The problems encountered in attempting full Lagrange and Hermite interpolation are severely multiplied when one attempts to construct smooth (i.e., continuous or differentiable) *piecewise* Lagrange, *piecewise* Hermite, and other spline interpolates on a random grid. One can easily demonstrate cases in which such smooth interpolates do not exist. The problems here arise from the location of the knots p_i of the grid and the matching of separate elements of the piecewise interpolate across boundaries. In all the well-known cases there must be some sort of regularity in the location of the knots before one arrives at smooth approximating functions. With this in mind, we briefly survey some of the approximates that have been constructed on rectangular and triangular grids, indicating directions in which these results have been generalized. Throughout, our discussions $C^k[\Omega]$ denotes the set of all functions $f(x,y)$ defined on R whose kth partials (mixed and otherwise) are continuous on Ω. The set $C[\Omega]=C^0[\Omega]$ is the set of all functions that are continuous on Ω; the Tchebycheff norm on $C[\Omega]$, $f\in C[\Omega]$, is defined by

$$\|f\|=\max_{p\in\Omega}|f(p)|.$$

The space $C^{m,n}[\Omega]$ is the set of functions f whose mth partials in the x direction and nth partials in the y direction are continuous on Ω. Throughout the book $D_x^k f(p)=(\partial^k f/\partial x^k)(p)$ and $D_y^k f(p)=(\partial^k f/\partial y^k)(p)$.

5.2 APPROXIMATES ON A RECTANGULAR GRID

Let Ω be the rectangle $R=\{(x,y): a\leqslant x\leqslant b \text{ and } c\leqslant y\leqslant d\}$. Let $\pi_x: a=x_0<x_1<\cdots<x_n=b$ and $\pi_y: c=y_0<y_1<\cdots<y_m=d$ be partitions of the intervals $[a,b]$ and $[c,d]$, respectively. The set of points

$$\pi=\{(x_i,y_j): 0<i<n;\ 0\leqslant j\leqslant m\}=\pi_x\cdot\pi_y$$

forms a *rectangular grid* partitioning the rectangle R. Let $h_x = \max\{x_{i+1} - x_i : 0 \leqslant i \leqslant n-1\}$ and $h_y = \max\{y_{j+1} - y_j : 0 \leqslant j \leqslant m-1\}$ be the mesh of the partitions π_x and π_y, respectively. The *mesh of* π is then given by $h = \max\{h_x, h_y\}$ and is easily seen to be the length of the longest side of all the rectangles $R_{ij} = [x_i, x_{i+1}] \times [y_j, y_{j+1}]$ formed by the partition π (see Figure 5.3).

To construct the analogs of one-dimensional Lagrange, piecewise Lagrange, and other piecewise polynomial interpolates on a rectangular grid such as that of Figure 5.3, one forms the so-called tensor products of the appropriate one-dimensional basis functions to derive a basis for the two-dimensional analog. Specific examples are provided next.

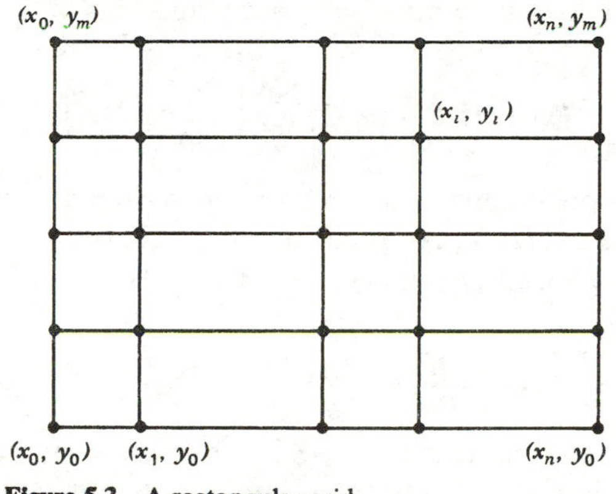

Figure 5.3 A rectangular grid.

Lagrange Polynomials

Let $\{l_i(x) : 0 \leqslant i \leqslant n\}$ and $\{l_j(y) : 0 \leqslant i \leqslant m\}$ be the Lagrange polynomials of degrees m and n, respectively, solving the interpolation problem

$$l_i(x_k) = \delta_{ik}, \qquad 0 \leqslant i,k \leqslant n$$

$$l_j(y_\mu) = \delta_{j\mu}, \qquad 0 \leqslant j,\mu \leqslant m.$$

Then letting

$$l_{ij}(x,y) = l_i(x) l_j(y),$$

it follows that

$$\boxed{P(x,y) = \sum_{\substack{0 \leqslant i \leqslant n \\ 0 \leqslant j \leqslant m}} f(x_i, y_j) l_{ij}(x,y)} \tag{1}$$

is a polynomial of degree n in x and m in y solving the interpolation problem

$$P(x_i,y_j)=f(x_i,y_j), \qquad 0\leqslant i\leqslant n;\ 0\leqslant j\leqslant m. \tag{2}$$

This follows simply by observing that

$$\delta_{ij}(x,y)=\delta_{ik}\delta_{j\mu}=\begin{cases}1 & \text{if } i=k, j=\mu\\ 0 & \text{otherwise.}\end{cases}$$

This establishes the *existence* of such a polynomials.

Moreover, $P(x,y)$ is the *unique* polynomial of degree n in x and m in y solving this problem. To see this, suppose $p(x,y)=\sum_{j=0}^{m}\sum_{i=0}^{n}a_{ij}x^iy^j$ is a polynomial of degree n in x and of degree m in y also solving (2). Let k be a fixed integer between 0 and n, let $x=x_k$, and let

$$p_k(y)=\sum_{j=0}^{m}b_{kj}y^j=\sum_{j=0}^{m}\left(\sum_{i=0}^{n}a_{ij}(x_k)^i\right)y^j.$$

Then $p_k(y)$ is a polynomial of degree m solving the interpolation problem $p_k(y_l)=f(x_k,y_l)$, $0\leqslant l\leqslant m$. As such, its coefficients b_{kj}, $0\leqslant j\leqslant m$ are unique. But the system of equations

$$b_{kj}=\sum_{i=1}^{n}a_{ij}(x_k)^i, \qquad k=0,1,\ldots,n$$

has a unique solution $(a_{0j},a_{1j},\ldots,a_{nj})$ for each $j=0,1,\ldots,m$, since the coefficient matrix $c_{ik}=(x_k^i)$, being a Vandermonde matrix, is nonsingular. That is, we have proved the following theorem.

THEOREM

There exists a unique polynomial of degree n in x and degree m in y solving the interpolation problem (2) on a rectangular grid.

One cannot in general lower the degree of $P(x,y)$. That is, there exist sets $\{(x_i,y_j):0\leqslant i\leqslant n \text{ and } 0\leqslant j\leqslant m\}$ of points in the plane such that no polynomial of the form

$$q(x,y)=\sum_{j=0}^{m'}\sum_{i=0}^{n'}a_{ij}x^iy^j, \qquad m'<m \quad\text{or}\quad n'<n$$

can solve

$$q(x_i,y_j)=t_{ij}, \qquad 0\leqslant i\leqslant n;\ 0\leqslant j\leqslant m,$$

where the t_{ij}'s are given real numbers (see Exercises).

The functions $l_{ij}(x,y)$ are popularly known as *shape functions* in engineering circles. To mathematicians they are simply a *basis* (and a rather nice one for computational purposes) for the Lagrange polynomials of degree n in x and m in y. Note that there are $(m+1)(n+1)$ l_{ij}'s. We prove in the next section that the l_{ij}'s are linearly independent on R; so that

THEOREM

The set $\mathcal{P}_{mn}$ of Lagrange polynomials of degree n in x and degree m in y on a rectangular grid is a *linear space* of dimension $(m+1)(n+1)$ with basis $\{l_{ij}: 0 \leqslant i \leqslant n \text{ and } 0 \leqslant j \leqslant m\}$.

The same arguments regarding choice of basis for ease of computation that applied to one-dimensional approximates apply to two-dimensional approximates. The shape functions $l_{ij}(x,y)$ are an especially nice choice for Lagrange interpolates. Graphs of some of these functions (Figure 5.4*) are helpful in visualizing the geometry of Lagrange polynomials in two variables on rectangular grids.

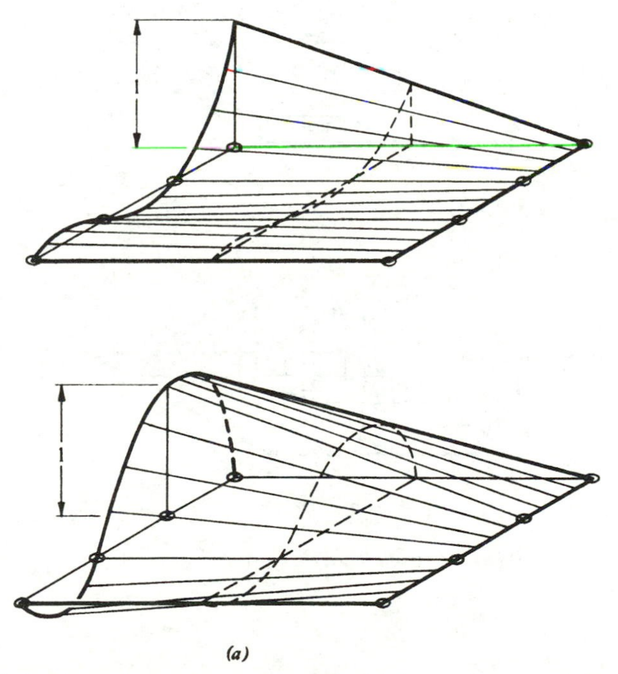

(*a*)

Figure 5.4 Typical basis Lagrangian polynomials. (*a*) Degree 1 in x and degree 3 in y. (*b*) Degree 2 in both x and y. (*c*) Degree 5 in x and degree 4 in y.

*The author is grateful to Professor O. C. Zienciewicz for permission to reproduce these figures from his McGraw-Hill text.

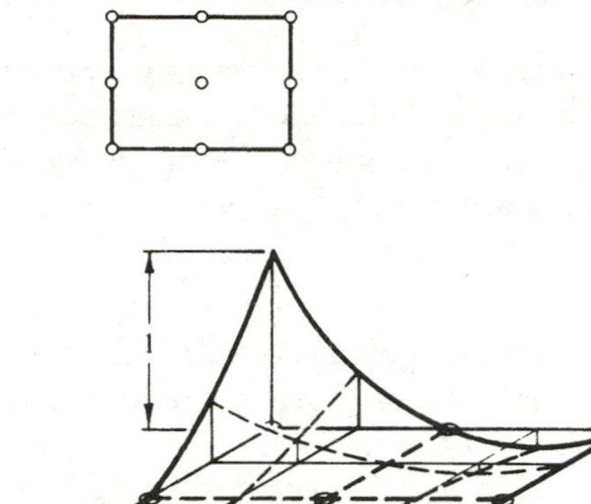

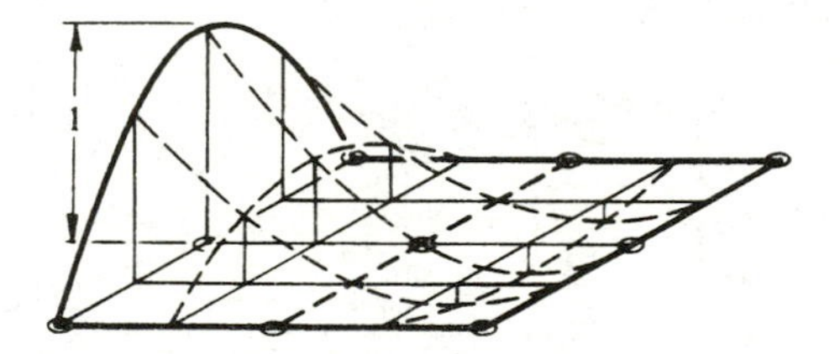

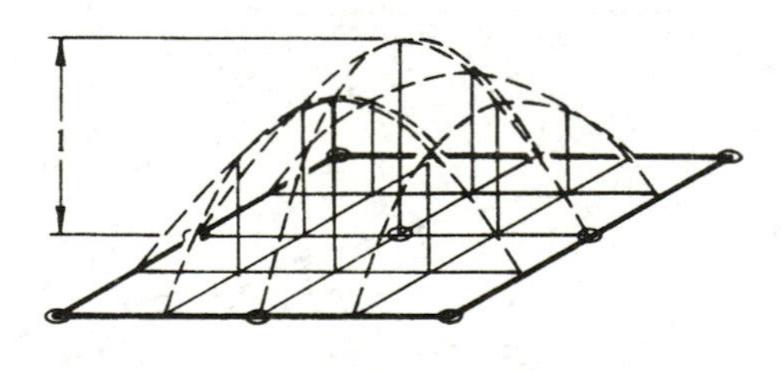

(b)

Figure 5.4 (*Continued*)

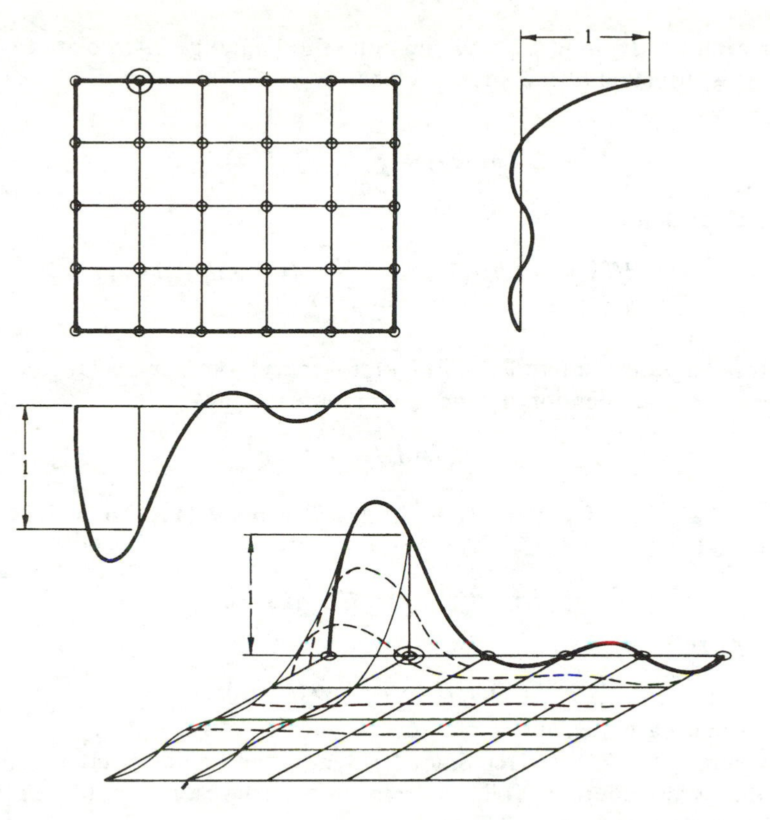

Figure 5.4 (*Continued*)

We can also estimate the error

$$\|f-P\| = \max_{P\in R} |f(p)-P(p)|$$

as well as $\|D^{|j|}f - D^{|j|}P\|$ for mixed partials. Specifically, let $j=(j_1,j_2)$ be a pair of positive integers, $0 \leqslant j_2 \leqslant m$ and $0 \leqslant j_1 \leqslant n$ and let $|j| = j_1 + j_2$. Then $D^{|j|}f = \partial^{|j|}f / \partial x^{j_1} \partial y^{j_2}$. Methods developed by Schultz (1969) for rectangular grids are useful in deriving these estimates, although the methods we choose for triangular grids are applicable as well. We first introduce some notation that considerably simplifies our proof. For each fixed $\bar{x}$ in $[a,b]$ let $I_{\bar{x}}f$ be the Lagrange interpolate to $f(\bar{x},y)$ in the y direction interpolating $f(\bar{x},y)$ at the knots of π_y. Specifically,

$$I_{\bar{x}}f(\bar{x},y) = \sum_{j=0}^{m} f(\bar{x},y_j) l_j(y).$$

For each fixed $\bar{y}$ in $[c,d]$, let $I_{\bar{y}}f(x,\bar{y})$ be the Lagrange interpolate to $f(x,\bar{y})$ in the x direction at the knots π_x. Then

$$I_{\bar{y}}f(x,\bar{y})=\sum_{i=0}^{n} f(x_i,\bar{y})l_i(x).$$

It is clear that

$$If(x,y)=I_xI_yf(x,y)=\sum_{\substack{0\leqslant i\leqslant n\\ 0\leqslant i\leqslant m}} f(x_i,y_j)l_i(x)l_j(y)$$

is the Lagrange interpolate $P(x,y)$ to $f(x,y)$ on our rectangular grid. Moreover, the operators I_x and I_y commute, giving

$$I_xI_yf=I_yI_xf=If=P. \tag{3}$$

More than that, if $D_y^kf=\partial^kf/\partial y^k$ exists at a point (x,y) in R, it is easily seen that

$$D_y^kI_yf(x,y)=I_yD_y^kf(x,y). \tag{4}$$

Similarly, we have

$$D_x^kI_xf(x,y)=I_xD_x^kf(x,y) \tag{5}$$

when that partial exists.

Assume $f\in C^{n+1,m+1}[R]$. Since $I_xf(x,y)$ is the Lagrange interpolate to f in the y direction, it follows from our error estimates for Lagrange interpolates (see Section 2.5) that

$$\|f-I_xf\|\leqslant\frac{\|D_y^{m+1}f\|}{4(m+1)}h_y^{m+1}. \tag{6}$$

Now invoke the triangle inequality, so that with $I_{xy}=I_xI_y$

$$\|f-If\|\leqslant\|f-I_xf\|+\|I_xf-I_{xy}f\|. \tag{7}$$

To estimate the second quantity on the right-hand side, use (3) to obtain

$$I_xf-I_{xy}f=(I_xf)-I_y(I_xf).$$

But $I_y(I_xf)$ is the Lagrange interpolate of degree $n+1$ in the x direction to $I_xf(x,y)$, which interpolates I_xf at the knots of π_x. It again follows from our Lagrange estimates that

$$\|I_xf-I_{xy}f\|\leqslant\frac{\|D_x^{n+1}(I_xf)\|}{4(n+1)}h_x^{n+1}, \tag{8}$$

considering $I_x f$ as a function of x. But from (5), $D_x^{n+1}(I_x f) = I_x(D_x^{n+1} f)$ is the Lagrange interpolate of degree m in y to $D_x^{n+1} f$ considered as a function of y with knots at π_y. Thus we have

$$\|D_x^{n+1} f - I_x(D_x^{n+1} f)\| \leq \frac{\|D_y^{m+1} D_x^{n+1} f\|}{4(m+1)} h_y^{m+1};$$

so that

$$\|D_x^{n+1}(I_x f)\| \leq \|D_x^{n+1} f\| + \frac{\|D_y^{m+1} D_x^{n+1} f\|}{4(m+1)} h_y^{m+1}. \tag{9}$$

Let

$$\boxed{\beta_{mn} = \max\left\{ \frac{\|D_y^{m+1} f\|}{4(m+1)}, \frac{1}{4(n+1)} \left[\|D_x^{n+1} f\| + \frac{\|D_y^{m+1} D_x^{n+1} f\|}{4(m+1)} \right] h_x^{n+1} \right\}.} \tag{10}$$

Combining (6), (7), (8), and (9) we find

$$\|f - If\| = \|f - P\| \leq \beta_{mn} \max\{h_x^{n+1}, h_y^{m+1}\}.$$

That is, we have proved

THEOREM

Let $f \in C^{n+1,m+1}[R]$ and let $P = If$ be the Lagrange interpolate of degree n in x and degree m in y to f on a rectangular grid π of R. Then

$$\boxed{\|f - P\| \leq \beta_{mn} \max\{h_x^{n+1}, h_y^{m+1}\},}$$

where β_{mn} is given by (10).

If $f \in C^n[R]$ and $m = n$, then clearly

$$\|f - P\| \leq \beta_n h^{n+1},$$

where $\beta_n = \beta_{nn}$.

To estimate the errors $\|D^{[j]} f - D^{[j]} P\|$, we simply use the estimates for the one-dimensional case $F' - p'$, $F'' - p''$, and so on (see Section 2.5) when $F(x)$ is a function of one variable and $p(x)$ is its Lagrange interpolate of degree m or n. For example,

$$\begin{aligned} \|D_x f - D_x P\| &= \|D_x f - D_x If\| \\ &\leqslant \|D_x f - D_x I_y f\| + \|D_x I_y f - D_x I_x I_y f\| \\ &\leqslant \|D_x f - D_x I_y f\| + \|D_x I_y f - I_x D_x I_y f\|, \end{aligned} \tag{11}$$

where the last line follows from (5). But once again invoking the estimates of Section 2.5, we write

$$\|D_x f - D_x(I_y f)\| \leqslant \frac{\|D_x^{n+1} f\|}{2} h_x^n \tag{12}$$

while

$$\|D_x I_y f - I_x D_x I_y f\| \leqslant \frac{\|D_y^{m+1} D_x I_y f\|}{4(m+1)} h_y^{m+1}, \tag{13}$$

since $I_x D_x I_y f$ is the Lagrange interpolate of degree m in the y direction to $D_x I_y f$. But

$$D_y^{m+1} D_x I_y f = D_x D_y^{m+1} I_y f = D_x I_y D_y^{m+1} f,$$

and from Section 2.5 we have

$$\|D_x D_y^{m+1} f - D_x I_y D_y^{m+1} f\| \leqslant \frac{\|D_x^{n+1} D_y^{m+1} f\|}{2} h_x^n.$$

Thus

$$\|D_y^{m+1} D_x I_y f\| \leqslant \frac{\|D_x^{n+1} D_y^{m+1} f\|}{2} h_x^n + \|D_x D_y^{m+1} f\|. \tag{14}$$

Combining (11), (12), (13), and (14), we find

$$\boxed{\|D_x f - D_x P\| \leqslant \gamma_{mn} \max\{h_x^n, h_y^{m+1}\},}$$

where

$$\gamma_{mn} = \max\left\{ \frac{\|D_x^{n+1} f\|}{2}, \frac{1}{4(m+1)} \left[\|D_x D_y^{m+1} f\| + \frac{\|D_x^{n+1} D_y^{m+1} f\|}{2} h_x^n \right] \right\}.$$

Waning sadism forbids us to go further. The reader can see that in general we have

THEOREM

If $f \in C^{n+1,m+1}[R]$, then

$$\boxed{\|D_x^j D_y^k f - D_x^j D_y^k P\| \leqslant \alpha_{jk} \min\{h_x^{n+1-j}, h_y^{m+1-k}\},} \tag{15}$$

where α_{jk} is a constant independent of h_x and of h_y but dependent on the Tchebycheff norm of mixed partials of f.

Appropriate decreases in the power of h are experienced with decrease in the smoothness of f just as in the one-variable case. The enthusiastic reader is referred to the Exercises. Note that $P \in C^k[R]$ for all k, which means that f can be differentiated an infinite number of times.

If we renumber the l_{ij}'s to obtain $\phi_1(x,y), \phi_2(x,y), \ldots, \phi_N(x,y)$, it follows from (1) and $P = f_N$ that

$$f_N(x,y) = c_1\phi_1(x,y) + c_2\phi_2(x,y) + \cdots + c_N\phi_N(x,y),$$

where $N = (n+1)(m+1)$. Viewed in this light, $P = f_N$ is simply an approximate to f from the N-dimensional subspace $W_N = \operatorname{span}\{\phi_1, \phi_2, \ldots, \phi_N\}$ of $X = C[R]$, which uniquely solves the interpolation problem

$$f_N(p_k) = f(p_k), \qquad 1 \leqslant i \leqslant N,$$

where the set $\{p_k : 1 \leqslant k \leqslant N\} = \{(x_i, y_j) : 0 \leqslant i \leqslant n,\ 0 \leqslant j \leqslant m\}$. Thus we are again looking at the problem of approximating from finite dimensional subspaces subject to pure interpolatory constraints (viz. Section 1.6).

Piecewise Lagrange Polynomials

The piecewise Lagrange polynomials are particularly popular among engineers and are formed by taking sums of products of piecewise basis Lagrange polynomials in the x direction with piecewise basis Lagrange polynomials in the y direction. If the degrees in the x and y directions are the same, we obtain piecewise bilinear, piecewise biquadratic, piecewise bicubic, and so forth, Lagrange polynomials on a rectangular grid. Figure 5.5 is the graph of a basis piecewise bilinear Lagrange polynomial $\phi_{ij}(x,y)$ (a shape function). Note that each such function vanishes outside at most every four rectangles of our grid. This feature (small compact support, in the parlance of the mathematician) makes them very attractive for comput-

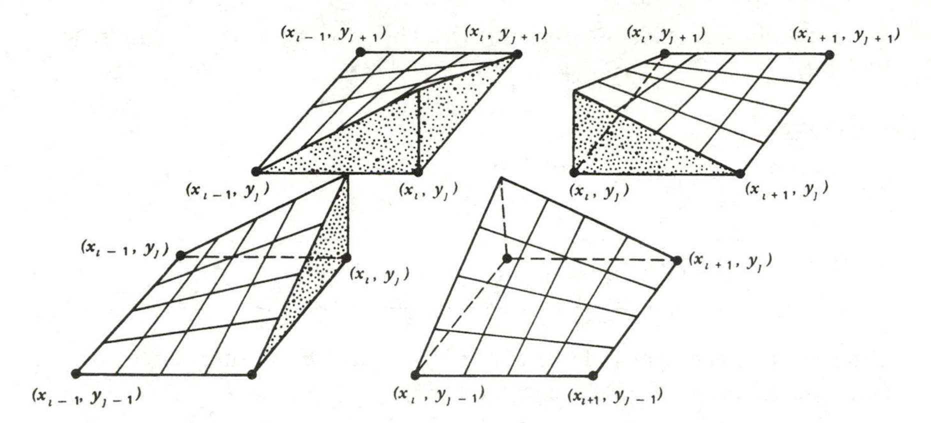

Figure 5.5 A canonical basis bilinear Lagrange polynomial over its region of support.

ing approximate solutions of partial differential equations. The error analysis for these approximates follows from the error analysis for the Lagrange polynomials on a rectangular grid, just as the error analysis for piecewise Lagrange polynomials in one dimension followed from that of the one-dimensional Lagrange polynomials. In particular, if $s_1(x,y)$ is the piecewise bilinear interpolate

$$s_1(x,y) = \sum_{\substack{0 \leqslant i \leqslant n \\ 0 \leqslant j \leqslant m}} f(x_i,y_j)\phi_{ij}(x,y),$$

where the ϕ_{ij}'s are the shape functions illustrated in Figure 5.5, then from our error analysis of the last section

$$\boxed{\|f - s_1\| \leqslant \beta_{11}h^2}$$

if $f \in C^2[R]$, and

$$\boxed{\|D_x f - D_x s_1\| \leqslant \alpha_{11}h}$$

$$\boxed{\|D_y f - D_y s_1\| \leqslant \alpha_{11}h.}$$

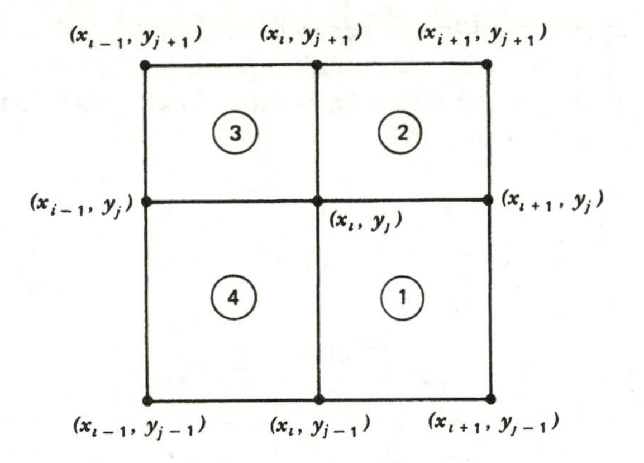

As the diagram indicates, the equation of $\phi_{ij}(x,y)$ is given by

$$\phi_{ij}(x,y)=\begin{cases} \dfrac{(x_{i+1}-x)}{(x_{i+1}-x_i)}\,\dfrac{(y-y_{j-1})}{(y_j-y_{j-1})} & \text{on } 1\\[2ex] \dfrac{(x_{i+1}-x)}{(x_{i+1}-x_i)}\,\dfrac{(y_{j+1}-y)}{(y_{j+1}-y_j)} & \text{on } 2\\[2ex] \dfrac{(x-x_{i-1})}{(x_i-x_{i-1})}\,\dfrac{(y_{j+1}-y)}{(y_{j+1}-y_j)} & \text{on } 3\\[2ex] \dfrac{(x-x_{i-1})}{(x_i-x_{i-1})}\,\dfrac{(y-y_{j-1})}{(y_j-y_{j-1})} & \text{on } 4\\[2ex] 0 & \text{otherwise.} \end{cases}$$

Piecewise Bicubic Hermites

Next we seek a piecewise bicubic polynomial $f_N(x,y)$ subject to the interpolatory constraints

$$f_N(x_i,y_j)=f(x_i,y_j),$$

$$\frac{\partial f_N}{\partial x}(x_i,y_j)=\frac{\partial f}{\partial x}(x_i,y_j),$$

$$\frac{\partial f_N}{\partial y}(x_i,y_j)=\frac{\partial f}{\partial y}(x_i,y_j),$$

$$\frac{\partial^2 f_N}{\partial x\partial y}(x_i,y_j)=\frac{\partial^2 f}{\partial x\partial y}(x_i,y_j), \qquad 0\leqslant i,j\leqslant n.$$

(Here we assume without loss of generality that $m=n$.) The solution to this problem is also easily written down after recalling the basic cubic Hermite polynomials $\phi_i(x)=\phi_{i0}(x)$ and $\psi_i(x)=\phi_{i1}(x)$, $0\leqslant i\leqslant n$, in one variable of Section 3.2. In particular

$$f_N(x,y)=\sum_{i,j=0}^{n} f(x_i,y_j)\phi_i(x)\phi_j(y)$$

$$+\sum_{i,j=0}^{n}\frac{\partial f(x_i,y_i)}{\partial x}\psi_i(x)\phi_j(y)$$

$$+\sum_{i,j=0}^{n}\frac{\partial f}{\partial y}(x_i,y_j)\phi_i(x)\psi_j(y)$$

$$+\sum_{i,j=0}^{n}\frac{\partial^2 f}{\partial x\partial y}(x_i,y_j)\psi_i(x)\psi_j(y),$$

which establishes the existence of f_N. Proof of uniqueness is left to the exercises and follows much the same line of reasoning of the Lagrange case.

Error estimates are also available and the proof of these estimates can be derived using the same techniques used for the Lagrange polynomials. In particular, one can prove (see Exercises) that when $f\in C^4[R]$

$$\boxed{\|f-f_N\|\leqslant\gamma_0 h^4}$$

$$\boxed{\gamma_0=\frac{1}{384}\left\{\|D_x^4 f\|+24\|D_x^2D_y^2 f\|+\|D_y^4 f\|\right\}}$$

while

$$\boxed{\max_{p\in R_{ij}}\left|D^{|j|}f(p)-D^{|j|}f_N(p)\right|\leqslant\gamma_{ij}\max\left\{h_x^{4-j_1},h_y^{4-j_2}\right\},}$$

where

$$j=(j_1,j_2)$$

$$|j|=j_1+j_2, \qquad 1\leqslant j_1,j_2\leqslant 4,$$

$$D^{|j|}f=\frac{\partial^{|j|}f}{\partial x^{j_1}\partial y^{j_2}}$$

and γ_{ij} is a constant dependent only on the Tchebycheff norm of the mixed partials of order k of f up through $k=4$. Once again the derivative estimates are important in solving partial differential equations by variational methods. The exercises on these estimates are strongly recommended. Corresponding lower order formulas, available when $f\in C^k[R]$, $0\leqslant k\leqslant 3$, follow from the one-dimensional estimates for these smoothness classes.

Note that $f_N(x,y)$ is once continuously differentiable in all directions, thus belonging to $C^1[R]$. However, it is not generally the case that $f\in C^2[R]$ since jump discontinuities in the normal derivatives exist across all the subrectangular interfaces.

Bicubic Splines

Bicubic Hermite polynomials are bicubic splines, but it is common practice to use the terminology *bicubic spline* to refer to piecewise cubic polynomials on a rectangular grid which are of smoothness class $C^2[R]$. In particular, the bicubic spline interpolate s_N to f is a piecewise bicubic polynomial belonging to $C^2[R]$ which solves the interpolation problem

$$\begin{aligned}
s_N(x_i,y_j)&=f(x_i,y_j), && 0\leqslant i,j\leqslant n\\
\frac{\partial s_N}{\partial x}(x_i,y_j)&=\frac{\partial f}{\partial x}(x_i,y_j), && 0\leqslant j\leqslant n; i=0,n\\
\frac{\partial s_N}{\partial y}(x_i,y_j)&=\frac{\partial f}{\partial y}(x_i,y_j), && 0\leqslant i\leqslant n; j=0,n\\
\frac{\partial^2 s_N}{\partial y\partial x}(x_i,y_j)&=\frac{\partial^2 f}{\partial y\partial x}(x_i,y_j), && i,j=0,n.
\end{aligned} \tag{16}$$

We can prove there exists a unique cubic spline

$$s_N(x,y)=\sum_{i,j=-1}^{n+1} c_{ij}B_i(x)B_j(y) \tag{17}$$

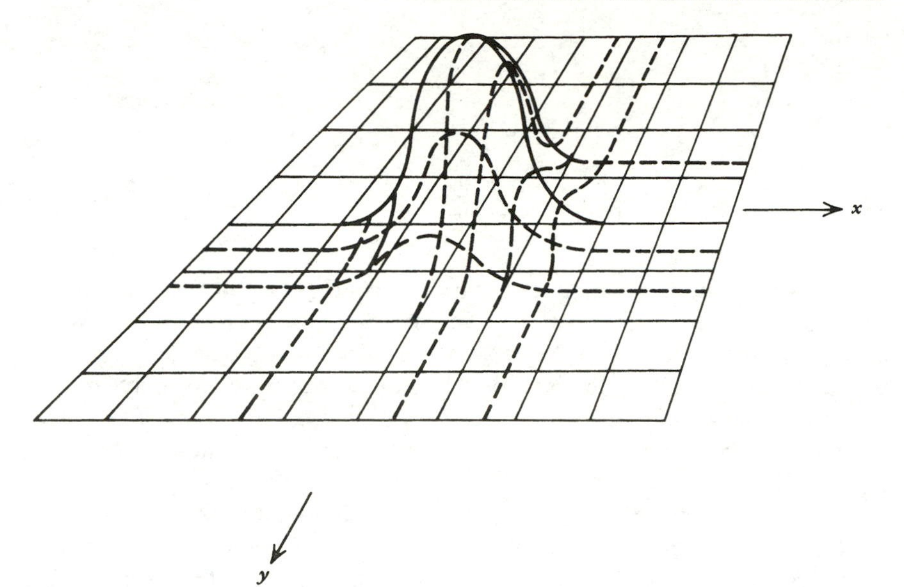

Figure 5.6 Typical cubic B spline $B_{ij}(x,y) = B_i(x)B_j(y)$.

solving (16) where the $B_i(x)$'s and $B_j(y)$'s are the cubic B splines. Figure 5.6 graphs a basis B spline $B_i(x)B_j(y)$.

Moreover, error estimates completely analogous to the one-dimensional case obtain, and we have

$$\boxed{\|f - s_N\| \leqslant \beta_0 h^4}$$

$$\boxed{\|D^{|1|}f - D^{|1|}s_N\| \leqslant \beta_1 h^3}$$

$$\boxed{\|D^{|2|}f - D^{|2|}s_N\| \leqslant \beta_2 h^2,}$$

with evenly spaced knots, where β_0, β_1, and β_2 depend only on the Tchebycheff norms of mixed partials of f up through the kth, $k = 0, 1, 2$ (see Exercises).

To establish the existence of s_N given by (17) solving (16), let $f(x,y)$ be a given function of (x,y) defined on R whose partials exist at boundary knots of R and whose mixed partials exist at the corner knots of R. Choose an $\bar{x}$ belonging to $[a,b]$, hold it fixed, and let $I_{\bar{x}}f(\bar{x},y)$ denote the unique cubic spline interpolate to the function $f(\bar{x},y)$ in the y direction. This spline exists and is unique by the one-dimensional theory. For fixed $\bar{y}$ in $[c,d]$ let $I_{\bar{y}}f(x,\bar{y})$ denote the unique cubic spline interpolate to $f(x,y)$ in the x direction. The reader will easily verify that $I_xI_yf(x,y)=I_yI_xf(x,y)$ and

$$s_N(x,y)=I_xI_yf(x,y).$$

Since for each x in $[a,b]$

$$I_xf(x,y)=\sum_{j=-1}^{m+1} a_j(x)B_j(y),$$

where the coefficients $a_j(x)$ change with changing x, we see that

$$I_yI_xf(x,y)=\sum_{j=-1}^{m+1} I_ya_j(x)B_j(y). \tag{18}$$

But

$$I_ya_j(x)=\sum_{i=-1}^{n+1} a_{ij}B_i(x). \tag{19}$$

Combining (18) with (19) we arrive at an expression (17) for $s_N(x,y)$. Proof of uniqueness, which is much the same as that demonstrated in the Lagrange case, utilizing (18) and (19) is called for in the Exercises.

Each example in this section is an instance of approximation from a tensor product space. We take up these ideas in the next section.

EXERCISES

1. Prove that the bicubic spline interpolate to a given function $f(x,y)$ on a rectangular grid is unique.
2. Prove that the bicubic Hermite interpolate to a given function $f(x,y)$ on a rectangular grid is unique.

3. Let $\{C_{-1}, C_0, C_2, \ldots, C_n, C_{n+1}\}$ be the unique cardinal bicubic splines solving the following interpolation problems:

$$C_i'(x_0)=0$$

$$C_i(x_j)=\delta_{ij}, \qquad 0 \leqslant i,j \leqslant n$$

$$C_i'(x_n)=0$$

$$C_{-1}'(x_0)=1$$

$$C_{-1}(x_i)=0, \qquad 0 \leqslant i \leqslant n$$

$$C_{-1}'(x_n)=0$$

$$C_{n+1}'(x_0)=0$$

$$C_{n+1}(x_i)=0, \qquad 0 \leqslant i \leqslant n$$

$$C_{n+1}'(x_n)=1.$$

Let $C_{-1}(y), C_0(y), \ldots, C_{n+1}(y)$ be the corresponding functions in the y direction. Write the unique bicubic spline interpolate to a given function $f(x,y)$ as a linear combination of products $C_i(x)C_j(y)$, $-1 \leqslant i,j \leqslant n+1$. Should we object for any reason to using these functions to compute $s_N(x,y)$? Explain.

4. Graph the cardinal basis bicubic Hermite polynomials.

5. Let $p_0, p_1, \ldots, p_5$ be any six distinct points in the plane.
   (a) Find a polynomial $q(p)=q(x,y)=\Sigma_{i,j=0}^{6} a_{ij} x^i y^j$ solving $q(p_k) = f(p_k)$, $0 \leqslant k \leqslant 5$, where $f(p)=f(x,y)$ is a given function.
   (b) Find a set of points at which this problem can be solved by a cubic in x and y but not by a quadratic.
   (c) What is the lowest possible degree polynomial you could have solving this problem and what constraints would be imposed on the location of the p_k's in this case? Explain.

6. Compute and graph the basis piecewise biquadratic Lagrange polynomials on a rectangular grid. Program your computer to compute the piecewise biquadratic Lagrange interpolate to $\cos x^2 y$ on $[0,\pi]\times[0,\pi]$. with 10 evenly spaced knots in the x and the y directions. Call out and plot $\cos x^2 y - s(x,y)$, where $s(x,y)$ is this piecewise Lagrange interpolate to $\cos x^2 y$.

7. Give a simpler proof than that in the text of the uniqueness of the Lagrange interpolate to $f(p)$, $p=(x,y)$, on a rectangular grid.

8. Let $s_3(x,y)$ be the bicubic Lagrange interpolate to $f \in C^k[R]$, $k=0$. 1, 2. Estimate

$$\|D^{|j|}f - D^{|j|}s_3\|, \qquad 0 \leqslant j \leqslant k.$$

(Hint: use the corresponding one-dimensional error estimates of Section 2.5.)

9. Let $s(x,y)$ be the bicubic spline interpolate to $f \in C^4[R]$. Prove

$$\|D^{|j|}f - D^{|j|}s\| \leqslant \beta_j h^{4-j}, \qquad j=0,1,2$$

and find an exact expression for each β_j.

10. Derive the error estimates

$$\|D^{|i|}f - D^{|i|}s_1\| \leqslant \beta_i h^{2-i}, \qquad i=0,1,2,$$

where s_1 is the piecewise bilinear Lagrange interpolate to $f \in C^2[R]$. Determine the β_i's exactly.

5.3 TENSOR PRODUCTS

All the approximates of the last section were special cases of approximations from finite dimensional tensor product spaces. To define this concept in a simple setting, suppose $X = C[a,b]$ and $Y = C[c,d]$. Let

$$X_N = \text{span}\{\phi_1, \phi_2, \ldots, \phi_N\} \qquad \text{and} \qquad Y_M = \text{span}\{\psi_1, \psi_2, \ldots, \psi_M\}$$

be N- and M-dimensional subspaces of X and Y, respectively. Then

DEFINITION $X_N \otimes Y_M$

The tensor product $X_N \otimes Y_M = \text{span}\{\phi_i(x)\psi_j(y) : 1 \leqslant i \leqslant N \text{ and } 1 \leqslant j \leqslant M\}$.

That is, $X_N \otimes Y_M$ is simply all possible linear combination of $\phi_i(x)$ from $\{\phi_1, \phi_2, \ldots, \phi_N\}$ with functions $\psi_j(y)$ from $\{\psi_1, \psi_2, \ldots, \psi_M\}$. Being defined as all possible linear combinations of these NM functions, $X_N \otimes Y_M$ is a linear space. Moreover, it is a subspace of the linear space $C[R]$, being composed of sums of products of continuous functions.

A number of interesting questions concerning these spaces come immediately to mind; some are quite easy to answer and others a bit more challenging. For example, we can easily settle the dimension question by way of the following theorem.

THEOREM

The set $\mathfrak{B} = \{\phi_i(x)\psi_j(y): 1 \leqslant i \leqslant N \text{ and } 1 \leqslant j \leqslant M\}$ is linearly independent and forms a basis for $X_N \otimes Y_M$. Thus the dimension of $X_N \otimes Y_M = = MN$.

Proof. Suppose $\sum_{i,j} a_{ij}\phi_i(x)\psi_j(y) = 0$ on $R = [a,b] \times [c,d]$. Let $c_j(x) = \sum_{i=1}^N a_{ij}\phi_i(x)$. It follows that

$$\sum_{j=1}^{M} c_j(\bar{x})\psi_j(y) = 0$$

on all of $[c,d]$ for each fixed $\bar{x}$ in $[a,b]$. But the ψ_j's are independent on the interval $[c,d]$. Thus $c_j(\bar{x}) = 0$ for each j and each $\bar{x}$ in $[a,b]$. Therefore $c_j(x) \equiv 0$ on $[a,b]$. But then for each j, $1 \leqslant j \leqslant M$, $\sum_{i=1}^N a_{ij}\phi_i(x) \equiv 0$, and by the independence of the ϕ_i's, $a_{ij} = 0$, $1 \leqslant i \leqslant N$, and $1 \leqslant j \leqslant M$. ■

A second pair of questions arises naturally in this setting: namely, is $X_N \otimes Y_M$ the same as $X_N \cdot Y_M = \{f(x)g(y): f \in X_N \text{ and } g \in Y_M\}$ and is $X_N \cdot Y_M$ a linear space? Since $X_N \cdot Y_M$ is another subset of $C[R]$, it would suffice to prove that the subset was closed under addition and scalar multiplication to prove it to be a linear space. Closure under scalar multiplication is easily verified. Closure under addition is not available readily, if at all (see Exercises). It is quite simple to prove $X_N \cdot Y_M \subset X_N \otimes Y_M$, but does the converse containment obtain? We leave some theorems on these matters to the Exercises, remarking that the algebraists have these matters well in hand. Thus the seriously interested reader is referred to any good algebraic exposition on tensor product spaces and multilinear algebras.

Approximation theorists and numerical analysts are principally concerned with these spaces because they want to determine the kind of approximations to functions of several variables that are provided from $X_N \otimes Y_M$ subject to pure interpolatory constraints, pure variational constraints, mixed constraints, orthogonality constraints, and so forth (viz. Section 1.6). All such approximates f are of the form

$$f(x,y) = \sum_{\substack{1 \leqslant i \leqslant N \\ 1 \leqslant j \leqslant M}} a_{ij}\phi_i(x)\psi_j(y).$$

The previous section gave three examples of such approximates, all subject to pure interpolatory constraints at the knots of our partition $\pi = \pi_x \cdot \pi_y$ of R. The reader can easily see how to build numerous comparable approximates such a biquintic Hermites, biquintic splines, and cubic by quintic splines. We assign some of these approximates as exercises. We also remark that is is *not* always necessary to confine the interpolatory constraints at the knots π of R. In fact, as illustrated in Section 3.3, there are times when this is precisely what must *not* be done. The matter is brought up again in Chapter 8. Also, not interpolating precisely at the knots of π yields greater versatility, as was illustrated in Section 4.2.

EXERCISES

1. Prove $X_N \cdot Y_M \subset X_N \otimes Y_M$.
2. (a) Let $X_N = \text{span}\{1, x, x^4\}$ and $Y_M = \text{span}\{1, \sin x, \sin y\}$. Does $X \cdot Y = X_N \otimes Y_M$? Prove your answer.
   (b) Let $X_N = \text{span}\{1, x^2, x^4\}$ and $Y_M = \text{span}\{1, y, y^2\}$. Does $X_N \cdot Y_M = X_N \otimes Y_M$? Prove your answer.
3. Let $X_N = S_3(\pi_x)$, the cubic splines with knots at π_x. Let $Y_M = L_2(\pi_y)$, the piecewise quadratic Lagrange polynomials with knots at π_y.
   (a) What unique interpolation problems with knots at $\pi = \pi_x \cdot \pi_y$ can you solve with approximates from $X_N \otimes Y_M$? Prove your answer. (Solve this problem in the greatest possible generality.)
   (b) What unique interpolates do you conjecture can be built when the interpolatory constraints are not confined to the knots of π? Explain. Can you prove your conjecture?
   (c) Under what conditions does $X_N \cdot Y_M = X_N \otimes Y_M$? Prove your answer.

5.4 APPROXIMATES ON TRIANGULAR GRIDS

The constructions we want to undertake next require the knots $\{p_0, p_1, \ldots, p_n\}$ of our partition π to generate a proper triangulation of Ω. Specifically, any set of nondegenerate triangles $\tau = \{T_0, T_1, \ldots, T_{n-1}\}$ such that:

1. the set of all vertices of triangles from τ is π,
2. each pair of triangles $T_i = T_j$ from τ either intersect at exactly one vertex or intersect on one complete side or do not intersect at all,
3. the union of the T_i's and their interiors is Ω,

is said to be a *proper triangulation of* Ω. It is easy to see that a given polygonally bounded domain in the plane can have several triangulations. These ideas are simply illustrated in Figures 5.7 and 5.8.

In any event, suppose Ω is a polygonally bounded domain together with a proper triangulation τ. We give two specific methods for constructing piecewise polynomials on Ω together with τ and indicate some directions in which these methods can and need be generalized. The actual triangulation of a given region Ω is a nontrivial programming problem to which engineers have devoted a good deal of attention under the title "automatic mesh generation." We take up such automatic mesh generation schemes in

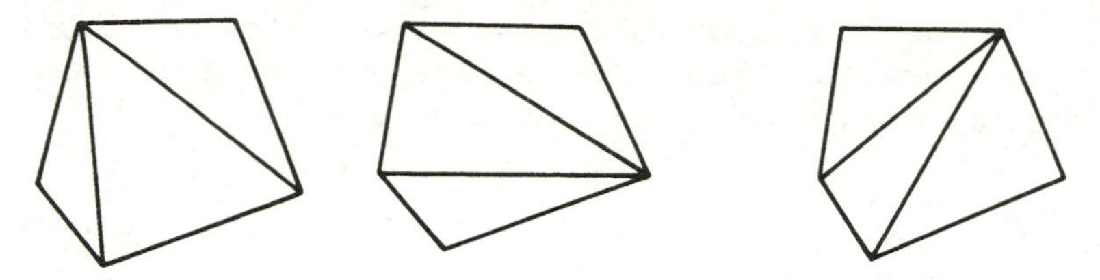

Figure 5.7 Three distinct triangulations of the same region Ω.

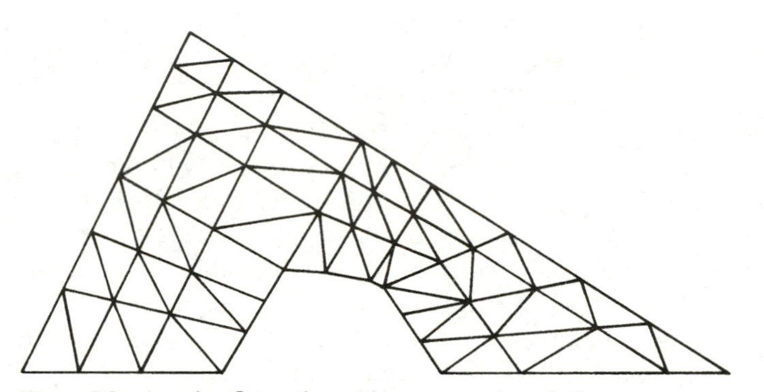

Figure 5.8 A region Ω together with a proper triangulation τ.

Section 5.5. In the meantime, we proceed as if the triangulation of our region is an accomplished fact. Our first example is a three-point scheme known popularly as the "method of plates"; our second scheme is a six-point formula. Both schemes are simple piecewise Lagrange polynomial fits over a triangular, rather than a rectangular, grid. Method I comprises piecewise linear fits, and Method II consists of piecewise quadratic fits.

Throughout our discussion, the *mesh h* of τ is the length of the edge of greatest length among the edges of all the triangles of τ.

The Method of Plates

The idea of the *method of plates* is that of approximating a function $f(x,y)$ by *piecewise planes* that interpolate f at the knots (or vertices) of some proper triangulation of the domain Ω of f. For example, suppose Ω is the region shown in Figure 5.8 together with its triangulation τ.

On each triangle T_i of Ω, we construct a plane $p_i(x,y)$

$$p_i(x,y) = a_i x + b_i y + c_i,$$

interpolating f at the vertices of T_i. We define our piecewise linear interpolate $s_N(x,y)$ to f by

$$s_N(x,y) = p_i(x,y)$$

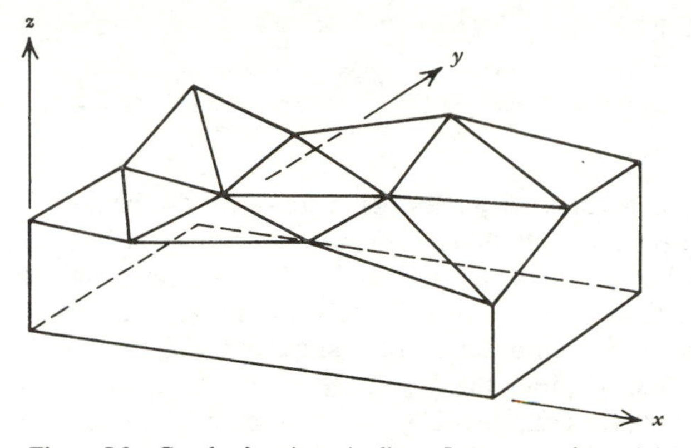

Figure 5.9 Graph of a piecewise linear Lagrange polynomial.

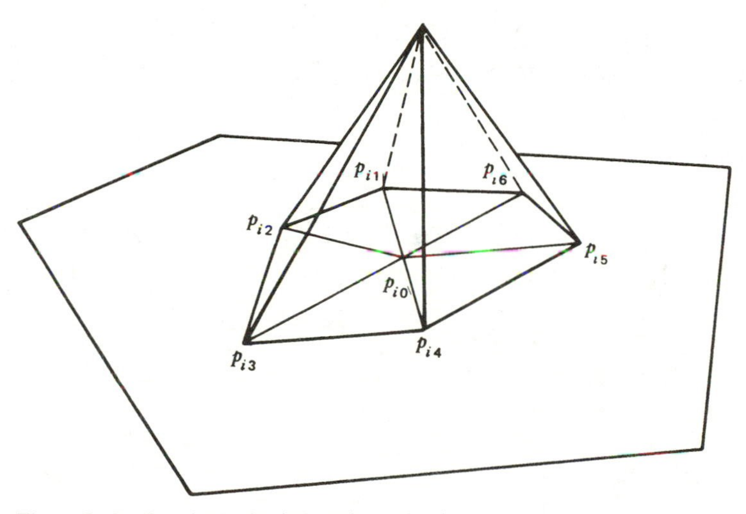

Figure 5.10 Graph of a basis function $\phi_i(x,y)$.

if (x,y) is on T_i or in its interior. The resulting function is a piecewise Lagrange polynomial of degree one whose graph is not unlike portions of the surface of a geodesic dome as illustrated in Figure 5.9. The canonical *basis* for computing $s_N(x,y)$ is the set of *shape functions* $\{\phi_0, \phi_1, \ldots, \phi_n\}$, which are the piecewise linear polynomials arising from the method of plates and which solve the interpolation problem

$$\phi_i(p_j) = \delta_{ij}, \qquad 0 \leqslant i,j \leqslant n.$$

The graph of a typical shape function $\phi_i(x,y)$ corresponding to an interior knot p_i has the shape given in Figure 5.10.

It follows that

$$s_N(x,y) = \sum_{i=0}^{n} f(p_i)\phi_i(x,y)$$

and that s_N is unique, its planes being uniquely determined by any three noncolinear points. Note that s_N is continuous on Ω, since $p_i(x,y) = p_j(x,y)$ along shared edges of adjacent triangles T_i and T_j, respectively. However, s_N may fail to be even once differentiable since jump discontinuities in normal derivatives can occur at all triangular interfaces.

Computing s_N requires the ϕ_i's. If we have a triangle T with vertices at $p_i = (x_i, y_i)$, $i = 0,1,2$ (Figure 5.11), the equation of the plane interpolating f at $(x_0, y_0, 1)$, $(x_1, y_1, 0)$, and $(x_2, y_2, 0)$ is

$$\phi(x,y) = \left[1 - \frac{(y_0 - y_2)}{(y_1 - y_2)} - \frac{(x_0 - x_1)}{(x_2 - x_1)}\right]^{-1}\left[1 - \frac{(y - y_2)}{(y_1 - y_2)} - \frac{(x - x_1)}{(x_2 - x_1)}\right].$$

It follows that

$$\phi_i(x,y) = \left[1 - \frac{(y_{0i} - y_{2i})}{(y_{1i} - y_{2i})} - \frac{(x_{0i} - x_{1i})}{(x_{2i} - x_{1i})}\right]^{-1}\left[1 - \frac{(y - y_{2i})}{(y_{1i} - y_{2i})} - \frac{(x - x_{1i})}{(x_{2i} - x_{1i})}\right]$$

p_2

p_0

p_1 **Figure 5.11**

on the triangle T_i and its interior (see Figure 5.12 where $p_i = (x_{0i}, y_{0i})$.

Error estimates for $f - s_N$ are also easily obtained. In particular, we can prove that

$$\| f - s_N \| \leqslant 4 M_2 h^2$$

when $f \in C^2[\Omega]$, where M_2 is a constant independent of h, which we define

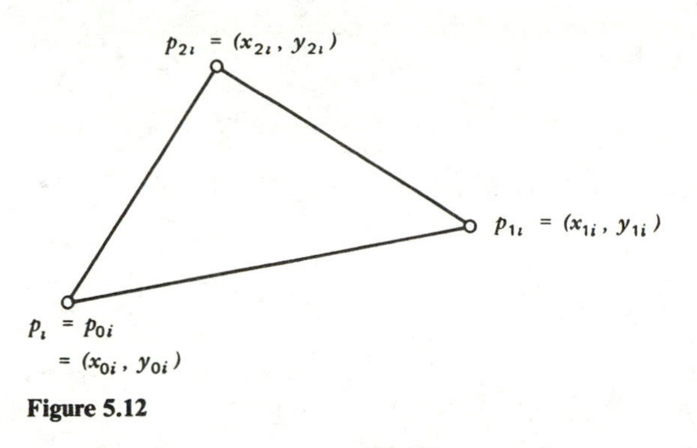

Figure 5.12

in a moment; and

$$\|f - s_N\| = \begin{cases} \beta_1 h & \text{if } f \in C^1[\Omega] \\ \beta_0 \omega(f,h) & \text{if } f \in C[\Omega], \end{cases}$$

where β_1 and β_0 are also constants independent of h and $\omega(f,h)$ is the modulus of continuity of f. Estimates for directional derivatives are also available with

$$|D_\beta f(p) - D_\beta s_N(p)| \leqslant 2M_2 h$$

on each triangle T_i of τ when $f \in C^2[\Omega]$.

For simplicity we work on a triangle T with vertices p_0, p_1, and p_2 as illustrated in Figure 5.11. Let $f \in C^2[\Omega]$. Then $s_N(x,y)$ on T is the unique plane interpolating f at each of the p_i's; thus we have

$$s_N(p_i) = f(p_i), \qquad i = 0, 1, 2.$$

For each ordered pair of integers $i = (i_1, i_2)$, $0 \leqslant i_1, i_2 \leqslant 2$, let $|i| = i_1 + i_2$, let $D^{|i|}f = \partial^{|i|}f / \partial x^{i_1} \partial y^{i_2}$, let

$$M_k = \max_{|i| = k} \|D^{|i|}f\|, \qquad k = 1, 2,$$

and let $M_0 = \|f\|$. For each unit vector $\beta = (\cos\theta, \sin\theta)$ and each function $g(x,y)$ defined on T, the *directional derivative* $D_\beta g(p)$

$$D_\beta g(p) = \lim_{t \to 0} \frac{g(p + t\beta) - g(p)}{t}$$
$$= D_x g(p) \cos\theta + D_y g(p) \sin\theta$$

when $g \in C^1[T]$; thus

$$D_\beta g(p) = \operatorname{grad} g(p) \cdot \beta.$$

With this in mind it is simple to prove

LEMMA

Let $f \in C^2[T]$. Then

1. $\max_{p \in T} |f(p) - s_N(p)| \leqslant 4M_2h^2$
2. $\max_{p \in T} |D_\beta f(p) - D_\beta s_N(p)| \leqslant 6M_2h$

for all unit vectors β pointing along or interior to T.

Proof. Let $g(p) = f(p) - s_N(p)$. Since both f and s_N belong to $C^2[T]$, $g \in C^2[T]$. Let p_0 be a vertex having an acute angle θ_0 at p_0, let p be any point belonging to the interior of T and let $\beta = p - p_0/|p - p_0|$, where $|p - p_0| = \sqrt{(x-x_0)^2 + (y-y_0)^2}$ and $p = (x,y)$, (see Figure 5.13). We expand g in a Taylor's expansion and p_0 in the direction β. In particular

$$g(p) = g(p_0) + D_\beta g(p_0)(p - p_0) + \frac{D_\beta^2 g(\tilde{p})}{2}(p - p_0)^2,$$

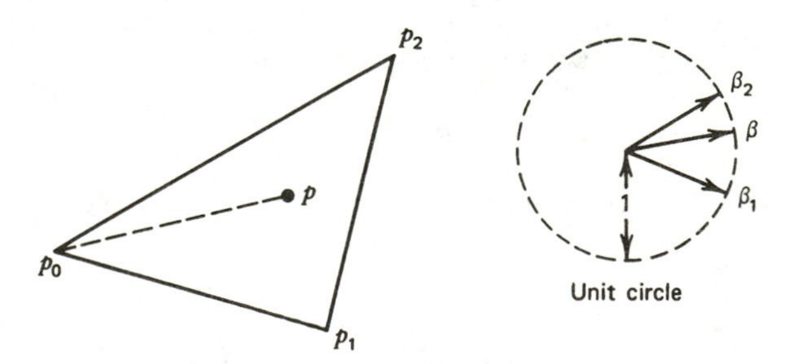

Figure 5.13

where $\tilde{p}$ is some point on the line segment $\overline{p_0 p}$ different from p and p_0. Since $s_N(x,y) = ax + bx + c$ on T, where a, b, and c are constraints, $D_\beta^2 g(\tilde{p}) = D_\beta^2 f(\tilde{p})$, second partials of s_N vanishing. Since $p \in T$,

$$|p - p_0| \leqslant h.$$

Thus

$$\frac{|D_\beta^2 g(\tilde{p})|}{2}|p-p_0|^2=\frac{|D_\beta^2 f(\tilde{p})|}{2}|p-p_0|^2\leqslant 2M_2h^2. \tag{1}$$

Let

$$\beta_1=\frac{p_1-p_0}{|p_1-p_0|} \qquad \text{and} \qquad \beta_2=\frac{p_2-p_0}{|p_2-p_0|}.$$

Since T is a proper triangle, β_1 and β_2 are linearly independent and there exist constants b_1 and b_2, $0\leqslant|b_1|,|b_2|\leqslant 1$, such that

$$\beta=b_1\beta_1+b_2\beta_2.$$

But then

$$\begin{aligned}D_\beta g(p_0)&=\operatorname{grad} g(p_0)\cdot\beta\\&=b_1\operatorname{grad} g(p_0)\cdot\beta_1+b_2\operatorname{grad} g(p_0)\cdot\beta_2\\&=b_1D_{\beta_1}g(p_0)+b_2D_{\beta_2}g(p_0).\end{aligned}$$

However, along the line segments $\overline{p_0p_1}$ and $\overline{p_0p_2}$, $s_N(p)$ reduces to the straight lines (Lagrange polynomials of degree 1) interpolating f at p_0,p_1, and p_0,p_2, respectively. It follows from our error estimates for Lagrange interpolates (Section 2.5) that

$$|D_{\beta_1}f(\bar{p})-D_{\beta_1}s_N(\bar{p})|\leqslant\frac{\|D_{\beta_1}^2 f\|}{2}h\leqslant M_2h$$

for all $\bar{p}$ on $\overline{p_0p_1}$; and

$$|D_{\beta_2}f(\bar{p})-D_{\beta_2}s_N(\bar{p})|\leqslant\frac{\|D_{\beta_2}^2 f\|}{2}h\leqslant M_2h$$

for all $\bar{p}$ on $\overline{p_0p_2}$. But p_0 lies on both these line segments. Thus

$$\begin{aligned}|D_\beta f(p_0)-D_\beta s_N(p_0)|&=|D_\beta g(p_0)|\\&=|b_1D_{\beta_1}g(p_0)+b_2D_{\beta_2}g(p_0)|\\&\leqslant|b_1|\,|D_{\beta_1}g(p_0)|+|b_2|\,|D_{\beta_2}g(p_0)|\\&\leqslant M_2h+M_2h\\&\leqslant 2M_2h.\end{aligned}$$

It follows that

$$|D_\beta g(p_0)||p-p_0| \leqslant 2M_2h^2. \tag{2}$$

Combining (1) and (2) with $s_N(p_0)=f(p_0)$, we find

$$|f(p)-s_N(p)| \leqslant 4M_2h^2$$

for all p in the interior of T. If p belongs to an edge, the Lagrange estimates apply directly. Similar reasoning proves 2. ■

Since the estimates of our lemma apply to arbitrary triangles T_i of τ, we have proved that if $f \in C^2[\Omega]$

$$\boxed{\|f-s_N\| \leqslant 4M_2h^2.}$$

On the other hand, $D_\beta f(p)-D_\beta s_N(p)$ may not exist along edges and at vertices but does exist at all interior points p of triangles of τ. Thus for all unit vectors β

$$\boxed{\operatorname{ess\,sup}|D_\beta f(p)-D_\beta s_N(p)| \leqslant 6M_2h,}$$

where in this case

$$\operatorname*{ess\,sup}_{p\in\Omega}|D_\beta f(p)-D_\beta s_N(p)| = \sup_{\substack{p\in \mathring{T}_i\\ T_i\in\tau}}|D_\beta f(p)-D_\beta s_N(p)|.$$

and $\mathring{T}_i$ is the interior of T.

A Six-Point Scheme on a Triangular Grid

The method of plates consisted of constructing piecewise linear Lagrange polynomials on a triangular grid. We now construct piecewise quadratic Lagrange polynomials on such a grid. Specifically, let $\tau=\{T_0,T_1,\ldots,T_{n-1}\}$ again be a triangulation of Ω. On each triangle T_i of τ construct the midpoints of each side. Our grid will consist of the vertices p_{i0}, p_{i2}, p_{i4} together with the midpoints p_{i1}, p_{i3}, p_{i5} of each triangle T_i of τ (see Figure 5.14).

On each triangle T_i compute the quadratic surface

$$P_i(x,y)=a_ix^2+b_ixy+c_iy^2+d_ix+e_iy+f_i,$$

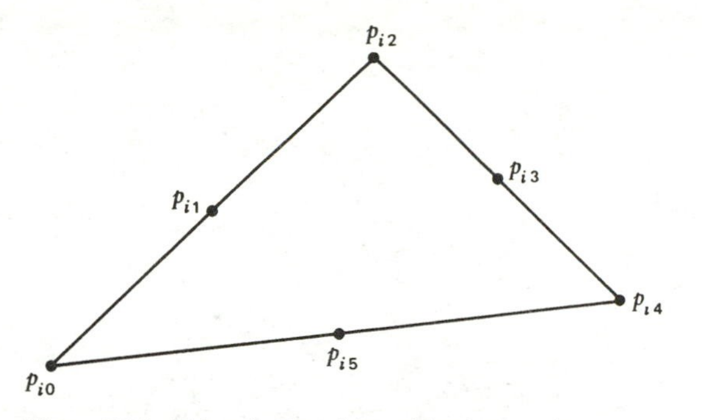

Figure 5.14 Triangle T_i and its grid points p_{ij}, $0 \leqslant j \leqslant 5$.

interpolating f at the points p_{ij}, $0 \leqslant j \leqslant 5$. That is,

$$P_i(p_{ij}) = f(p_{ij}), \qquad 0 \leqslant j \leqslant 5.$$

The piecewise quadratic polynomial interpolate $s_N(x,y)$ to $f(x,y)$ with respect to this grid is defined by

$$s_N(x,y) = P_i(x,y)$$

if (x,y) belongs to triangle T_i or its interior. It follows that the graph of $s_N(x,y)$ is a piecewise parabolic surface rather like a distended egg carton in which one can both deposit and suspend eggs (see Figure 5.15.) We prove existence and uniqueness of these functions by simply computing them (i.e., producing an algorithm).

The preferred method of computation in engineering circles (see the books of Zienkewicz (1967) and of Desai and Abel (1972) is again via the

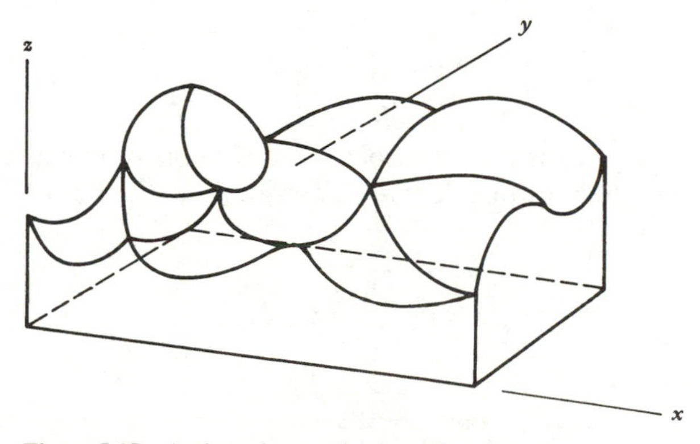

Figure 5.15 A piecewise quadratic surface.

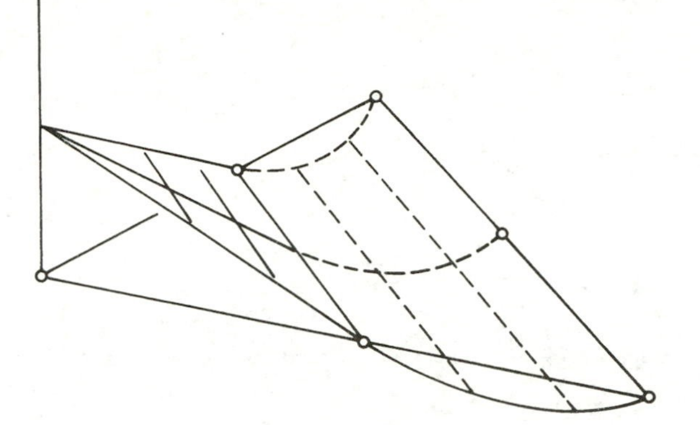

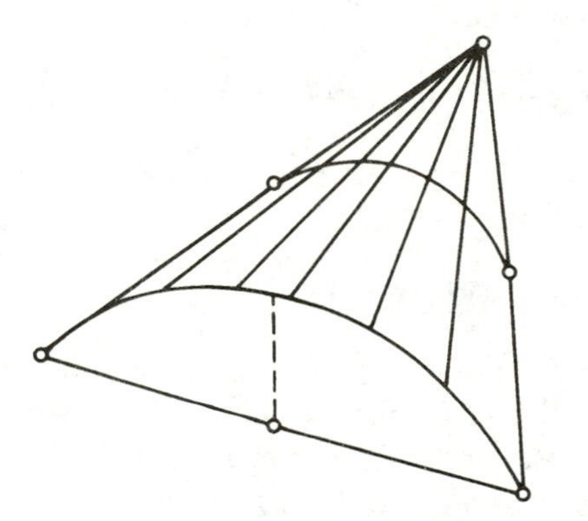

Figure 5.16 Basis quadratic Lagrange polynomials.

canonical basis or shape functions $\phi_i(x,y)$ solving the interpolation problem

$$\phi_i(p_j)=\delta_{ij}, \qquad 0\leqslant i,j\leqslant N. \tag{3}$$

where $\{p_0,p_1,\dots,p_N\}$ is the total collection of knots of our grid. Graphs of typical shape functions on triangle T_i are given in Figure 5.16.* It follows that

$$s_N(x,y)=\sum_{i=0}^{n} f(x_i,y_i)\phi_i(x,y). \tag{4}$$

*The author is grateful to Professors Desai and Abel and to Van Nostrand Reinhold for permission to reproduce these figures.

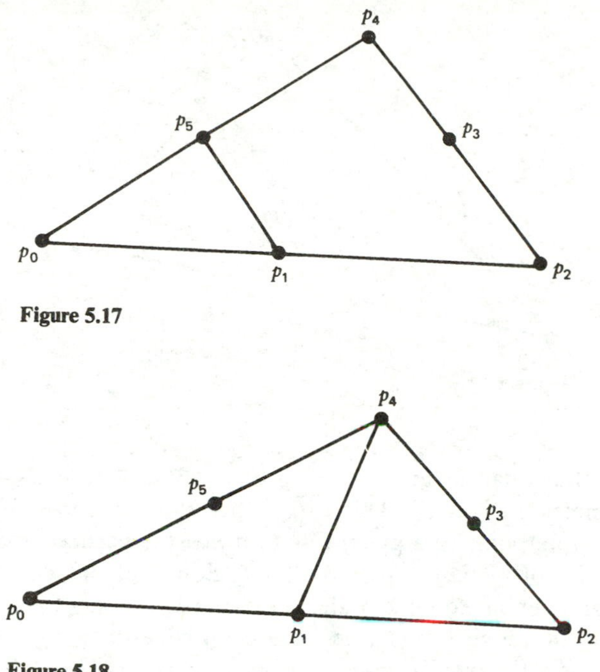

Figure 5.17

Figure 5.18

Computation of the ϕ_i's is easily achieved as products of planes. For example, if triangle T_0 has vertices and midpoints $p_0, p_1, \ldots, p_5$ as illustrated in Figure 5.17, and if Q_0 is the unique plane interpolating 1, 0, and 0 data at p_0, p_5, and p_1, respectively, and $\tilde{Q}_0$ is the unique plane interpolating 1, 0, and 0 data at p_0, p_4, and p_2 respectively, it is evident that

$$\phi_0(x,y) = Q_0(x,y) \cdot \tilde{Q}_0(x,y)$$

in triangle T_0. It is given by other products of planes on triangles of τ different from T_0 sharing vertex p_0. Moreover, ϕ_0 vanishes identically outside all triangles τ sharing vertex p_0. Now partition T_0 in a somewhat different manner (Figure 5.18). To compute ϕ_1 let Q_1 be the unique plane interpolating 1, 0, and 0 data at p_1, p_0, and p_4, respectively, and let $\tilde{Q}_1$ be the unique plane interpolating 1, 0, and 0 data at p_1, p_2, and p_4, respectively. It is then clear that the shape function

$$\phi_1(x,y) = Q_1(x,y) \cdot \tilde{Q}_1(x,y)$$

on triangle T_0, whereas it is given by a different product of planes on

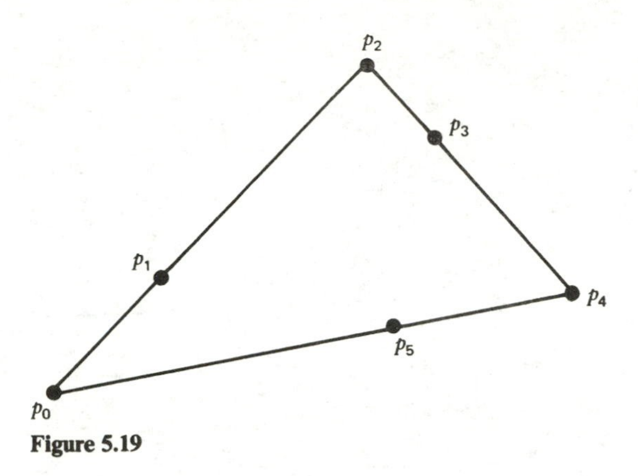

Figure 5.19

triangle $\tilde{T}_0$ of τ adjacent to T_0 sharing edge $\overline{p_0p_1p_2}$. Clearly ϕ_1 vanishes outside triangles T_0 and $\tilde{T}_0$. This establishes the *existence* of a piecewise quadratic s_N interpolating a given function f at the vertices and midpoints of the triangles of a triangulation τ of a region Ω of the x–y plane. Note that it is not essential to choose the interior edge points to be midpoints, any three interior edge points, one from each edge sufficing. Formulas (3) and (4) constitute an *algorithm* for computing $s_N(x,y)$. To establish uniqueness of such piecewise quadratics, we prove a theorem, letting p_0, p_2, and p_4 be the vertices of any nondegenerate triangle T in the x–y plane and $f(x,y)$ be any function defined on T. On each edge of T choose exactly one point distinct from the vertices of T and label these edge points p_1, p_3, and p_5 as illustrated in Figure 5.19. Then

THEOREM

There exists exactly one quadratic polynomial

$$q(x,y) = ax + bxy + cy + dx + ey + f$$

solving $q(p_i) = f(p_i)$, $0 \leqslant i \leqslant 5$, where $p_i(x_i, y_i)$.

Proof. Existence is already established by our algorithm (3) coupled with (4) restricted to a single triangle. For a simple proof of uniqueness, let $R = [a,b] \times [c,d]$ be any rectangle interior to the triangle T. The set $\mathcal{A} = \{1, x, x^2\}$ is independent on $[a,b]$ and the set $\mathcal{B} = \{1, y, y^2\}$ is independent on $[c,d]$. Let $A = \operatorname{span} \mathcal{A}$ and $B = \operatorname{span} \mathcal{B}$. It follows that the tensor product $A \otimes B$ has dimension 9 and is spanned by $\{1, x, y, x^2, xy, y^2, x^2y, xy^2, x^2y^2\}$. Thus the subset $\{1, x, y, x^2, xy, y^2\}$ is also independent on R, hence is independent on the larger set consisting of T and its interior. Let $W = \operatorname{span}\{1, x, y, y^2, xy, x^2\}$ and let $Z = \operatorname{span}\{\phi_0, \phi_1, \ldots, \phi_5\}$, where the ϕ_i's are the shape functions associated with triangle T. The dimension both of W and Z is 6, the former by the independence of $\{1,x,y,y^2,xy,x^2\}$ and the latter because the matrix $(\phi_i(p_j))$, $0 \leqslant i \leqslant 5$, is the identity matrix. But each ϕ_i belongs to W, being a product of planes; hence $Z \subset W$. Since the dimensions of these two spaces are the same, the spaces are equal. Since there is a unique q in Z solving $q(p_i)=f(p_i)$, the result follows trivially. ■

Note that the function $s_N(x,y)$ is continuous on all of Ω, since $s_N(x,y)$ reduces to a single parabola along adjacent edges of intersecting triangles $T_i = T_j$. However, s_N in general fails to be once continuously differentiable because the normal derivatives of s_N along adjacent triangular elements usually experience a *jump discontinuity*.

We now want to estimate the error $\|f-s_N\|$. We begin by proving a lemma, letting T be a nondegenerate triangle with vertices p_0, p_1, p_2 and angles θ_0, θ_1, θ_2 (see Figure 5.20). Assume θ_0 is the smallest angle of T, and let

$$\beta_1 = \frac{p_1 - p_0}{|p_1 - p_0|}$$

$$\beta_2 = \frac{p_2 - p_0}{|p_2 - p_0|}$$

$$\beta_3 = \frac{p_1 - p_2}{|p_1 - p_2|}$$

be unit vectors in the direction of the sides of T. Let

$$\beta = \frac{p - p_0}{|p - p_0|},$$

where p is any point in the interior of T. Then

LEMMA

There exist constants b_1, b_2, c_1, c_3, d_3, and d_2, all smaller than $1/\sin\theta_0$, in absolute value, such that

$$\begin{aligned}\beta &= b_1\beta_1 + b_2\beta_2\\ &= c_1\beta_1 + c_3\beta_3\\ &= d_3\beta_3 + d_2\beta_2.\end{aligned}$$

Proof. The linear independence of each pair from $\beta_i : i = 1,2,3$ implies the existence of these constants. It is clear that $|b_1|$, $|b_2| \leqslant 1$, since θ_0 is an acute angle and β lies on a ray passing through p_0 interior to angle θ_0. Consider either other vertex of T, for example, p_2. Construct a unit circle centered at p_2 and construct the triangle $\tilde{T}$ with vertices p_2, p_3, and p_4 as in Figure 5.20, where $p_4 - p_2 = \beta$, $p_3 - p_4 = d_3\beta_3$, and $p_3 - p_2 = d_2\beta_2$. Let θ_3, θ_4, and θ be the angles of $\tilde{T}$. Clearly $\theta_3 = \theta_2$, whereas $0 < \theta_1 < \theta_4$, and $0 < \theta_2 < \theta_0$. Noting that $|\beta| = 1$ and that

$$\frac{|d_3|}{\sin\theta} = \frac{1}{\sin\theta_2} = \frac{|d_2|}{\sin\theta_4},$$

we find $|d_3| = \sin\theta/\sin\theta_2 < 1/\sin\theta_0$ and $|d_2| = \sin\theta_4/\sin\theta_2 \leqslant 1/\sin\theta_0$. A similar argument pertains to vertex p_1. ∎

THEOREM ERROR ESTIMATE

Let $f \in C^3[\Omega]$ and let s_N be the piecewise quadratic polynomial interpolate to f over a triangular grid τ. Let θ be the smallest angle among the angles of triangles of τ and suppose $\theta \geqslant \theta_0$. Then

$$\text{(a)} \quad \|f - s_N\| \leqslant \frac{8M_3}{\sin\theta} h^3$$

$$\text{(b)} \quad \text{ess sup}\,|D_\beta f(p) - D_\beta s_N(p)| \leqslant \frac{12M^3}{\sin\theta} h^2$$

$$\text{(c)} \quad \text{ess sup}\,|D_\beta^2 f(p) - D_\beta^2 s_N(p)| \leqslant \frac{20M_3}{\sin\theta} h$$

for all unit vectors β where

$$M_3 = \max_{|i|=3} \{\|D^{|i|}f\|\}, \qquad i = (i_1, i_2);\ 0 \leqslant i_1, i_2 \leqslant 3.$$

Proof. The proof of these estimates is only slightly more complicated than that given for the method of plates. Once again, it suffices to work on a single triangle T of τ and to note that both f and s_N are three times continuously differentiable in all directions on T (see Figure 5.21). Choose p belonging to the interior of T and let $g(p) = f(p) - s_N(p)$. Let

$$\beta_5 = \frac{p_5 - p_0}{|p_5 - p_0|}, \qquad \beta_1 = \frac{p_1 - p_0}{|p_1 - p_0|}, \qquad \beta_3 = \frac{p_3 - p_2}{|p_3 - p_2|}$$

be the unit vectors parallel to edges of T, and let

$$\beta = \frac{p - p_0}{|p - p_0|}.$$

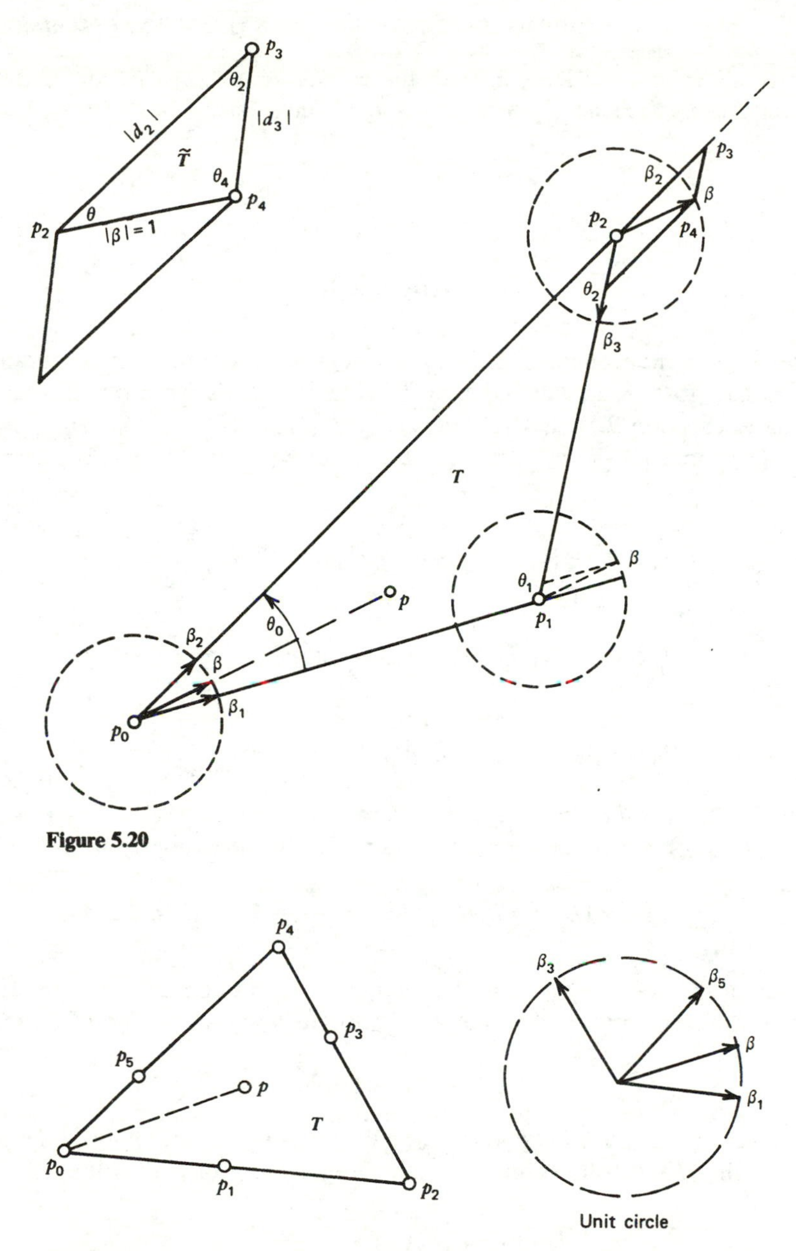

Figure 5.20

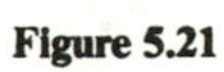

Figure 5.21

It follows from our lemma that there exist constants b_i, c_i, and d_i, $0<|b_i|,|c_i|,|d_i|<1/\sin\theta_0$, where $i=1,2$, or 3 such that

$$\beta=b_1\beta_1+b_5\beta_5$$

$$\beta=c_1\beta_1+c_3\beta_3$$

$$\beta=d_3\beta_3+d_5\beta_5,$$

since T is nondegenerate. But s_N reduces to the quadratic Lagrange interpolate to f along each edge of T. Thus it follows from our Lagrange estimates (Section 2.5) that $|D_{\beta_1}g(p_0)|$, $|D_{\beta_5}g(p_0)|$, $|D_{\beta_1}g(p_2)|$, $|D_{\beta_3}g(p_2)|$, $|D_{\beta_3}g(p_4)|$, and $|D_{\beta_5}g(p_4)|$ are all less than or equal to $2M_3h^2$. But then

$$D_\beta g(p_0)=|b_1D_{\beta_1}g(p_0)+b_5D_{\beta_5}g(p_0)|\leqslant\frac{4M_3}{\sin\theta_0}h^2$$

$$D_\beta g(p_2)=|c_1D_{\beta_1}g(p_2)+c_3D_{\beta_3}g(p_2)|\leqslant\frac{4M_3}{\sin\theta_0}h^2$$

$$D_\beta g(p_4)=|d_3D_{\beta_3}g(p_4)+d_5D_{\beta_5}g(p_4)|\leqslant\frac{4M_3}{\sin\theta_0}h^2.$$

However, since $s_N(p)$, $p=(x,y)$ is a quadratic $ax^2+by^2+cxy+dx+ey+f$ on T, $D_\beta s_N(p)$ is a plane $Ax+By+C$ on T with the property that

$$|D_\beta g(p_i)|=|D_\beta f(p_i)-D_\beta s_N(p_i)|\leqslant\frac{4M_3}{\sin\theta_0}h^2,\qquad i=0,2,4$$

at each of the vertices p_0, p_2, p_4. Let $\tilde{s}(p)$ be the unique plane interpolating $D_\beta f(p)$ at these vertices. Then by our estimates for the method of plates

$$|D_\beta f(p)-\tilde{s}(p)|\leqslant 6M_3h^2 \tag{5}$$

for all p in T. Since $|\tilde{s}(p_i)-D_\beta s_N(p_i)|=|D_\beta f(p_i)-D_\beta s_N(p_i)|=|D_\beta g(p_i)|$ $\leqslant 4M_3/\sin\theta_0 h^2$, it follows simply from the geometry of planes that

$$|\tilde{s}(p)-D_\beta s_N(p)|\leqslant\frac{6M_3}{\sin\theta_0}h^2 \tag{6}$$

for all directions β and all points p of T. Combining (5) and (6) with the triangle inequality, we find

$$|D_\beta f(p)-D_\beta s_N(p)|\leqslant\frac{12M_3}{\sin\theta_0}h^2. \tag{7}$$

This proves part (b) of the theorem.

Next expand g in a Taylor's expansion in the direction β about p_0. Since $g(p_0)=0$, we have

$$g(p)=g(p_0)+D_\beta g(\tilde{p})|p-p_0|=D_\beta g(\tilde{p})|p-p_0|,$$

where $\tilde{p}$ is some point on the line segment $\overline{p_0 p}$. But then from (5)

$$\begin{aligned}|g(p)|&=|f(p)-s_N(p)|\\&\leqslant|D_\beta g(\tilde{p})||p-p_0|\\&\leqslant|D_\beta g(\tilde{p})|h=|D_\beta f(\tilde{p})-D_\beta s_N(\tilde{p})|h\\&\leqslant\frac{12M_3}{\sin\theta_0}h^3,\end{aligned}$$

as was to be proved. Part (c) is left as an exercise. ■

As usual, the results can be extended to accommodate the case $f\in C^k[\Omega]$, $0\leqslant k<3$, using the Jackson type estimates of Section 2.5 (see Exercises).

Numerous other approximation schemes exist for building piecewise polynomial approximates over a triangular grid. Among these are higher order piecewise Lagrange interpolates such as piecewise cubics, quartics, and quintics. Methods I and II comprises the piecewise linear and piecewise quadratic fits, respectively. Piecewise cubics can be obtained by introducing four knots on each side of each triangle of τ including vertices plus a midpoint; whereas piecewise quartics come from introducing 5 knots on each side of each triangle of τ including vertices 3 interior knots. In each case we seek a piecewise Lagrange polynomial $s(p)$ of a given degree on each triangle of τ solving

$$s(p_i)=f(p_i),\qquad 1\leqslant i\leqslant N,$$

where $f(p)$ is a given function defined on Ω and $\{p_i:1\leqslant i\leqslant N\}$ is collection of vertices, interior and interior edge knots of triangles of τ. To prove existence and uniqueness of such interpolates, it suffices to construct a cardinal basis $\mathfrak{B}=\{\phi_i(p):1\leqslant i\leqslant N\}$ solving

$$\phi_i(p_j)=\delta_{ij},\qquad i\leqslant i,j\leqslant N;$$

for then

$$s(p)=\sum_{i=1}^{N}f(p_i)\phi_i(p).$$

is the unique interpolate to f from span $\mathfrak{B}$. Such bases are easily constructed recursively from products of lower dimensional basis Lagrange polynomials. For example, a basis for the piecewise cubics $L_3(\tau)$ is built triangle by triangle by taking products of planes $P = ax + by + c$ with quadratic surfaces $Q = Ax^2 + By^2 + Cxy + Dx + Ey + F$. In each case P and Q are themselves cardinal basis functions. The reader is left to exercise his geometric ingenuity in constructing these functions (see Exercises) and the spaces $L_k(\tau)$, $k = 0, 1, 2, \ldots$.

Moreover, $L_3(\tau)$ is not the only space of piecewise cubics one can construct over τ. There also exist spaces of piecewise cubics that

(a) prescribe the values of the polynomial and its first derivatives at the vertices of each triangle and the value of the function at the center of gravity of each triangle as well as cubics that

(b) prescribe the values of the polynomial and its first partials at the vertices, normal derivatives and values at midpoints of edges, and the value of p at the center of gravity,

to mention two other possibilities. In all these cases error estimates are available either by a generalization of the methods given for $L_1(\tau)$ and $L_2(\tau)$ or by other means (see in particular Strang and Fix (1973) and Zlamal 1968, 1969). This proliferation of cubics is already excessive, and the demography of higher degree piecewise polynomial fits becomes more appalling the larger the degree becomes. Despite this, however, the situation in two-dimensional approximation over triangular grids is far from satisfactory. In particular there are no *simple* algorithms for constructing piecewise approximates to $C^1[\Omega]$, $C^2[\Omega]$, and smoother functions on triangular grids and on nonregular grids. In all our algorithms (and there are some), the size of the basis becomes larger and larger (i.e., more and more undetermined coefficients per triangle) as more and more smoothness is built in. Examples of such constructions are given in the papers of Zenicek (1970), Mansfield, Birkhoff and Mansfield (1974) C. A. Hall (1969), and Hulme (1968), as well as in the vast engineering literature on this subject. From the pure approximation viewpoint, we do not even know the analog of the polynomial and other splines (if they exist) and whether such approximates would be of practical value. Work on nonpolynomial approximates yielding well-conditioned matrices is in its beginnings [see, viz., Birkhoff and Mansfield (1974) for some rational approximates].

It is only appropriate to point out that the engineers have been the innovators in the field of two-dimensional and higher dimensional interpolation over triangular and curvilinear grids mathematicians being latter day converts, so to speak. (see Zienciewicz (1967, 1971), de Veubeke (1968), and Ergatoudis et. al. (1968) for a small sample. It is only in recent years that more than ε, ε small but positive, mathematicians have seriously turned their attention to these matters. Indeed some have reaped an impressive mathematical harvest in placing the engineers' innovations on a firm mathematical footing (to the mathematicians' way of thinking if not to the engineers'; see, e.g., the papers of Zlamal (1968,1969,1970)). And, of course, the field of mathematics has been enriched as well. Despite all this current activity, there remains much work to be done, the aforementioned problems being a representative but by no means exhaustive list.

EXERCISES

1. (a) Construct the cardinal basis for the piecewise cubic Lagrange polynomials $L_3(\tau)$ over a triangulation τ or Ω.
   (b) Let $s(p) \in L_3(\tau)$ be the piecewise Lagrange interpolate to $f \in C^4[\Omega]$. Derive and prove error estimates for

$$\|f-s\|$$

   and for

$$\operatorname{ess\,sup}|D_\beta^k f(p) - D_\beta^k f(p)|, \qquad k=1,2,3.$$

2. (a) Prove that there exists a unique piecewise cubic polynomial interpolating f and its first partials at the vertices of each triangle T of τ and interpolating f at the center of gravity of each triangle of τ.
   (b) Prove $s \in C[\Omega]$. Does $s \in C^1[\Omega]$? Why?
   (c) Try to prove $\|f-s\| \leqslant \gamma_3 M_3 h^4$ where $f \in C^4[\Omega]$ and γ_3 is a constant

5.5 AUTOMATIC MESH GENERATION AND ISOPARAMETRIC TRANSFORMS

We have now built several (finite element) approximating schemes over triangulations τ of polygonally bounded domains Ω in the x–y plane. To compute with such approximates in solving partial differential equations, or even in straightforward approximation problems, we must perform such a proper triangulation in a systematic way. This would be a formidable task if we could not bring the computer to our aid, particularly for triangulations involving a great number of triangles. (Partitions of Ω into several hundred triangles are not uncommon in large-scale computations—in some cases they are the rule rather than the exception.) Computer-aided triangulations and other geometric partitions of a domain Ω are known as *automatic mesh generating schemes.*

PROBLEM 1 AUTOMATIC MESH GENERATION

Given a closed polygonally bounded and connected domain of the real plane, how do we program the computer to partition Ω into a proper triangulation that allows us to use the method of plates and other piecewise Lagrange polynomials over such triangulations?

Two such triangulations are illustrated in Figures 5.22 and 5.23, along with a triangulation of a nonpolygonally bounded region (Figure 5.24).*

Our second problem is to decide how to find finite element approximates (piecewise polynomials) over nonpolygonally bounded domains. In particular, what do we do about *curved boundaries*? We would still like to work over proper triangulations of a region Ω or an approximate to Ω_N as illustrated in Figures 5.22, 5.23, and 5.24. This problem has been very nicely solved in part by Zienkiewicz and Phillips (1971), B. M. Irons (1970), and Ergatoudis, Irons, and Zienkiewicz (1968), using what are known in the literature as *isoparametric transformations*. If the boundary of

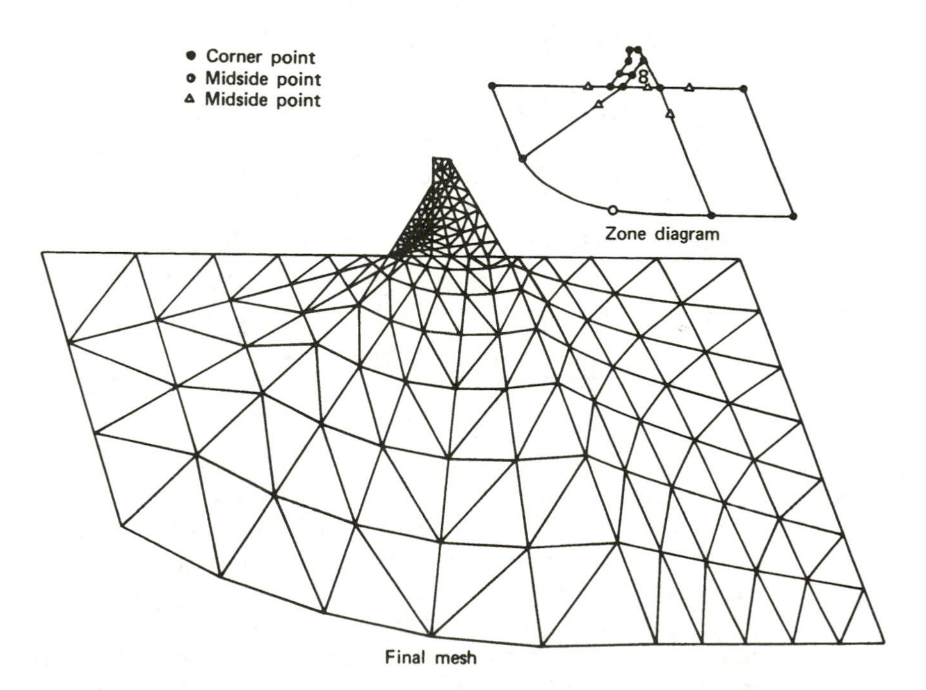

Figure 5.22 Automatic mesh generation of a polygonally bounded region.

*The author is grateful to professors Zienkiewicz and Phillips for permitting her to use these diagrams.

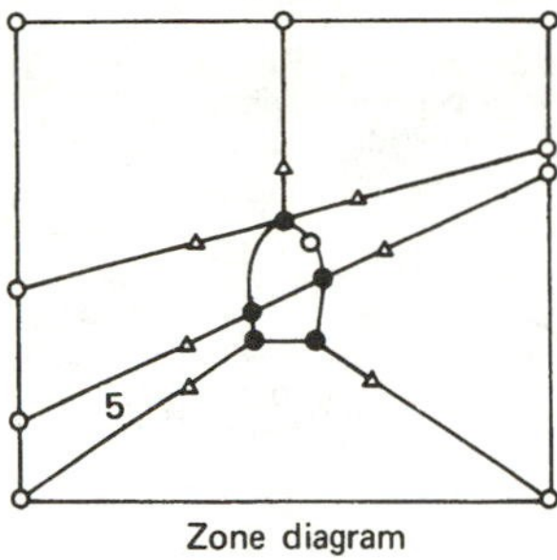

Zone diagram

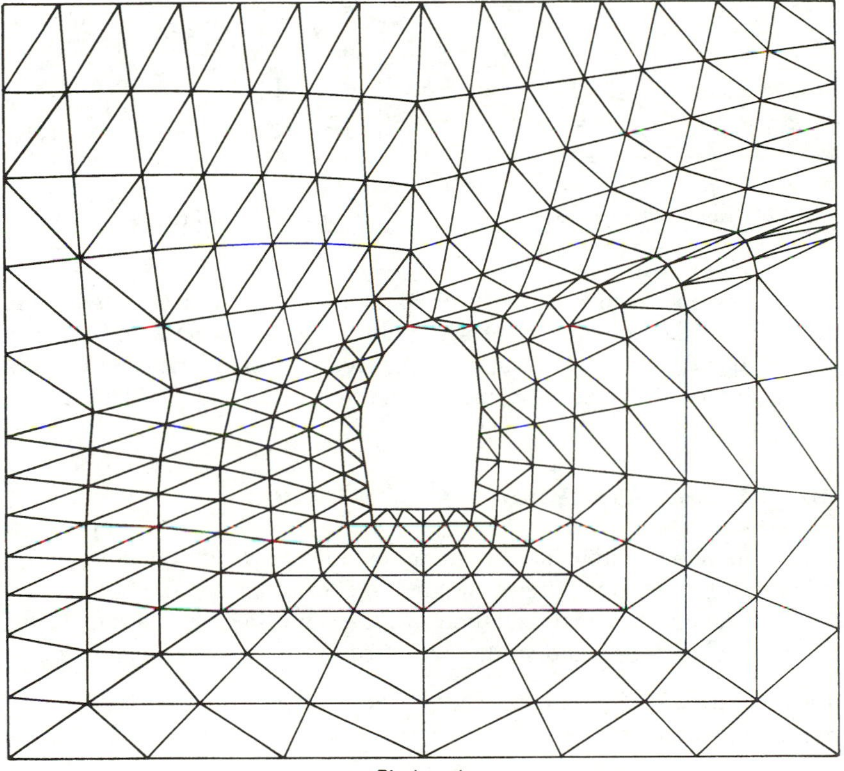
Final mesh

Figure 5.23 Automatic mesh generation of a polygonally bounded region.

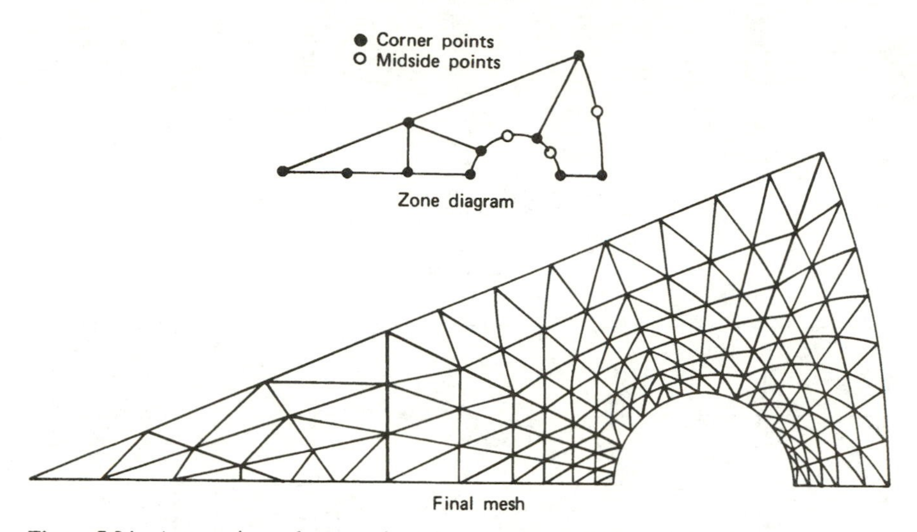

Figure 5.24 Automatic mesh generation of a nonpolygonally bounded region.

Ω consists of polynomial segments or spline segments, we try to use splines of exactly the same degree parametrically, to map a square onto Ω such that boundaries go exactly or approximately onto boundaries; and so that the resulting transformation is one to one. So our second problem is

PROBLEM 2 ISOPARAMETRIC TRANSFORMATIONS

Given a domain Ω with curved boundaries how do we get the computer to automatically generate a "triangulation" of Ω or of an approximation to Ω? What are the analogs of the method of plates and the piecewise quadratic Lagrange polynomials, and so on, over "triangles" for such a partition?

In the interest of good pedagogy, we approach these problems by looking at a simple example.

EXAMPLE

Let $I=[-1,1]$ and let $\mathbb{S}=[-1,1]\times[-1,1]$ be the square graphed in Figure 5.25. Let Ω be the "quadrilateral" in Figure 5.25 whose edges are two parabolas and two straight lines. We want to find a $1:1$ mapping T from $\mathbb{S}$ *onto* Ω such that *boundaries map onto boundaries*. To this end let $p_1,p_2,p_3,\ldots,p_6,p_i=(x_i,y_i)$, be vertices and midpoints of edges of Ω as illustrated and let $q_1,q_2,\ldots,q_6,q_i=(s_i,t_i)$, be the corresponding points of $\mathbb{S}$. We define the mapping T of $\mathbb{S}$ onto Ω *parametrically* (i.e., by parametric equations). In particular, we let

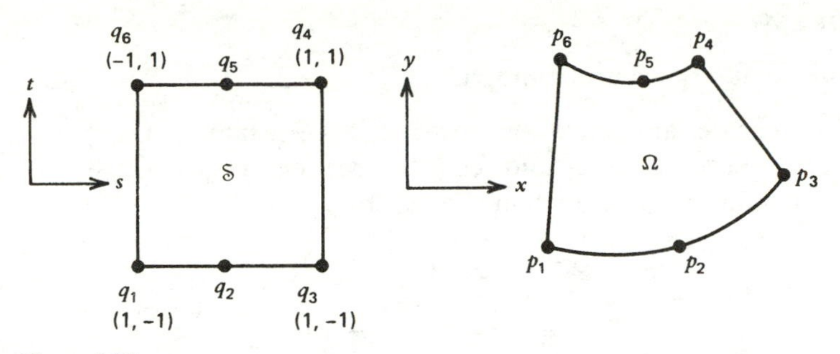

Figure 5.25

$$\begin{pmatrix} x \\ y \end{pmatrix} = T(s,t) = \begin{pmatrix} \sum_{i=1}^{6} x_i L_i(s,t) \\ \sum_{i=1}^{6} y_i L_i(s,t) \end{pmatrix}, \tag{1}$$

where the L_i's are the unique polynomials quadratic in s and linear in t solving the interpolation problem

$$L_i(q_j) = \delta_{ij} \qquad 1 \leqslant i,\ j \leqslant 6. \tag{2}$$

Clearly $T(s_i, t_i) = T(q_i) = p_i$. To obtain the L_i's, let

$$s_1 = -1, \qquad s_2 = 0, \qquad s_3 = 1$$

$$t_1 = -1, \qquad t_6 = 1.$$

Let the $l_i(s)$'s be the quadratic basis Lagrange polynomials solving

$$l_i(s_\mu) = \delta_{i\mu}, \qquad 1 \leqslant i, \mu \leqslant 3,$$

and let $l_j(t)$ be the linear basis Lagrange polynomials in the t direction solving

$$l_j(t_k) = \delta_{jk}, \qquad j,k = 1,6.$$

Then the six functions $\{l_i(s)l_j(t) : 1 \leqslant i \leqslant 3,\ j = 1,6\}$ give us a set $\{L_1(q), L_2(q), \ldots, L_6(q)\}$ of six linearly independent quadratic polynomials in $q = (s,t)$ solving the interpolation problem (2).

We now must prove that

1. T is one to one from $\mathbb{S}$ onto Ω.
2. T maps boundaries onto boundaries.

THEOREM

T maps boundaries onto boundaries.

Proof. Consider any one of the edges of Ω. For example, take $\widehat{p_1p_2p_3}$. This edge is a parabola and thus can be written as $y=Ax^2+Bx+C$. A parametrization of this function is given by

$$x=s=\phi_1(s)$$

$$y=As^2+Bs+C=\phi_2(s).$$

But $T(s,-1)=(T_1(s,-1),T_2(s,-1))^T$ is a two-tuple of parabolas (quadratics) which interpolates the ordered pair $(\phi_1(s),\phi_2(s))^T$ of quadratics at $-1,0$, and 1. Since any parabola is uniquely determined by three points, it follows that $\phi_1(s)=T_1(s,-1)$ and $\phi_2(s)=T_2(s,-1)$. Thus $\{T(s,-1): -1\leqslant s\leqslant 1\}$ is $\widehat{p_1p_2p_3}$. A similar argument obtains along the remaining edges. ∎

THEOREM

T is one to one.

To prove the foregoing theorem, it suffices to observe that for fixed s, $-1\leqslant s\leqslant 1$, line segments $l_s=\{(s,t): -1\leqslant t\leqslant 1\}$ in $\mathbb{S}$ parallel to the t-axis map one to one onto line segments λ_s in Ω, that λ_s does not intersect $\lambda_{\bar{s}}$ when $s\neq\bar{s}$, and that each point in Ω lies on one and only one such segment λ_s (see Exercises). Figure 5.26 illustrates l_s and λ_s. In particular, T generates a coordinate system for Ω, coordinate lines being the curve $s=$constant and $t=$constant in the $x-$ plane.

Although the proof of the one-to-oneness of T is rather easy in this simple case, proofs that T is one to one and onto, with boundaries mapping onto boundaries, can become rather complicated in the case of more general *parametric transformations* T. In fact, not all such parametric coordinate transforms are one to one (see Exercises). One-to-one para-

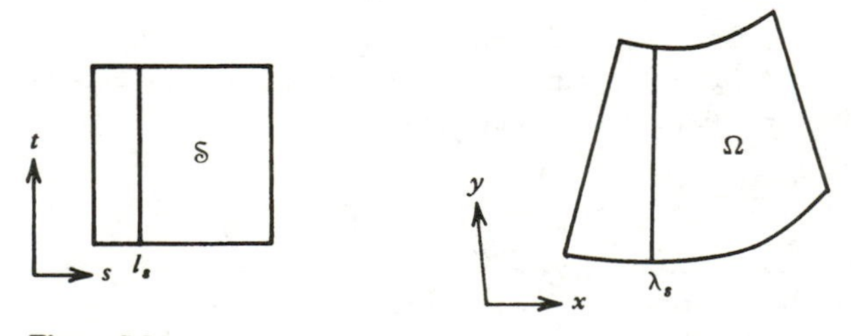

Figure 5.26

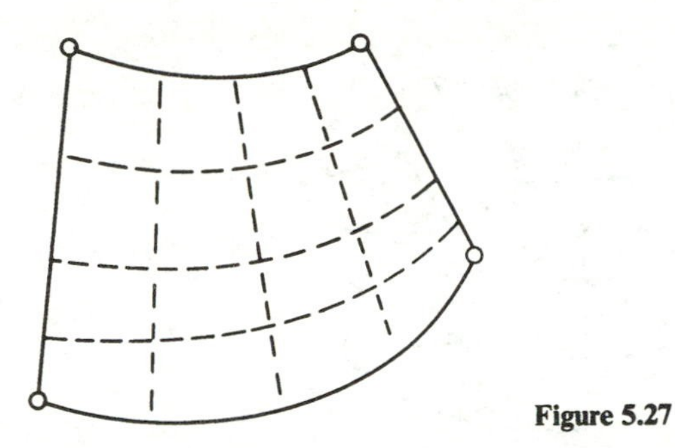

Figure 5.27

metric transformations such as T from a square $\mathbb{S}$ in the s–t plane onto a region $\tilde{\Omega}$ of the x–y plane and their higher dimensional analogs are referred to as *isoparametric transformations*. From the applied point of view we are interested in isoparametric transformations for which $\tilde{\Omega}$ is a good approximate to Ω. The T we are working with does have these properties, and it can be used to automatically generate a triangulation of Ω or of an approximation to Ω. One automatic mesh *generation algorithm* arising from T is the following.

AUTOMATIC MESH GENERATION ALGORITHM I: CURVED TRIANGULAR ELEMENT

STEP 1. Let $-1=s_0 \leqslant s_1 \leqslant \cdots \leqslant s_n = 1$ and $-1=t_0 \leqslant t_1 \leqslant \cdots \leqslant t_m = 1$ be two partitions of the interval $[-1,1]$ in the s and t directions, respectively.

STEP 2. Let $\{T(s_i,t): -1 \leqslant t \leqslant 1\}$ and $\{T(s,t_j): -1 \leqslant s \leqslant 1\}$, where $0 \leqslant i \leqslant n$ and $0 \leqslant j \leqslant m$ be the set of coordinate curves in Ω appearing in Figure 5.27.

STEP 3. Step 2 partition Ω into mn quadrilaterals R_{ij} with curved boundaries. Draw in the shortest diagonal of each R_{ij}. The result (Figure 5.28) is a triangulation of Ω in which triangles can have curved boundaries.

Using this algorithm to compute approximates to a given function $f(x,y)$ requires a basis or set of shape functions $\phi_i(x,y)$ for the piecewise linear, quadratic, and so forth, Lagrange polynomials on such a triangular grid. If all the sides of Ω were straight lines, the ϕ_i's would simply be the shape functions of the previous section. However, some of the boundaries are curved; thus we need a set of shape functions built smoothly over these curves edges. This would require a specific formula for T^{-1}. In particular, the natural shape functions ϕ_i for the "triangulation" described by

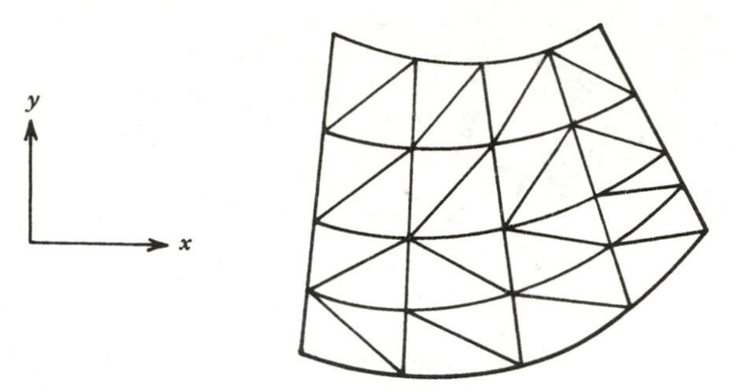

Figure 5.28

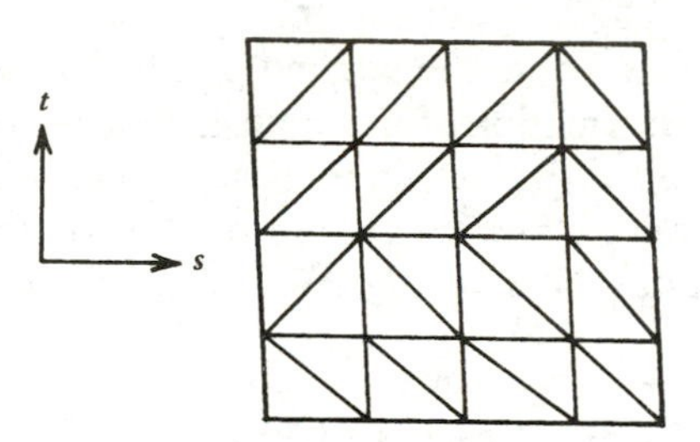

Figure 5.29 Analogous triangulation of $\mathbb{S}$.

Algorithm I are the functions

$$\phi_i(x,y) = \psi_i(T^{-1}(x,y)),$$

where the ψ_i's are the shape function over the analogous triangulation of $\mathbb{S}$ in the s–t plane (see Figure 5.29).

The problem with this approach is that we usually lack an explicit expression for T^{-1}. To overcome this difficulty we straighten out all the curves to have true triangular elements. The price to be paid is that we are no longer triangulating Ω but an approximate Ω_N to Ω. However, this approach allows us to use the approximates of the previous section directly, without having to bother with T^{-1}. In this case our automatic mesh generation scheme is slightly modified.

AUTOMATIC MESH GENERATION SCHEME II: NONCURVED TRIANGULAR ELEMENTS

STEP 1. Same as Step 1 of Algorithm I.

STEP 2 Let $\lambda_{ij} = T(s_i, t_j) : 0 \leqslant i \leqslant n$ and $0 \leqslant j \leqslant m$. Then $\{(x_i, y_j) = T(s_i, t_j)\}$ is our *grid*.

STEP 3. Connect all successive pairs of points (x_i, y_j) and (x_{i+1}, y_j), and

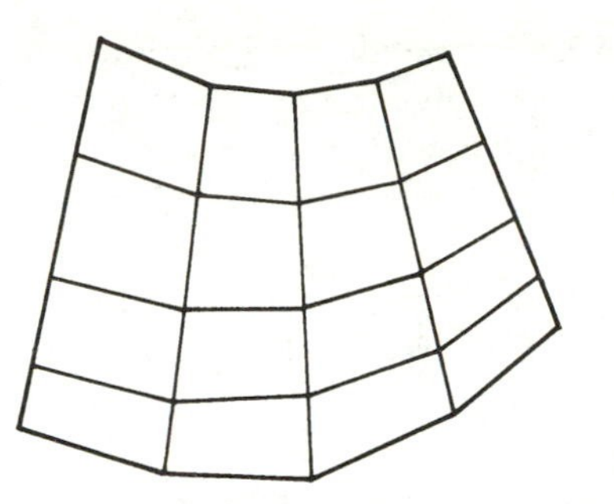

Figure 5.30

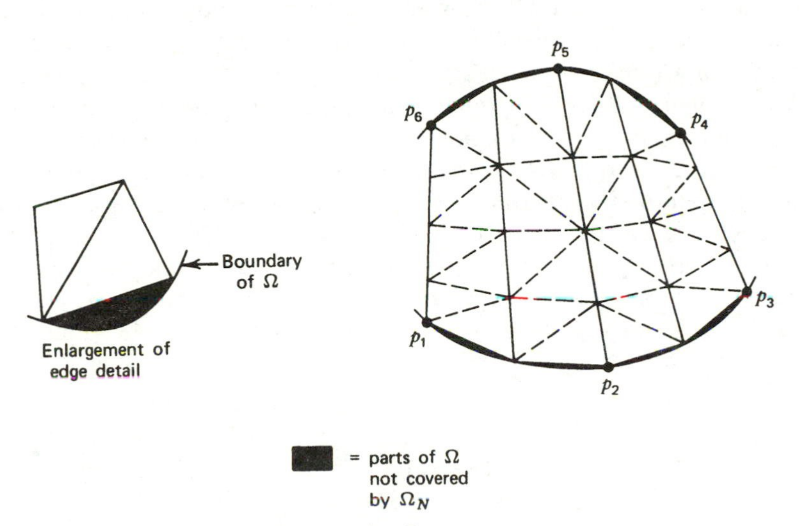

Figure 5.31 A triangulation of a region Ω.

pairs of points (x_i,y_j) and (x_i,y_{j+1}) by line segments. The result is a partition consisting of a collection of quadrilaterals R_{ij} (Figure 5.30), some of which may fail to be contained in Ω (this can happen when (x_i,y_j) is a boundary knot).

STEP 4. Draw in the shortest diagonal each R_{ij}.

The result of this algorithm is a proper triangulation τ_N of a region $\Omega_N = \cup\{R_{ij} : 0 \leqslant i \leqslant n-1 \text{ and } 0 \leqslant j \leqslant m-1\}$ (the region covered by all the quadrilaterals R_{ij}), which approximates Ω when Ω has curved boundaries, as illustrated in Figure 5.31 for a different Ω.

Now suppose $f(x,y)$ is a function defined on Ω which we wish to approximate using the methods of Section 5.4. We use the functions of Section 5.4 over the triangulation τ_N of Ω_N. The method of plates will always be well defined, since the vertex of each triangle of Ω_N belongs to Ω. However, the piecewise quadratic Lagrange polynomials might not be defined along boundary edges of Ω_N at midpoints of edges of triangles that lie outside Ω. At such points we can always force the approximate f_N to f to interpolate the piecewise linear fit to f at the midpoint; the resulting approximate will be at worst $o(h^2)$ near edges while remaining $o(h^3)$ on the interior of Ω when $f \varepsilon C^3[\Omega]$ (see Exercises). The resulting approximate, unless continuously extended, will fail to be defined at points of Ω that do not lie in Ω_N such as p^* shown in Figure 5.32.

The method can be extended to more general curved boundaries using other choices of $l_i(s,t)$. For example, let Ω be some closed, connected, and bounded region of the plane as appears in Figure 5.33. To triangulate Ω (actually an approximation to Ω), partition Ω into "foursided" regions as shown in Figure 5.33 by introducing knots along the boundary. In this case we have introduced nine such boundary knots. We triangulate each of the regions I, II, III and IV by approximating the four bounding edges of each region by parametric splines or parametric piecewise polynomials of a fixed degree on each edge, making sure to use the same degree along edges forming interfaces between adjoining regions. For example, we might decide to fit cubic splines to each edge by introducing n knots p_i, $i \leqslant i \leqslant n$, including end knots on each edge. This gives us a total of $4n$ knots along each quadrilateral. For example, let $p_i = (x_i, y_i)$, $i \leqslant i \leqslant 4n$, on quadrilateral I. Define $T: \mathbb{S} \to \mathrm{I}$ by

$$\begin{pmatrix} x \\ y \end{pmatrix} = T(s,t) = \begin{pmatrix} T_1(s,t) \\ T_2(s,t) \end{pmatrix},$$

$$T_1(s,t) = \sum_{i=1}^{4n} x_i C_i(s,t)$$

$$T_2(s,t) = \sum_{i=1}^{4n} y_i C_i(s,t).$$

The functions $C_i(s,t) = C_\lambda(s) C_k(t)$, where $C_\lambda(s)$ and $C_k(t)$ are the cardinal cubic splines solving

$$C_\lambda(s_\mu) = \delta_{\lambda\mu}, \qquad 1 \leqslant \lambda, \mu \leqslant n$$

$$C_k(t_j) = \delta_{jk}, \qquad 1 \leqslant j,k \leqslant n,$$

and $-1 = s_1 < s_2 < \cdots < s_n = 1$ and $-1 = t_1 < t_2 < \cdots < t_n = -1$. T maps $\mathbb{S}$ onto some approximate to I with boundary knots going to boundary knots.

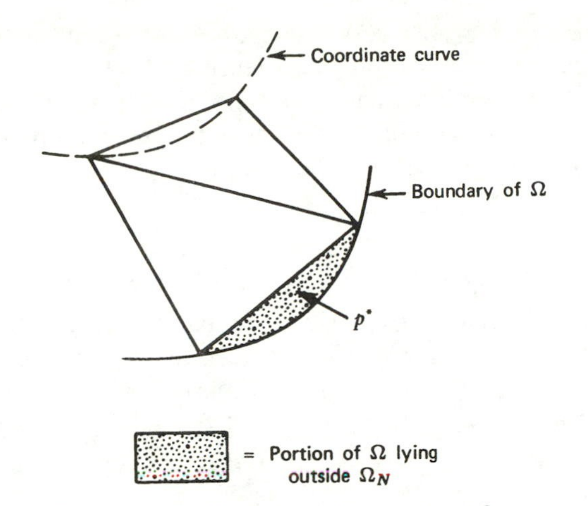

Figure 5.32 Section of Ω_n compared with corresponding section of Ω.

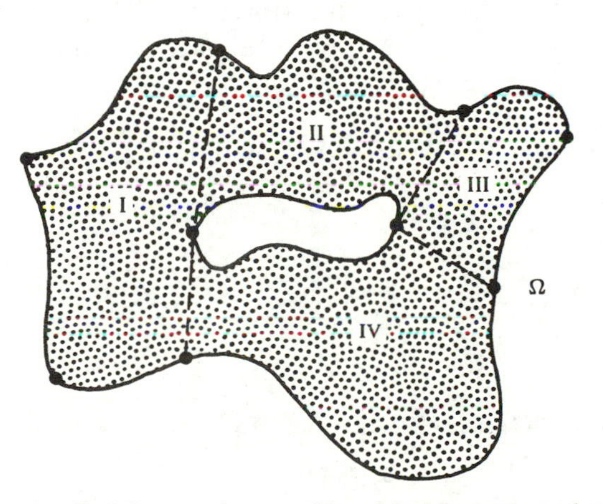

Figure 5.33 A region Ω of the x–y plane partitioned for automatic mesh generation.

Repeat the process on the other pieces II, III, and IV. Although boundaries *may not* map onto boundaries exactly, boundaries will map into approximate boundaries. If the map is one to one, we can proceed just as in the previous case to generate a triangulation of an approximate Ω_N to Ω. The real problem is to establish whether T is one to one. To the very applied individual this is tantamount to calling out a visual display of the coordinate curves $s=$constant and $t=$constant in the x–y plane generated by T and to see whether distinct "coordinate curves" $s_1=$constant and $s_2=$constant, $s_1 \neq s_2$, or $t_1=$constant and $t_2=$constant, $t_1 \neq t_2$ intersect. If they do, T is not a proper coordinate transformation and is inadmissible for automatic mesh generation (see Figure 5.34). The mathematical problem here is to determine what class of piecewise parametric polynomial transformations T of $\mathbb{S}$ both generates a proper coordinate system for Ω and yields tight approximates to the boundary of Ω?

EXERCISES

1. Prove that the mapping T given in the text example is one to one.
2. Prove that the parametric piecewise quadratic Lagrange interpolates to a given function f defined on Ω over the triangulation τ_N of Ω_N as suggested in the text is $o(h^3)$ on the interior of Ω and at worst $o(h^2)$ near the boundary of Ω when $f \in C^3[\Omega]$.
3. Let $\Omega = \{(x,y): 1 \leqslant x^2 + y^2 \leqslant 9\}$ be the annulus in the x–y plane illustrated in Figure 5.35.
   (a) Program your computer to compute and graph (plot) a triangulation of Ω by first partitioning $\mathbb{S}$ into two parts (Figure 5.35a) and then let T parametrically transform $\mathbb{S}$ to I and then to II, using the tensor product of piecewise bilinear Lagrange polynomials in the s and t directions, respectively with $n = 9$ and $m = 1$ as described in Exercise 4.
   (b) Repeat the problem using the partition of Figure 5.35b.
4. Let $\mathbb{S} = [-1,1] \times [-1,1]$ and number the edges of $\mathbb{S}$ through 4′ as illustrated in Figure 5.36.

$$\pi_s = -1 = s_0 < s_1 < \cdots < s_n = 1$$

$$\pi_t = -1 = t_0 < t_1 < \cdots < t_m = 1.$$

Let $l_i(s)$ and $l_j(t)$ be the hat functions (piecewise linear Lagrange polynomials) solving

$$l_i(s_k) = \delta_{ik}, \qquad 0 \leqslant i, k \leqslant n$$

$$l_j(t_\mu) = \delta_{j\mu}, \qquad 0 \leqslant j, \mu \leqslant m.$$

Let Ω be a "quadrilateral" plus interior (as in Figure 5.36) in the x–y plane and number successive sides of Ω 1 through 4. Introduce $n+1$ distinct points, including end points on each of sides 1 and 3 and $m+1$ such points on each of sides 2 and 4. Let $l_{ij}(s,t) = l_i(s) l_j(t)$. Define a parametric transformation of $\mathbb{S}$ into the x–y plane using the l_{ij}'s that map the knots of edge i' onto the corresponding knots of edge i. Is T always one to one and onto? Explain.

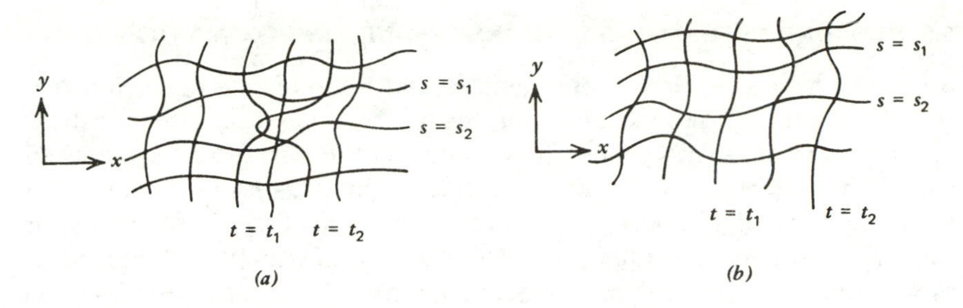

Figure 5.34 (*a*) Improper and (*b*) proper coordinate transformation.

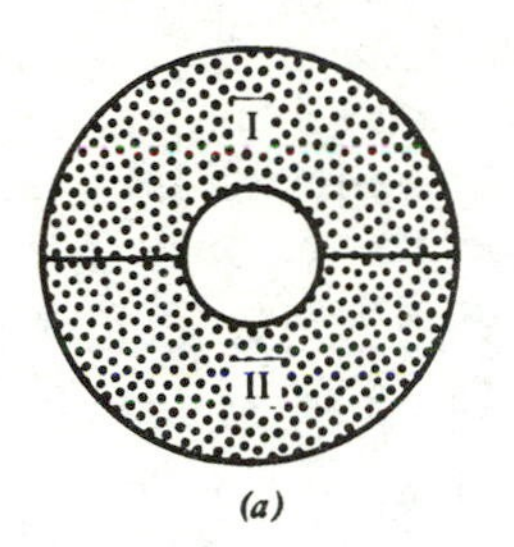

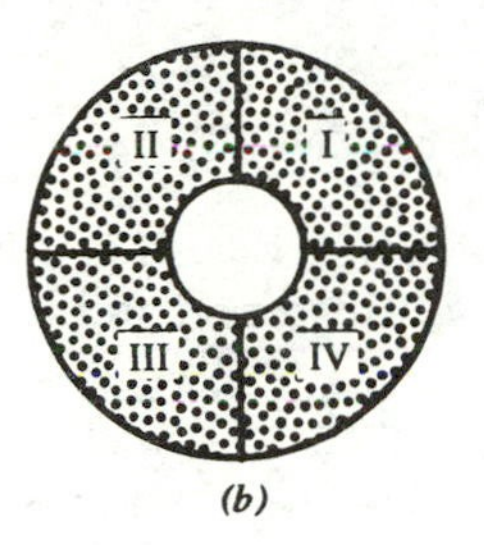

Figure 5.35

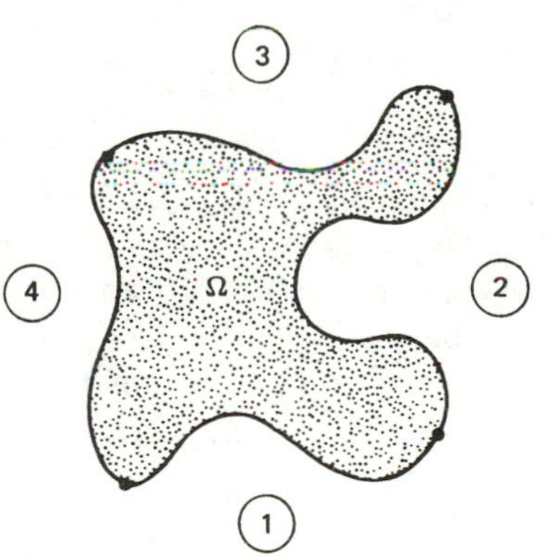

Figure 5.36

5.6 BLENDED INTERPOLATES AND SURFACE APPROXIMATION

So far we have considered approximates that interpolate a given function $f(x,y)$ at a given set of points in the plane. One can also construct approximates that interpolate $f(x,y)$ along entire curves or segments of curves in the plane. Such approximates are known as *blended interpolates* after the work of Birkhoff and Gordon (1968) and Gordon (1969), or as transfinite elements after the work of Gordon and Hall (1972). These ideas as well of as those of the previous section, also apply to the approximation of surfaces. The simple extension of isoparametric transformations to three-dimensional surface approximation are sometimes called *Coons's patches* after the work of Coons (1968).

A simple example of a blended interpolate is the following. Let $\pi = \{(x_i,y_j): 0 \leqslant i \leqslant n \text{ and } 0 \leqslant j \leqslant m\}$ be a rectangular grid, $x_0 < x_1 < \cdots < x_n$, and $y_0 < y_1 < \cdots < y_m$, contained in some region Ω of the x–y plane (see Figure 5.37). Let $l_i(x)$ and $l_j(y)$ be the cardinal basis Lagrange polynomials of degrees n and m in the x and y directions, respectively. The function

$$I_x f(x,y) = s(x,y) = \sum_{i=0}^{n} f(x_i,y) l_i(x)$$

coincides with $f(x,y)$ along each of the lines $\{(x_i,y): -\infty < y < \infty\} \cap \Omega$, $0 \leqslant i \leqslant n$. Similarly,

$$I_y f(x,y) = \bar{s}(x,y) = \sum_{j=0}^{m} f(x,y_j) l_j(y)$$

interpolates $f(x,y)$ along each of the lines $\{(x,y_j): -\infty < x < \infty\} \cap \Omega$, $0 \leqslant j \leqslant m$. The surfaces associated with $s(x,y)$ and $\bar{s}(x,y)$ are called blended interpolating surfaces or transfinite interpolates because they coincide with the surface associated with $f(x,y)$ along entire curves in 3-space. Furthermore, one can easily estimate the errors $\|f - s\|$ and $\|f - \bar{s}\|$ when f is confined to the rectangle $[x_0,x_n] \times [y_0,y_m]$ using the methods of Section 5.2. In particular

$$\|f^{(j)} - s^{(j)}\| = o\left(h_x^{n+1-j}\right), \qquad 0 \leqslant j \leqslant m,$$

when $f \in C^{n+1,0}[R]$, and

$$\|f^{(j)} - \bar{s}^{(j)}\| = o\left(h_y^{m+1-j}\right), \qquad 0 \leqslant j \leqslant m,$$

when $f \in C^{0,m+1}[R]$. The reader will easily see how to construct various piecewise polynomial and other blended interpolates over rectangular

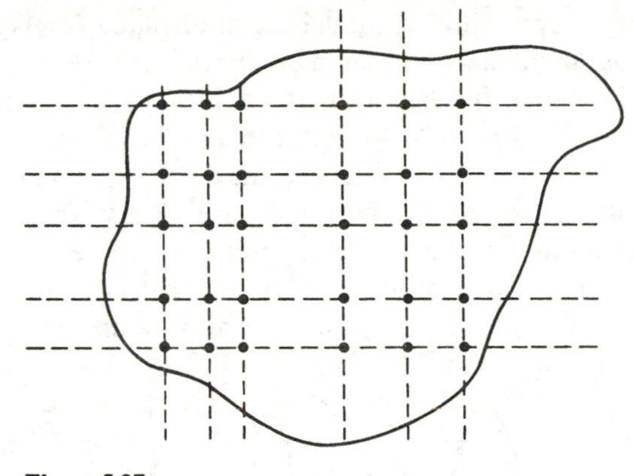

Figure 5.37

grids; error estimates dependent on the smoothness of f can also be provided.

Let Bf be the linear combination

$$Bf = I_x f + I_1 f - I_x I_y f$$

where $I_x f$ and $I_y f$ are blended interpolates to f and $I_x I_y f$ is the ordinary tensor product interpolate to f given in Section 5.2. This combination has come to be known as *the blended spline approximate* to f. It has the advantage over $I_x I_y f$, $I_x f$, and $I_y f$ of greatly increased accuracy. In particular, if $R_x = f - I_x f$, and $R_y = f - I_y f$, then

$$\begin{aligned} f - Bf &= f - [I_x f + I_y f - I_x I_y f] \\ &= \{I - [(I - R_x) + (I - R_y)] + (I - R_x)(I - R_x)\} f \\ &= (R_x f)(R_y f). \end{aligned}$$

To illustrate the order of convergence of such blended spline approximates, let $I_x I_y f$ be the piecewise bicubic polynomial interpolate to $f \in C^4[\Omega]$. Then $\|R_x f\| = \|f - I_x f\|$, $\|R_y f\| = \|f - I_y f\|$, and $\|f - I_x I_y f\|$ are all $o(h^4)$. However,

$$\|f - Bf\| = \|R_x f\| \|R_y f\| = o(h^8),$$

is a dramatic improvement in accuracy.

The extension of isoparametric transformations to three dimensions is equally simple. The construction of such approximates is fairly common in engineering design. For example, you might wish to construct a 1:1000 scale, automatic machine-tooled model of an airplane fuselage to test in a wind tunnel. You might start by placing a curvilinear coordinate system on the surface of the actual fuselage (a surface in 3-space) and measuring the coordinates (x_i, y_i, z_i), $1 \leq i \leq N$, of all the points of intersection of the coordinate curves. Some of the coordinate curves might coalesce, as illustrated in Figure 5.38 by the points P_0 and P_N, which are counted in this case as many times as there are distinct coordinate curves emanating from them. To approximate the surface Ω, we construct a transformation T of the square $\mathbb{S} = [-1,1] \times [-1,1]$ onto an approximating surface Ω_N. In particular we can take

$$\begin{pmatrix} x \\ y \\ z \end{pmatrix} = \tilde{T}(s,t) = \begin{pmatrix} \tilde{T}_1(s,t) \\ \tilde{T}_2(s,t) \\ \tilde{T}_3(s,t) \end{pmatrix},$$

where

$$\tilde{T}_1(s,t) = \sum_{i=1}^{N} x_i l_i(s,t)$$

$$\tilde{T}_2(s,t) = \sum_{i=1}^{N} y_i l_i(s,t)$$

$$\tilde{T}_3(s,t) = \sum_{i=1}^{N} z_i l_i(s,t),$$

where the l_i's are, say, bilinear Lagrange polynomials or bicubic splines solving

$$l_i(s_j, t_j) = \delta_{ij}, \qquad 1 \leq i,j \leq N,$$

and (s_i, t_j), $1 \leq i,j \leq N$, are points of a rectangular grid on $\mathbb{S}$. The transformation $\tilde{T}$, so defined, maps $\mathbb{S}$ onto a surface Ω_N, which hopefully is a good approximate to the original surface Ω (see Figure 5.39). To scale $\tilde{T}$, let $T = 1/100\ \tilde{T}$.

The T just defined is given parametrically and is an isoparametric transformation if and only if $\tilde{T}$ is one to one (no two coordinate lines intersect more than once) and maps boundaries onto boundaries. Sometimes such approximating surfaces have to be built piecewise, as in the case of automatic mesh generation routines, by partitioning the surface to be approximated into separate regions and mapping $\mathbb{S}$ onto each approximate to these regions separately, so that continuity is preserved across

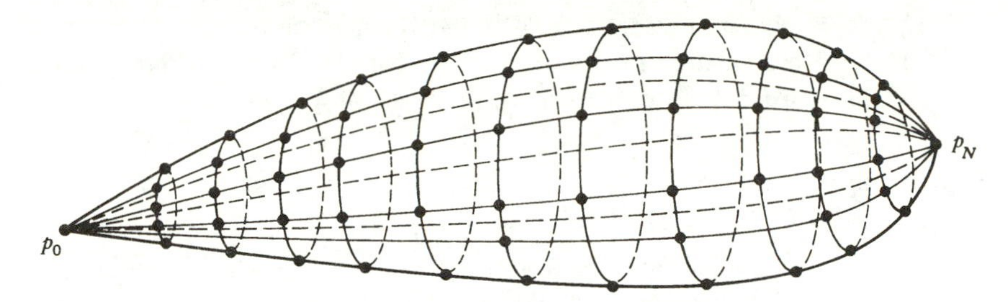

Figure 5.38 Three-dimensional surface to be approximated, showing grid points $(x_i, y_i, z_i)=0$ of an introduced coordinate system.

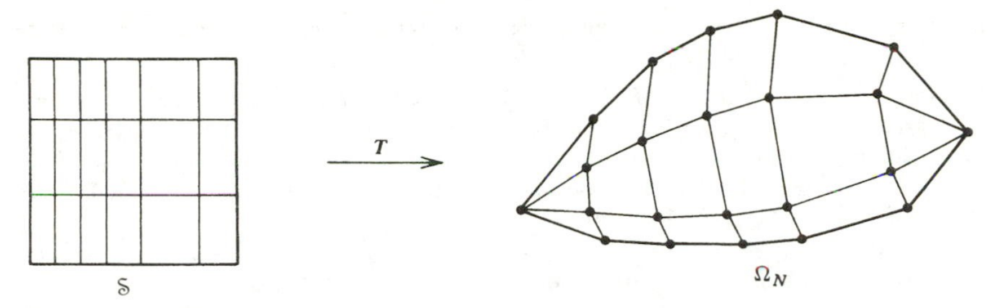

Figure 5.39 A piecewise bilinear approximate to half a fuselage.

regional interfaces. Such surfaces constitute Coons patches. Assuring that T is one to one can be difficult, and you may wish to resort to a visual display as an initial check. If the mesh spacing is small enough on the grid in the s–t plane, difficulties (T not one to one) will not usually be encountered, provided the surface that is being approximated is not too erratic (see Exercises).

The foregoing presentation was deliberately sketchy. It should, however, be clear enough to enable the interested reader to experiment with such surface fits and to recognize improper approximates to surfaces.

EXERCISES

1. Let Ω be the surface of a torus (doughnut) of external radius 9 and internal radius 1. Construct an isoparametric mapping of the square $\mathbb{S}=[-1,1]\times[-1,1]$ onto Ω.
2. Let T be the parametric piecewise bilinear approximate to a given surface Ω in 3-space described in the text. Then T changes with changing n and m. Let h be the mesh spacing of the rectangular grid on $\mathbb{S}$, and suppose $h \to 0$ as m and $n \to \infty$.
   (a) Is it true that T is always one to one? Why?

(b) Is T one to one for all h sufficiently small? Explain.
(c) Give a measure of error $|\Omega - T[\mathbb{S}]|$ for $\Omega - T[\mathbb{S}]$. In terms of this measure, estimate $|\Omega - T[\mathbb{S}]|$ in terms of h.

REFERENCES

Barnhill, R. E. and Lois Mansfiled. Error bounds for smooth interpolation over triangles. Technical report, University of Utah, 1971.

Birkhoff, G. Piecewise bicubic interpolation. In *Approximation with special emphasis on spline functions*, I. J. Schoenberg, ed. Academic Press, New York, 1969, pp. 185–221.

Birkhoff, G. and W. J. Gordon, the draftsman's and related equations. *J. Approx. Theory*, **1** (1968), 191–208.

Birkhoff, G. and Lois Mansfield. Compatible triangular finite elements. To appear, *J. Math. Anal. Appl.*, **47** (1974), pp. 531–553.

Birkhoff, G., M. H. Schultz, and R. S. Varga. Piecewise Hermite interpolation in one and two variables with application to partial differential equations. *Numer. Math.*, **11** (1968), pp. 232–256.

Bramble, J. H. and M. Zlamal. Triangular elements in the finite element method. *Math. Comp.*, **24**, no. 112, (1970), pp. 809–820.

Carlson, R. E. and C. A. Hall. Bicubic spline interpolation and approximation in right triangles. WAPD-T-2488, Bettis Atomic Power Laboratory, Technical Report, 1970.

Carlson, R. E. and C. A. Hall. Bicubic spline interpolation in rectangular polygons. WAPD-T-2391, Bettis Atomic Power Laboratory Technical Report, 1970.

Ciarlet, P. G. and C. Wagschal. Multipoint Taylor formulas and applications to the finite element method. *Numer. Math.*, **17** (1971), pp. 84–100.

Coons, S. A. Surfaces for computer-aided design of space forms. Project MAC, Design Division, Department of Mechanical Engineering, MIT, 1968. Available from: Clearinghouse for Federal Scientific–Technical Information, National Bureau of Standards, Springfield, Va.

de Veubeke, B. Fraeijs. A comforming finite element for plate bending. *Int. J. Solid Struct.*, **4** (1968), pp. 95–108.

Ergatoudis, I., B. M. Irons, and O. C. Zienkiewicz. Curved isoparametric quadrilateral elements for finite element analysis. *Int. J. Solid Struct.*, **4** (1968), pp. 31–42.

Gordon, William J. Spline-blended surface interpolation through curved networks. *J. Math Mech.*, **18**, (1969), pp. 931–952.

Gordon, W. J. and C. A. Hall. Geometric aspects of the finite element method: Construction of curvilinear coordinates systems and their application to mesh generation. Research Report GMR-1286, General Motors Research Laboratories, Warren, Mich., Spetember 1972.

Hall, C. A. Bicubic interpolation over right triangles. *J. Math. Mech.*, **19**, no. 5 (1969), pp. 1–11.

Hulme, B. L., Interpolation by Ritz approximation. *J. Math. Mech.*, **18**, no. 4 (1968), pp. 337–341.

Hulme, Bernie L. A new bicubic interpolation over right triangles. *J. Approx. Theory*, **5**, no. 1 (1972), pp. 66–73.

Irons, B. M., A frontal solution programme for finite element analysis. *Int. J. Numer. Meth. Eng.*, **2** (1970), pp. 5–32.

Mansfield, Lois. Higher order compatible triangular finite elements. To appear, *Numer. Math.*

Prenter, P. M., Lagrange and Hermite interpolation in Banach spaces. *J. Approx. Theory*, **4**, no. 4 (1971), pp. 419–432.

Ritter, Klaus. Two-dimensional splines and their extremal properties. *Z. Angew. Math.*, **49**, pp. 597–608.

Schultz, M. H. *L*-Multivariate approximation theory. *SIAM J. Numer. Anal.*, **6** (1969), pp. 161–183.

Desai C. S. and J. F. Abel. *Introduction to the Finite Element Method*, Van Nostrand Reinhold, New York, 1972.

Schultz, M. H. Multivariate *L*-spline interpolation. *J. Approx. Theory*, **2** (1969), pp. 127–135.

Strang, G. and G. Fix. *Analysis of the finite element method.* Prentice-Hall, Englewood Cliffs, N. J., 1973.

Zenisek, A. Interpolation polynomials in the triangle. *Numer. Math.*, **15** (1970), pp. 283–296.

Zienkiewicz, O. C. *The finite element method in structural and continuum mechanics*. McGraw-Hill, London, 1967.

Zienkiewicz, O. C. and D. V. Phillips. An automatic mesh generation scheme for plane and curved surfaces by isoparametric coordinates. *Int. J. Numer. Meth. Eng.*, **3** (1971), pp. 519–528.

Zlamal, Milos. On the finite element method. *Numer. Math.*, **12** (1968), pp. 394–409.

Zlamal, Milos. On some finite element properties for solving second-order boundary value problems. Numer. Math., **14** (1969), pp. 42–48.

Zlamal, Milos. A finite element procedure of the second-order accuracy. *Numer. Math.*, **14** (1970), pp. 396–402.

6

FUNDAMENTALS FOR VARIATIONAL METHODS

6.1 VARIATIONAL METHODS

Recall that in the very first paragraph of this book we posed a problem. We were given the differential equation

$$x''(t)+a(t)x'(t)+b(t)x(t)=f(t), \qquad a\leqslant t\leqslant b$$

subject to the boundary conditions

$$x(a)=\alpha \qquad \text{and} \qquad x(b)=\beta,$$

where a, b, and f were given continuous functions of t, and α and β were given real numbers. We assumed that this problem had a unique solution $x(t)$ and that $x''(t)$ was a continuous function. Our initial problem was to find $x(t)$ exactly. We decided that this was impossible for most choices of $a(t)$, $b(t)$ and $f(t)$ and that a reasonable compromise would be to look for a function $\tilde{x}(t)$ that was in some sense a good approximate to $x(t)$. This in turn led us into a rather lengthy dialog on the approximation of functions. In fact, the book to this point has been devoted almost exclusively to this topic. However, we have yet to answer our original question.

PROBLEM

What functions $\tilde{x}(t)$ make good "approximate solutions" to our differential equation?

We are finally ready to address ourselves to this problem. The answer is that *spline functions $\tilde{x}(t)$ obtained by variational methods make very good approximates to $x(t)$.*

Continuing with the same model, we cast it in a normed linear space setting. There are many ways of doing this. One simple possibility is to let $X=C^2[a,b]$ and $Y=C[a,b]$ with the Tchebycheff norm. We define the function (operator) A from X to Y by

$$Ax(t)=x''(t)+a(t)x'(t)+b(t)x(t), \qquad a<t<b.$$

Then $y(t)=Ax(t)$ is a continuous function; thus Ax belongs to $C[a,b]=Y$ whenever $x\in C^2[a,b]=X$. Our original problem, then, can be described as follows.

PROBLEM

Given $y\in Y$, find an $x\in X$ such that

$$Ax=y,$$

where $x(a)=\alpha$ and $x(b)=\beta$.

A variational method applied to this problem seeks a good approximate to a solution x of $Ax=y$ from a finite dimensional subspace X_N of X.

Thus suppose $B_N=\{\phi_1,\phi_2,\dots,\phi_N\}$ is a linearly independent set of functions from X and let

$$X_N=\operatorname{span}\{\phi_1,\phi_2,\dots,\phi_N\}.$$

For example, we might let

$$B_N=\{1,t,\dots,t^n\},$$

or we might choose

$$B_N=\{B_{-1}(x),B_0(x),B_1(x),\dots,B_{n+1}(x)\},$$

where B_i's are cubic B splines corresponding to the knots $a=x_0<x_1<\cdots<x_n=b$ of some partition π of $[a,b]$. In the first case $X_N=P_n[a,b]$, the linear space of all polynomials of exact degree n or less defined on $[a,b]$. In the second case, $X_N=S_3(\pi,2)$ is the set of all cubic splines with knots at points of π. Note that $X_N=P_n[a,b]$ is an $(n+1)$-dimensional subspace of

$C^2[a,b]=X$; thus $N=n+1$; whereas $X_N=S_3(\pi)$ is an $(n+3)$-dimensional subspace of $X=C^2[a,b]$, and $N=n+3$. A *variational method* seeks functions from X_N approximating the solution x. That is, we want to find

$$x_N=c_1\phi_1+c_2\phi_2+\cdots+c_N\phi_N$$

belonging to X_N which makes $\|x-x_N\|$ small.

DEFINITION VARIATIONAL METHOD

Given the equation

$$Ax=y,$$

where A maps the normed linear space X with norm $\|\ \|_X$ into the normed linear space Y with norm $\|\ \|_Y$ and a finite dimensional subspace $X_N=\text{span}\{\phi_1,\phi_2,\ldots,\phi_N\}$ of X, a *variational method* is simply a numerical algorithm for finding a function

$$x_N=c_1\phi_1+c_2\phi_2+\cdots+c_N\phi_N$$

belonging to X_N which makes

$$\|Ax_N-y\|_Y+\|x_N-x\|_X$$

small.

Most variational methods fit this description. Among the many approaches are

1. Least squares methods, including
   (a) The method of least squares
   (b) The Rayleigh-Ritz or finite element method
2. Collocation methods
3. Galerkin methods
4. Finite difference methods
5. The method of lines

All these can be viewed in a more global setting as *projective methods*, since solution by any of these methods is usually (but not always) tantamount to projecting the solution or an approximate thereof onto a finite dimensional subspace of X. Rather than attempt to given an in depth analysis of all such methods, we focus our attention on least squares methods, finite element methods, and collocation methods. Before examining any of these, however, we need a few more definitions and theorems from functional analysis.

6.2 LINEAR OPERATORS

Variational methods are especially simple to work with when the differential or integral equation whose solution (or solutions) we are attempting to approximate forms a linear operator equation. Such operators are easily defined.

DEFINITION LINEAR OPERATOR

Let X and Y be linear spaces. A function T mapping X into Y is said to be a *linear operator* or *linear transformation* provided

1. $T(x+y)=Tx+Ty$
2. $T(\alpha x)=\alpha Tx$

for all scalars α and vectors x and z from X.

Examples abound.

EXAMPLE 1 $m\times n$ MATRICES

Let $X=E^n$ and $Y=E^m$. Any $m\times n$ matrix $A=(a_{ij})$, $1\leqslant i\leqslant m$ and $1\leqslant j\leqslant n$, is a linear operator from X to Y.

EXAMPLE 2 LINEAR INTEGRAL EQUATIONS

Let $X=Y=C[0,1]$. Let $k(s,t)$ be any continuous function of two variables defined on the unit square $[0,1]\times[0,1]$. Let $x(t)$ belong to $C[0,1]$. Define Kx by

$$Kx(s)=x(s)+\int_0^1 k(s,t)x(t)\,dt.$$

If $y(s)=Kx(s)$, $0\leqslant s\leqslant 1$, it is simple to prove that $y(s)$ is a continuous function on $[0,1]$. Thus K maps X into X. Moreover, K is linear, since

$$\begin{aligned}[K(x+y)](s)&=[x(s)+y(s)]+\int_0^1 k(s,t)[x(t)+y(t)]\,dt\\&=\left[x(s)+\int_0^1 k(s,t)x(t)\,dt\right]+\left[y(s)+\int_0^1 k(s,t)y(t)\,dt\right]\\&=Kx(s)+Ky(s)\end{aligned}$$

and for each real number α, $[K(\alpha x)](s)=\alpha[Kx(s)]$.

EXAMPLE 3 ORDINARY DIFFERENTIAL EQUATIONS

Let $X = C^2[a,b]$ and $X = C[a,b]$. For each $x(t) \in X$, define Lx by

$$Lx(t) = x''(t) + (\sin \pi t)x'(t) + tx(t).$$

Since the derivative of a sum of two functions is the sum of the derivatives, and since the sum and product of continuous functions are continuous, it follows that L is indeed a linear operator from X to Y.

EXAMPLE 4 PARTIAL DIFFERENTIAL EQUATIONS

Consider the Dirichlet problem on a closed, bounded region D of E^2. Let $X = C^2[D]$ and for each $u(x,y) \in X$ define Δu by

$$\Delta u(p) = \left(\frac{\partial^2 u}{\partial x^2} + \frac{\partial^2 u}{\partial y^2} \right)(p)$$

at all points p of D. It follows by analogy with Example 3 that Δ is a linear operator from X to $Y = C[D]$.

If X and Y are normed linear spaces, we are able to define a *bounded linear operator* from X to Y.

DEFINITION BOUNDED LINEAR OPERATOR

Let X and Y be normed linear spaces where $\| \ \|_X$ denotes the norm in X and $\| \ \|_Y$ denotes the norm in Y. A linear operator T on X to Y is said to be a *bounded linear operator* if there exists a constant $M > 0$ with the property that

$$\|Tx\|_Y \leqslant M \|x\|_X$$

for each $x \in X$.

Most often we suppress the X and Y and simply write $\| \ \|_X = \| \ \|$ and $\| \ \|_Y = \| \ \|$, since it is clear from context which is which. It is important to note that the boundedness of a linear operator is highly sensitive to the choice of norm. We look at our previous examples in light of this definition. In each case we must define a norm on both X and Y.

EXAMPLE 1′ $m \times n$ MATRICES

For simplicity, let $m = n$. Then $X = Y = E^n$. If we let X carry the usual Euclidean norm

$$\|x\| = \sqrt{x_1^2 + x_2^2 + \cdots + x_n^2}$$

for each $x=(x_1,x_2,\ldots,x_n)\in X$ and let $A=(a_{ij})$, $1\leqslant i,j\leqslant n$, be an $n\times n$ matrix, A is a bounded linear operator on X into X. In particular recall Schwarz's inequality $|(x,z)|\leqslant\|x\|\cdot\|z\|$. Then for each i, $1\leqslant i\leqslant n$,

$$\left|\sum_{j=1}^{n} a_{ij}x_j\right|\leqslant\left(\sum_{j=1}^{n} a_{ij}^2\right)^{1/2}\left(\sum_{j=1}^{n} x_j^2\right)^{1/2}=\left(\sum_{j=1}^{n} a_{ij}^2\right)^{1/2}\|x\|.$$

It follows that

$$\|Ax\|=\left(\sum_{i=1}^{n}\left[\sum_{j=1}^{n} a_{ij}x_j\right]^2\right)^{1/2}\leqslant\left(\sum_{i=1}^{n}\sum_{j=1}^{n} a_{ij}^2\right)^{1/2}\|x\|.$$

Letting $M=(\Sigma_{i=1}^{n}\Sigma_{j=1}^{n}a_{ij}^2)^{1/2}$, we see that $\|Ax\|\leqslant M\|x\|$; thus A is a bounded linear operator.

EXAMPLE 2′ INTEGRAL EQUATION

Let $X=Y=C[0,1]$, let $k(s,t)$ be continuous of the unit square $[0,1]\times[0,1]$. For each $x(t)\in C[0,1]$, define

$$(Kx)(s)=x(s)+\int_0^1 k(s,t)x(t)\,dt.$$

Let $X=Y$ carry the Tchebycheff norm and let $\gamma=\max_{0\leqslant s,t\leqslant 1}|k(s,t)|$. Then

$$\begin{aligned}|Kx(s)|&\leqslant|x(s)|+\left|\int_0^1 k(s,t)x(t)\,dt\right|\\&\leqslant|x(s)|+\int_0^1|k(s,t)x(t)|\,dt\\&\leqslant\|x\|+\gamma\|x\|\int_0^1 1\,dt\\&=(1+\gamma)\|x\|.\end{aligned}$$

Thus $\|Kx\|\leqslant(1+\gamma)\|x\|$, and K is a bounded linear operator from X to X.

EXAMPLE 3′ DIFFERENTIAL OPERATOR

The case of the differential operator is more delicate than the previous two examples. Let $X=C^2[0,1]$ and $Y=C[0,1]$. Define L from X to Y by

$$Lx(t)=x''(t)+(\sin\pi t)x'(t)+tx(t),\qquad 0\leqslant t\leqslant 1.$$

We know that L is a linear operator. For each function $y(t) \in Y$, define

$$\|y\| = \|y\|_{\infty} = \max_{0 \leqslant t \leqslant 1} |y(t)|.$$

CASE A. Suppose we let X carry the Tchebycheff norm so that

$$\|x\| = \|x\|_{\infty} = \max_{0 \leqslant t \leqslant 1} |x(t)|.$$

Then we have

$$\begin{aligned} \|Lx\| &= \max_{0 \leqslant t \leqslant 1} |Lx(t)| \\ &= \max_{0 \leqslant t \leqslant 1} |x''(t) + (\sin \pi t)x'(t) + tx(t)|. \end{aligned}$$

If $x_n(t) = t^n$, then

$$\begin{aligned} \|Lx_n\| &= \max_{0 \leqslant t \leqslant 1} |n(n-1)t^{n-2} + n(\sin \pi t)t^{n-1} + t^{n+1}| \\ &\geqslant n(n-1) + 1 = [n(n-1)+1]\|x_n\|, \end{aligned}$$

since $\|x_n\| = \max_{0 \leqslant t \leqslant 1}|t^n| = 1$ and $Lx_n(1) = n(n-1)+1$. Since n can be chosen arbitrarily, it is clear that L *is unbounded when* X *carries the Tchebycheff norm.*

Now let us choose a different norm for $X = C^2[0,1]$.

CASE B. For each $x(t) \in X$ define the norm of x by

$$\|x\| = \|x\|_{\infty} + \|x'\|_{\infty} + \|x''\|_{\infty}.$$

In this case we have

$$\|Lx\| \leqslant \|x''\|_{\infty} + \|x'\|_{\infty} + \|x\|_{\infty} = \|x\|.$$

Letting $M = 1$, we see that

$$\|Lx\| \leqslant M\|x\|, \qquad M = 1,$$

for all $x \in C^2[0,1]$. Thus L *is a bounded linear operator from* X *to* Y.

Once we have an operator T that is bounded, it is important to consider the smallest constant M for which

$$\|Tx\| \leqslant M\|x\|.$$

Such a smallest M exists and is called $\|T\|$. That is:

$$\|Tx\| \leqslant \|T\|\|x\|,$$

and one can prove that

$$\begin{aligned}\|T\| &= \inf\{M : \|Tx\| \leqslant M\|x\|, \qquad \text{for all } x \in X\}\\ &= \sup\{\|Tx\| : \|x\| = 1\}\\ &= \sup\left\{\frac{\|Tx\|}{\|x\|} : x \in X, x \neq 0\right\}.\end{aligned}$$

It is possible for example, to prove that in Example 1 $\|A\| = (\sum_{i=1}^{n}\sum_{j=1}^{n}a_{ij}^2)^{1/2}$. We seldom compute $\|T\|$ but we take recourse to its existence quite often.

EXERCISES

1. Choose normed linear spaces X and Y such that the integral differential operator

$$Ax(t) = x'(t) + \int_0^1 k(s,t)x(t)\,dt$$

is a bounded linear operator from X to Y. Assume that $k(s,t)$ is continuous.

2. Let $A = (a_{ij})$, $1 \leqslant i, j \leqslant n$, be an $n \times n$ matrix. Let $X = E^n = \{(x_1, x_2, \ldots, x_n) : x_i \text{ is real}\}$. For each $x \in X$ define

$$\|x\|_2 = \sum_{i=1}^{n} x_i^2$$

$$\|x\|_1 = |x_1| + |x_2| + \cdots + |x_n|$$

$$\|x\|_\infty = \max_{1 \leqslant i \leqslant n} |x_i|.$$

Prove that A is a bounded linear operator from X into X for all three norms. Find $\|A\|$ in each case.

3. Let $X = C[0,1]$. Suppose that $k(s,t,v)$ is continuous on $[0,1] \times [0,1] \times X$.

X. For each $x \in X$ define

$$Kx(s)=\int_0^1 k(s,t,x(t))\,dt.$$

Does K map X into X? Prove your answer. Is K linear? Bounded? Why?

6.3 INNER PRODUCT SPACES

Least squares methods, including the method of least squares and the Rayleigh-Ritz method, are imbedded in an inner product space setting as are so-called Galerkin methods. To define these simple spaces we must generalize the definition of a *dot or inner product* $x \cdot y=(x,y)$ of two vectors to an abstract linear space setting. Recall that the ordinary dot product of elementary physics is a real-valued function satisfying the following properties:

1. $(x,x)>0 \quad$ unless $\quad x=0$
2. $(x,y)=(y,x)$
3. $(x,\alpha y+\beta z)=\alpha(x,y)+\beta(x,z)$

for any vectors x, y, z and real numbers α and β. Any real-valued function defined on pairs of vectors from a linear space X that obeys these three postulates is said to form an *inner product on* X.

DEFINITION INNER PRODUCT SPACE

An inner product space is a linear space on which there is defined an inner product.

EXAMPLE 1. n-DIMENSIONAL EUCLIDEAN SPACE

The simplest example is the space

$$E^n=\{(x_1,x_2,\ldots,x_n):x_i \text{ real}\}$$

of n-tuples of real numbers, with (x,y) defined by

$$(x,y)=x_1y_1+x_2y_2+\cdots+x_ny_n,$$

where $x=(x_1,x_2,\ldots,x_n)$ and $y=(y_1,y_2,\ldots,y_n)$. The reader can easily verify that when $X=E^2$, $(x,y)=\|x\|\|y\|\cos\theta$ where θ is the angle between x and y.

EXAMPLE 2 THE SPACE $C^2[0,1]$

Let $X = C^2[0,1]$ and let $x(t)$ and $y(t)$ be any two functions from X. There are numerous integral inner products. We define four of them, as follows, and leave it to the reader to prove that they are indeed inner products.

1. $(x,y) = \int_0^1 x(t)y(t)\,dt$
2. $(x,y) = \int_0^1 [x(t)y(t) + x'(t)y'(t)]\,dt$
3. $(x,y) = \int_0^1 [x(t)y(t) + x'(t)y'(t) + x''(t)y''(t)]\,dt$
4. $(x,y) = \int_0^1 \sigma(t)x(t)y(t)\,dt$, where $\sigma(t) > 0$ is a given continuous function on $[0,1]$.

All the foregoing integrals are Riemann integrals: we have occasion to use many inner products of this sort in the sequel.

Another example is the space $L_2[a,b]$.

EXAMPLE 3 $X = L_2[a,b]$

$X = L_2[a,b]$ and $f,g \in X$, we define

$$(f,g) = \int_a^b f(t)g(t)\,dt.$$

It follows from the Lebesgue theory of integration that this integral does indeed form an inner product.

Finally, note that if X is a linear space and $(\ ,\ )_i$, $1 \leqslant i \leqslant n$, are any n inner products defined on X, their sum is again an inner product. Specifically

$$(x,y) = (x,y)_1 + (x,y)_2 + \cdots + (x,y)_n$$

is an inner product.To define a large class of other inner products, we need what are known as positive definite, symmetric linear operators.

DEFINITION POSITIVE DEFINITE AND SYMMETRIC

A linear operator A mapping an inner product space X into itself is said to be *positive definite* (p.d.), provided

$$(Ax,x) > 0 \qquad \text{unless} \quad x = 0.$$

The operator A is said to be *symmetric* if

$$(Ax,y)=(x,Ay) \qquad \text{for all } x \text{ and } y \text{ in } X.$$

EXAMPLE 4 INNER PRODUCTS DERIVING FROM A P.D. SYMMETRIC MATRIX

Let $X=E^n$ and let $A=(a_{ij})$ be an $n\times n$ positive definite, real symmetric matrix. Then A maps X into X. Moreover, the function

$$(x,y)_A=(Ax,y)$$

for all x and y in E^n, where (,) denotes usual inner product of Example 1, forms an inner product for E^n.

There are many continuous analogs to the previous example, as the next chapter reveals. For the time being we will be content with one simple example.

EXAMPLE 5 AN INNER PRODUCT DERIVING FROM A P.D., SYMMETRIC DIFFERENTIAL OPERATOR

Let $Y=C[0,1]$ with the integral inner product $(f,g)=\int_0^1 f(t)g(t)\,dt$. Define $Ax(t)=-x''(t)$, and let $X=\{x(t)\in C^2[0,1]: x(0)=x(1)=0\}$. Then X is a linear subspace of Y and $Ax\in Y$ for all $x\in X$ (i.e., $x''(t)$ is continuous). Moreover, for all x, $z\in X$, $(Ax,z)=-\int_0^1 x''(t)z(t)\,dt$. Integrating this expression by parts and recalling that $x(0)=x(1)=z(0)=z(1)=0$, we find that

$$(Ax,z)=\int_0^1 x'(t)z'(t)\,dt=(x,Az);$$

thus A is *symmetric*. Moreover, A is positive definite. To see this, observe that

$$(Ax,x)=\int_0^1 [x'(t)]^2\,dt=0$$

implies $x'(t)\equiv 0$, since $x'(t)$ is continuous. But then $x(t)=k$, k a constant. Since $x(0)=0$, $k=0$. Thus $x(t)\equiv 0$ and $(Ax,x)>0$, unless $x\equiv 0$. We can now use the operator A to define a new inner product

$$(x,z)_A=(Ax,z)$$

on X. Such inner products are often referred to as *energy inner products*, since they can be used to minimize the potential energy of appropriate physical systems (see Mikhlin's book) and are the underlying inner product for the finite element method.

Thus we see from the previous two examples that

THEOREM

Every positive definite, symmetric linear operator A from a linear space X to a linear space Y having an inner product $(,)$ gives rise to a second inner product $(,)_A$ defined by

$$(x,y)_A = (Ax,y)$$

for each $x,y \in X$, whenever $X \subset Y$.

There is also a converse to this theorem. See if you can figure it out and prove it. Start in a finite dimensional setting.

Another interesting feature of inner products is that each inner product $(\, , \,)$ defined on a linear space X gives rise to a *norm* $\| \; \|$ defined by

$$\|x\| = \sqrt{(x,x)} \; .$$

Moreover, norms arising in this fashion satisfy

SCHWARZ'S INEQUALITY

For each x and y belonging to the linear space X

$$|(x,y)| \leqslant \|x\| \cdot \|y\|.$$

Proof. Let x and y belong to X and let t be a real number. Then

$$(x+ty, x+ty) = \|x\|^2 + 2t(x,y) + t^2\|y\|^2 \geqslant 0.$$

Thus $t^2\|y\|^2 + 2t(x,y) + \|x\|^2$ is a quadratic equation in t which is always nonnegative. But this can occur only if the discriminant $b^2 - 4ac = 4|(x,y)|^2 - 4\|x\|^2\|y\|^2 \leqslant 0$. The desired inequality now follows. ■

Note that in Examples 1 and 2 this implies

1. $$\left| \sum_{i=1}^{n} x_i y_i \right| \leqslant \left(\sum_{i=1}^{n} x_i^2 \right)^{1/2} \left(\sum_{i=1}^{n} y_i^2 \right)^{1/2}$$

and

2. $$\left| \int_a^b f(t) g(t)\, dt \right| \leqslant \left(\int_a^b [f(t)]^2 dt \right)^{1/2} \left(\int_a^b [g(t)]^2 dt \right)^{1/2}.$$

We use inequality 2 repeatedly in the sequel.

The notion of inner product also suggests perpendicularity or orthogonality. Two vectors x and y from an inner product space X with an inner product (,) are said to be *orthogonal* provided

$$(x,y)=0.$$

For example, in the linear space $C[0,\pi]$ with inner product $(f,g)=\int_0^\pi f(t)g(t)dt$, the functions $f(t)=\sin t$ and $g(t)=\cos t$ are orthogonal, as is well known to anyone who has worked with Fourier series.

EXERCISES

1. Let $X=E^n$ and let (,) be an inner product defined on X. Prove that (,) determines a linear operator A from X into X. Is such an A uniquely determined? Is A nonsingular? Prove your answers.
2. Let $X=\{u\in C^2[0,1]: u(0)=u(1)=0\}$. Define A from X into $Y=C[0,1]$ by

$$Au(t)=u''(t)+u(t).$$

Let $(u,v)=\int_0^1 u(t)v(t)dt$. Is A symmetric? Positive definite? Explain.
3. Let $X=C^2[D]$ and let $Y=C[D]$, where D is any closed, bounded, connected set in the real plane (e.g., a circle and its interior, or a rectangle and its interior). Let Δ be the Laplacian

$$\Delta u=u_{xx}+u_{yy}.$$

Is Δ positive definite and symmetric on X into Y where

$$(u,v)=\int\int_D uv?$$

Explain.
4. Let $X=C[0,1]$ and let

$$(u,v)=\int_0^1 u(t)v(t)\,dt,$$

for all u, $v\in X$. Find a set of polynomials $\{p_0(t),p_1(t),\ldots,p_3(t)\}$ of degree 3 or less having the property

$$(p_i,p_j)=\begin{cases}1 & \text{if} \quad i=j\\ 0 & \text{if} \quad i\neq j,\end{cases}$$

Such polynomials are known as *orthogonal polynomials*.

5. Let $X=\{u\in C^2[0,1]: u(0)=u(1)=0\}$, and let $Y=C[0,1]$. Define A by

$$Au(t)=-u''(t)+u(t).$$

Let $(u,v)=\int_0^1 u(t)v(t)dt$. Prove that A is symmetric and positive definite. Find four functions in X that are mutually orthogonal in the energy norm $(,)_A$.

6. Let $X=\{u\in C^1[0,1]: u(0)=u(1)=0\}$. For each $u\in X$ let

$$Au(t)=-u''(t)+\sin t\cdot u(t).$$

For each $u,v\in X$ define

$$\langle u,v\rangle_A=\int_0^1[u'v'+\sin t\cdot uv].$$

Is $\langle\ ,\ \rangle_A$ an inner product on X? Explain.

6.4 NORMS, CONVERGENCE, AND COMPLETENESS

We have seen in Chapter 1 that the choice of norm dictates the distance between two functions. For example, when $X=C[0,1]$,

$$\|f\|_\infty=\max_{0\leq t\leq 1}|f(t)|$$

and

$$\|f\|_2=\left(\int_0^1[f(t)]^2dt\right)^{1/2}$$

define two quite different norms on X. This was illustrated by taking $g(t)\equiv 0$ on $[0,1]$ and letting $f^2(t)$ be the function whose graph is the "roof function" shown in Figure 6.1. In this case, $\|f-g\|_\infty=1$ and $\|f-g\|_2\leq\sqrt{\varepsilon}$. Thus for very small $\varepsilon>0$, f is close to y in the $\|\ \|_2$ norm, and f is quite far from y in the $\|\ \|_\infty$ norm. For this reason what is meant by a "good approximation" depends critically on the choice of norm.

Another property that is totally dependent (for the same reasons) on choice of norm is convergence of an infinite sequence in a normed linear space. In particular suppose X is a normed linear space and (x_n), $n=1,2,\ldots$, is an infinite sequence of vectors from X. We need to know what is meant by $\lim_{n\to\infty}x_n=x$, where x is some other vector from X. A correct definition must be geometric in its intent. Suppose, for example, you have an ordered and countable collection (an infinite sequence) $x_1,x_2,\ldots,$ of

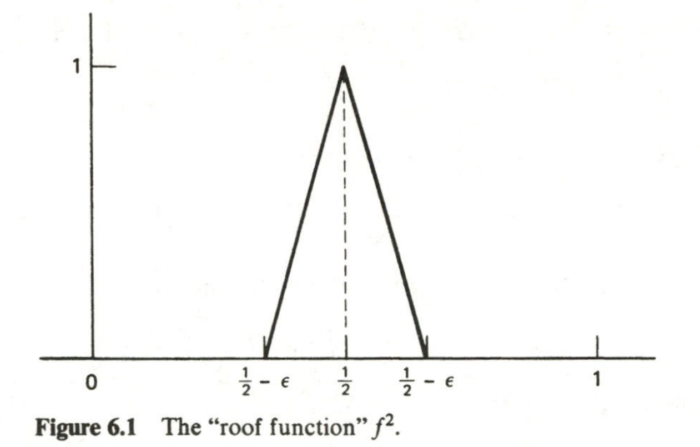

Figure 6.1 The "roof function" f^2.

points or vectors x_n from the real plane E^2 (see Figure 6.2) and a fixed vector x from E^2. Clearly the statement "x_n converges to x as n grows larger and larger" means simply that the distance from x_n to x becomes smaller and smaller as n goes to infinity. But the distance from x to x_n is precisely

$$\|x-x_n\|_2=\sqrt{(x_1-x_1^n)^2+(x_2-x_2^n)^2}\ ,$$

where $x=(x_1,x_2)$ and $x_n=(x_1^n,x_2^n)$. Thus it seems reasonable to define convergence in an arbitrary normed linear space as follows.

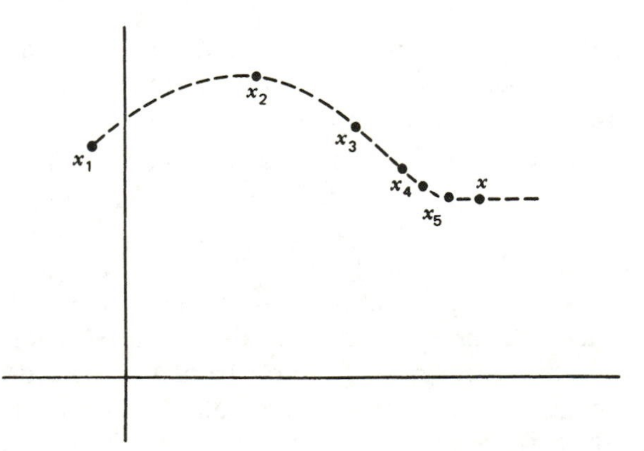

Figure 6.2 A sequence of vectors $x_1, x_2, \ldots$ converging to the vector x.

DEFINITION (STRONG) CONVERGENCE

Let X be a normed linear space with norm $\| \ \|$. A sequence (x_n), $n = 1, 2, \ldots$, of vectors from X is said to *converge (strongly)* to a vector $x \in X$ if

$$\lim_{n \to \infty} \|x - x_n\| = 0.$$

The adjective "strong" is used because there are other types of convergence in certain linear spaces. Among these are weak convergence and convergence in norm. We use the term convergence exclusively to refer to strong convergence. Thus x_n *converges to* x if the distance $\|x_n - x\|$ from x_n to x goes to zero as n goes to infinity. Since the choice of norm dictates the distance, the choice of norm also dictates the character of the convergence. Resorting again to $C[0,1]$ with the two norms $\| \ \|_\infty$ and $\| \ \|_2$, we see that the members of the sequence of roof functions $f_n(t)$, $n = 1, 2, \ldots$, the graph of whose squares appear in Figure 6.3, converge in the L_2 norm to the function $f(t) \equiv 0$. That is,

$$\|f - f_n\|_2 = |f_n\|_2 = \left(\int_0^1 [f_n(t)]^2 \, dt \right)^{1/2} = \frac{1}{\sqrt{n}},$$

which goes to zero as n goes to infinity. However,

$$\|f - f_n\|_\infty = \|f_n\|_\infty = 1$$

for all n. Thus $f_n(t)$ does *not* converge to $f(t)$ in the Tchebycheff norm.

The notion of convergence in a normed linear space leads to yet another fruitful concept—that of *completeness*. A normed linear space X is said to

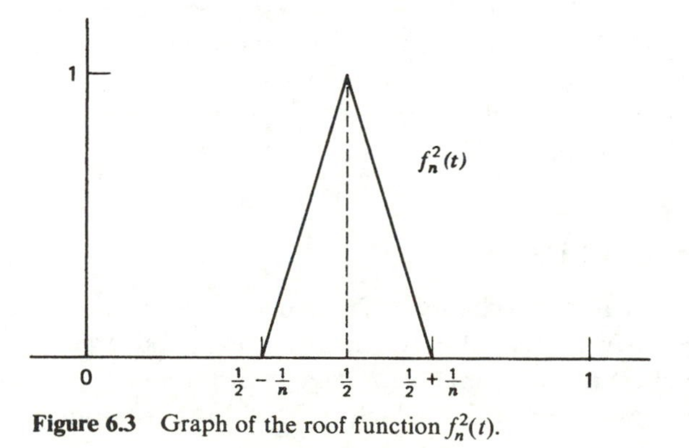

Figure 6.3 Graph of the roof function $f_n^2(t)$.

be *complete* if and only if every Cauchy sequence $(x_n : n = 1,2,\ldots,)$ in X converges to a vector $x \in X$. A sequence (x_n) in X is *Cauchy* if $\lim_{m,n\to\infty} \|x_m - x_n\| = 0$. For example, the sequence $(f_n(t))$ of roof functions just given is a Cauchy sequence in $C[a,b]$ with the L_2 norm $\| \ \|_2$, since for $m \geqslant n$, $\|f_m - f_n\|_2 \leqslant \|f_n\|_2 < 2/\sqrt{n}$, which goes to zero as n goes to infinity. The reader should check to see whether the same sequence of functions is Cauchy in the Tchebycheff norm. Inner product spaces that are complete are special; they are called *Hilbert spaces*. Similarly, normed linear spaces that are complete are specially named; they are called *Banach spaces*. Since every inner product space is a normed linear space with $\|x\| = \sqrt{(x,x)}$, every Hilbert space is a Banach space. However, the converse would not be true unless every norm were generated by an inner product. For example, just try and find an inner product on $C[a,b]$ that would give the Tchebycheff norm!

We have already discussed one instance of the concept of completeness playing an important role—recall our heuristic description of the space of Lebesgue square integrable functions $L_2[a,b]$ given in Chapter 1. We pointed out that although $\|f\|_2 = (\int_a^b [f(t)]^2 dt)^{1/2}$ defined a norm on $C[a,b]$, $C[a,b]$ failed to be complete in this norm, whereas the space $L_2[a,b]$ was complete in the same norm. This was, in fact, one of the principal reasons for bothering with a Lebesgue integral at all. The completeness of $L_2[a,b]$ is very important in showing the so-called equivalence of certain norms that we deal with in approximating solutions to differential equations. For this reason, we discuss the notion of equivalent norms in next.

EXERCISES

1. Let W be the set of all piecewise linear Lagrange polynomials defined on $[0,1]$ together with the Tchebycheff norm. Is W complete? Explain. Answer the same question when X carries the L_2 norm. Explain.
2. Let X be the set of all functions $f(t)$ that are piecewise continuous on $[0,1]$ and square integrable on $[0,1]$ (i.e., $\int_0^1 f^2(t)\,dt < \infty$). Is X complete? Explain. Does the square integrability have anything to do with the answer? Explain.

6.5 EQUIVALENT NORMS

We have seen that the Tchebycheff norm $\| \ \|_\infty$ and the L_2 norm $\| \ \|_2$ on $C[a,b]$ give us two very different measures of good approximation. One (the Tchebycheff norm) is much *stronger* than the other. In fact, if f and g are two functions from X for which $\|f-g\|_\infty < \epsilon$ for some small $\epsilon > 0$, it automatically follows that $\|f-g\|_2<\epsilon\sqrt{b-a}$. In particular

$$\|f-g\|_2=\left(\int_a^b [f(t)-g(t)]^2\,dt\right)^{1/2} \leqslant\left(\int_a^b \|f-g\|_\infty\,dt\right)^{1/2}=\|f-g\|_\infty\sqrt{b-a}\ .$$

Thus whenever $\|f-g\|_\infty$ is small, $\|f-g\|_2$ is automatically small unless $b-a$ is very large. Suppose we had some other norm $\|\ \|$ on $C[a,b]$ and two constants $M>0$ and $m>0$ such that for all f and g in $C[a,b]$

$$m\|f-g\|_\infty \leqslant \|f-g\| \leqslant M\|f-g\|_\infty.$$

In this case, whenever $\|f-g\|_\infty$ is small, $\|f-g\|$ would also be small *and vice versa*, provided M and m were not too large. More important yet, if (f_n) was an infinite sequence in $X=C[a,b]$ and f was a fixed vector or function from X, then $\lim_n\|f-f_n\|=0$ if and only if $\lim_n\|f-f_n\|_\infty=0$. That is, *convergence in one norm automatically implies convergence in the other norm as well.* Such norms are important and are known as *equivalent norms*.

DEFINITION EQUIVALENT NORMS

Let X be a linear space with two different norms $\|\ \|$ and $|\ |$. The two norms are said to be *equivalent* if there exist constants $M>0$ and $m>0$ such that

$$m|x| \leqslant \|x\| \leqslant M|x|$$

for all x in X.

As a simple example, look at $X=E^2$ with the three norms

$$\|x\|_2=\sqrt{x_1^2+x_2^2}\ ,$$

$$\|x\|_1=|x_1|+|x_2|$$

$$\|x\|_\infty=\max\{|x_1|,|x_2|\},$$

where $x=(x_1,x_2)$ is an arbitrary vector in the plane. Then $\|x\|_2=(x_1^2+x_2^2)^{1/2}\leqslant(2\|x\|_\infty^2)^{1/2}=\sqrt{2}\ \|x\|_\infty$.On the other hand, $\|x\|_\infty=\max\{|x_1|,|x_2|\}\leqslant\sqrt{(x_1^2+x_2^2)}^{\,1/2}=\|x\|_2$. Letting $M=\sqrt{2}$ and $m=1$, we see that

$$\|x\|_\infty \leqslant \|x\|_2 \leqslant \sqrt{2}\ \|x\|_\infty;$$

thus these two norms are equivalent. The proof that any pair of these norms is equivalent is equally simple.

This concept is important in the sequel since it enables us to replace a generalized L_2 convergence or convergence in a Sobolev norm (see Section 7.12) either by L_2 convergence or by a much preferable generalized Tchebycheff type of convergence in many important cases.

EXERCISES

1. Let $X=C^1[a,b]$. For each $x \in X$ define

$$\|x\| = \|x\|_\infty$$

$$|x| = \|x\|_\infty + \|x'\|_\infty.$$

   (a) Prove that both $\| \;\|$ and $| \;|$ are norms for X.
   (b) Let (x_n) be a sequence in X and let $x \in X$. Does $\lim_n |x - x_n| = 0$ imply $\lim_n \|x - x_n\| = 0$? Prove your answer. What about the converse? Prove your answer.
   (c) Are $\| \;\|$ and $| \;|$ equivalent? Prove your answer.
   (d) Prove that $C^1[a,b]$ is complete in the norm $| \;|$.
2. Let $\| \;\|$ and $| \;|$ be two equivalent norms on a linear space X. Prove that a sequence (f_n) in X converges to $f \in X$ in the norm $\| \;\|$ if and only if (f_n) converges to f in the norm $| \;|$.
3. Let $X = \{u \in C^2[0,1] : u(0) = u(1) = 0\}$. For each $x \in X$, define

$$|x| = \|x\|_\infty + \|x'\|_\infty + \|x''\|_2,$$

   where $\|u\|_2 = (\int_0^1 u^2)^{1/2}$. Let

$$|x| = \|x''\|_2 + \|x\|_2.$$

   Are $| \;|$ and $| \;|_*$ norms on X? Prove your answer. If so, are they equivalent? Prove it.

6.6 BEST APPROXIMATIONS

The idea behind least squares methods (including Rayleigh-Ritz, finite element, and the method of least squares) is a simple one; it derives from the geometric notion of finding a point x_0 (if it exists) in some hyperplane M which is closest to a given fixed point p_0 (see Figure 6.4). Least squares methods merely generalize this idea to find approximate solutions to various types of operator equations including matrix equations and differential and integral equations, as well as systems of such equations. Other variational methods are also related to this notion, but none so directly as least squares methods. There are many least squares methods, but *whatever approach is chosen, all least squares methods derive from the initial notion of finding the closest point x_0 in a finite dimensional subspace M of an inner product space X to a fixed point p_0 of X.* The importance of this single key notion leads us to examine a small but critical portion of approximation theory. In particular we introduce the idea of a *best approximation* to a fixed point p_0 of a normed linear space X by a point or points of some fixed subset M of X that may or may not be a subspace (hyperplane) of X.

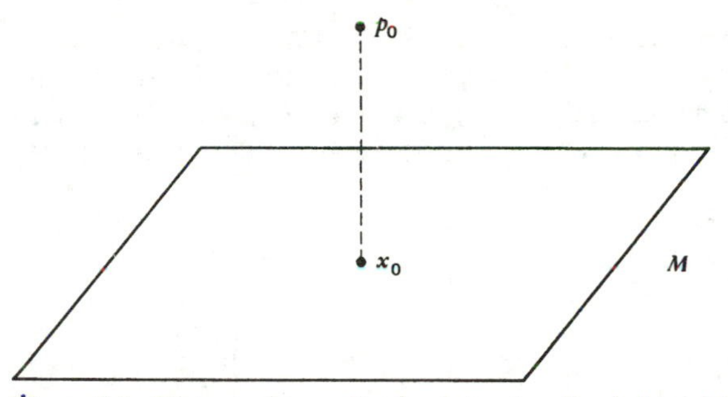

Figure 6.4 Distance from a fixed point p_0 to a fixed plane M.

DEFINITION BEST APPROXIMATION

Let X be a normed linear space with norm $\| \ \|$. Let $M \subset X$ be a subset of X. Choose a point $p \in X$. A point $y_0 \in M$ is said to be a *best approximation to p from M* provided

$$\|p-y_0\| \leqslant \|p-y\| \qquad \text{for all} \quad y \in M.$$

That is, y_0 is that point in M which is closest to p. Best approximations need not exist, as the following example illustrates.

EXAMPLE 1 NO BEST APPROXIMATION

Let $X = E^2$, the real plane, and let $M = \{(x,0) : x \neq 0\}$. Let $p = (0,1)$. Here there is no point of M closest to y_0 (see Figure 6.5).

It is also simple to see that even though best approximations may exist, they need not be unique.

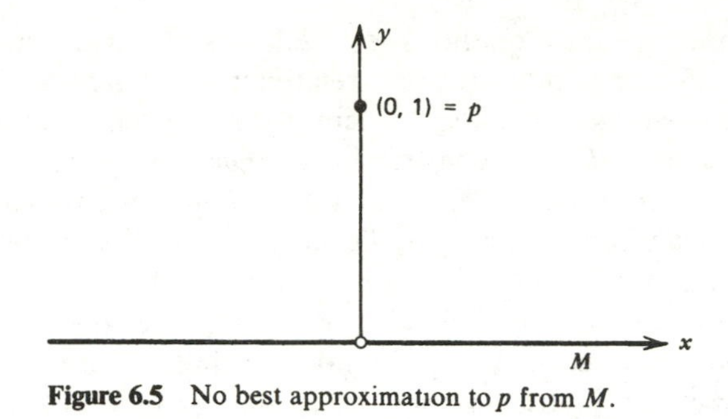

Figure 6.5 No best approximation to p from M.

EXAMPLE 2 NONUNIQUE BEST APPROXIMATES

Again let $X=E^2$, and this time (Figure 6.6) let $M=\{(1,y):y \text{ is real}\} \cup \{(-1,y):y \text{ is real}\}$. That is, M is two straight lines parallel to the y axis, one passing through $(1,0)$ and the other through $(-1,0)$. Let $p=(0,0)$, the origin. In this case there are clearly *two* best approximates, $z_0=(-1,0)$ and $z_1=(1,0)$ to p from M.

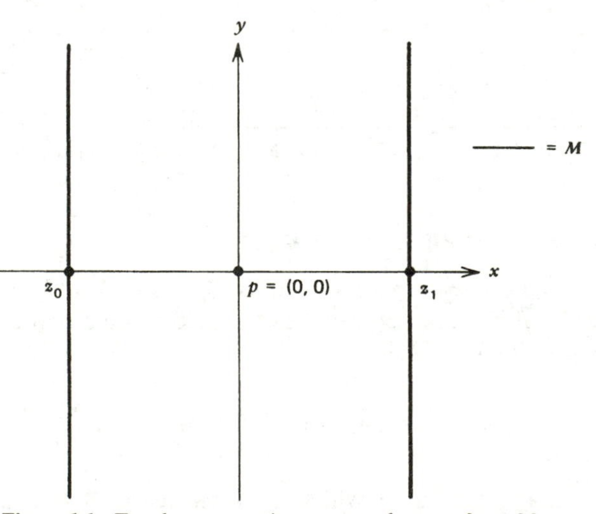

Figure 6.6 Two best approximates z_0 and z_1 to p from M.

It is only natural to ask when, if ever, best approximations do exist. The following important theorem illustrates that matters become considerably less gloomy when M is a finite dimensional subspace. The proof is easily understood by those knowing a bit of advanced calculus. The result is easily understood by anyone.

THEOREM EXISTENCE OF BEST APPROXIMATES

Let X be a normed linear space X with norm $\| \ \|$ and let X_N be a finite dimensional subspace of X. Then *each point* $x \in X$ *has a best approximation* $x_N \in X_N$; thus

$$\|x - x_N\| = \min_{y \in X_N} \|x - y\|.$$

Proof. Let $\bar{z} \in X_N$ and let $d = \|x - \bar{z}\|$. Let

$$K = \{ z \in X_N : \|x - z\| \leqslant d \}.$$

Then K is closed and bounded, since $\|x\|$ is a continuous function of x (see Exercises in Section 1.3). Being a closed, bounded subset of a finite dimensional space, K is compact. Let $g(z) = \|x - z\|$, $z \in K$. Then g is a continuous function of z. Since K is compact, g assumes its minimum at some point $x_N \in K$. Thus $\|x - x_N\| = \min_{y \in K} \|x - y\|$, as was to be proved. ∎

To see exactly what this powerful theorem implies in terms of the type of problem that interests us, let us return to our old friend, the differential equation

$$x''(t) + a(t)x'(t) + b(t)x(t) = f(t), \qquad a \leqslant t \leqslant b,$$

where $a(t)$, $b(t)$, and $f(t)$ are given continuous functions of t. We wish to approximate the unique solution $x(t)$ to this equation which satisfies the boundary conditions $x(a) = \alpha$ and $x(b) = \beta$. Suppose we decide to try to find an approximate solution $x_N(t)$ of the form

$$x_N(t) = a_1\phi_1(t) + a_2\phi_2(t) + \cdots + a_N\phi_N(t),$$

where the functions $\phi_1, \phi_2, \ldots, \phi_N$ are all given. For example, they might be cubic B splines with knots at $a = x_0 < x_1 < \cdots < x_n = b$ or basis piecewise Hermite polynomials of sufficient smoothness (what "sufficient smoothness" constitutes is discussed in detail later). Then $X_N = \text{span}\{\phi_1, \phi_2, \ldots, \phi_N\}$ is a finite dimensional subspace of some L_2 inner product space X, and our theorem *tells us there is an* $x_N(t)$ *from* X_N *which is closest to* $x(t)$*, the solution of our boundary value problem. Moreover, our error analysis of the previous chapters indicates just how close* $x_N(t)$ *is to* $x(t)$ *(i.e.,* $\|x - x_N\|$). Thus it remains only to find a way of computing $x_N(t)$. *Least squares algorithms are no more than such methods of computation for various choices of norm* $\| \ \|$.

In our zeal to proselytize on the marvels of such x_N's we have let one small detail slip by—*uniqueness*. Even though we have proved existence of best approximates from finite dimensional subspaces, we may not have uniqueness. For uniqueness, certain convexity properties are required (see Cheney (1966) and Meinardus (1964)). It happens that inner product spaces possess the appropriate convexity properties, and we can prove uniqueness of best approximations from finite dimensional subspaces X_N of an inner product space X without direct recourse to these properties. Letting $\|x\|=\sqrt{(x,x)}$, the key to uniqueness in this setting is simply that $\|x+y\|=\|x\|+\|y\|$ if and only if $(x,y)=\|x\|\cdot\|y\|$ and $(x,y)=\|x\|\cdot\|y\|$ if and only if $y=\lambda x$ for some real number λ. To obtain the first equivalence, note that $\|x+y\|^2=\|x\|^2+2(x,y)+\|y\|^2\leqslant\|x\|^2+2\|x\|\|y\|+\|y\|^2=(\|x\|+\|y\|)^2$. To obtain the second, let $y=\lambda x+z$, where $\lambda=\|y\|/\|x\|$. Suppose $(x,y)=\|x\|\cdot\|y\|$. Then $\|z\|^2=\|y-\lambda x\|^2=\|y\|^2-2\lambda(x,y)+|\lambda|^2\|x\|^2=\|y\|^2-2\|y\|^2+\|y\|^2=0$. Thus $z=\odot$. Note now that $\|x+y\|=\|x\|+\|y\|$ if and only if $y=\lambda x$. These observations enable us to prove

THEOREM UNIQUENESS OF BEST APPROXIMATIONS

Best approximations from (closed) finite dimensional subspaces of an inner product space are unique.

Proof. Let X be an inner product space and let M be a closed subspace of X (if M is finite dimensional, it is automatically closed). Let $x\in X$, and suppose x_0 and y_0 belonging to M are best approximations to x from M. Then $\|x-x_0\|=\|x-y_0\|=d$. Clearly $d=0$ implies $x_0=y_0=x$; so we assume $d>0$. Invoking the triangle inequality, we have

$$\|\tfrac{1}{2}(x_0+y_0)-x\|\leqslant\tfrac{1}{2}\|x-x_0\|+\tfrac{1}{2}\|x-y_0\|=d.$$

Thus $\frac{1}{2}(x_0+y_0)$ is also a best approximation, since $\frac{1}{2}(x_0+y_0)$ belongs to M. Thus

$$\|\tfrac{1}{2}(x_0+y_0)-x\|=\tfrac{1}{2}\|x-x_0\|+\tfrac{1}{2}\|x-y_0\|.$$

But this is the triangle inequality with equality. Thus $x-y_0=\lambda(x-x_0)$ for some real number λ. If $\lambda=1$, we are done. Suppose $\lambda\neq 1$. Then

$$x=\frac{y_0-\lambda x_0}{1-\lambda}$$

and $x\in M$, contrary to our initial assumption. Thus $\lambda=1$, and we are done. ■

Best approximations from subspaces of a Hilbert or inner product spaces are known as *least squares fits*. We examine such fits in detail in the next section.

EXERCISES

1. Find a linear space X that is not an inner product space and in which best approximations exist and are unique.
2. Let $X = C[0,1]$ with the Tchebycheff norm. Let $X_n = \text{span}\{1, t, t^2, \ldots, t^n\}$. Let $x \in X$. Prove that there exists a unique best approximate to x from X_n.

6.7 LEAST SQUARES FITS

We saw in the last section that if X is an inner product space and X_N is a finite dimensional subspace spanned by the vectors $\phi_1, \phi_2, \ldots, \phi_N$, then given any vector $x \in X$, there is a unique vector $x_N \in X$ which is closest to x in the norm $\|x\| = \sqrt{(x,x)}$. That is, x_N solves the minimization problem

$$\|x - x_N\| = \min_{\bar{x} \in X_N} \|x - \bar{x}\|,$$

and is known as the least squares fit to x from X_N. We saw, for example, that if $x(t)$ was the unique solution to the differential equation

$$x''(t) + a(t)x'(t) + b(t)x(t) = f(t), \qquad a \leqslant t \leqslant b,$$

subject to the boundary values $x(a) = \alpha$ and $x(b) = \beta$, and if $X_N = \text{span}\{B_0, B_1, \ldots, B_n\}$, where the $B_i(t)$'s are B splines with knots at $a = t_0 < t_1 < \cdots < t_n = b$, there is a unique spline function

$$x_N(t) = c_0 B_0(t) + c_1 B_1(t) + \cdots + c_n B_n(t)$$

which is closest to $x(t)$ in some norm deriving from an inner product. In general, one's choice of X and of inner product depends on the problem to be solved and the desired variational methods. That is, each choice of inner product gives rise to a distinct least squares method. For example, we see later that the inner product $(x,y)_A = (Ax,y)$, where A is a positive definite, symmetric linear operator and

$$(f,g) = \int_a^b f(t)g(t)\,dt$$

leads to the Rayleigh-Ritz method. Let us, however, postpone the details of choosing X and an inner product (,) until later, concentrating now on computing the least squares fit x_N to x from X_N in an abstract setting. The problem is surprisingly simple. In particular, we prove

THEOREM ALGORITHM FOR COMPUTING LEAST SQUARES FITS

Let X be an inner product space with inner product (,). Let $X_N = \text{span}\{\phi_1, \phi_2, \ldots, \phi_N\}$ be an N-dimensional subspace of X so that $\{\phi_1, \phi_2, \ldots, \phi_N\}$ is linearly independent. Let x be any point of X. The unique least squares fit x_N to x from X_N is given by

$$x_N = c_1\phi_1 + c_2\phi_2 + \cdots + c_N\phi_N,$$

where $c = (c_1, c_2, \ldots, c_N)'$ is the unique solution to the linear system

$$\sum_{j=1}^{N} (\phi_1, \phi_j)c_j = (x, \phi_i), \qquad 1 \leqslant i \leqslant N.$$

Moreover, the *Gram matrix* $(a_{ij}^N) = (\phi_i, \phi_j)$, $1 \leqslant i,j \leqslant N$ is nonsingular.

Proof. Let $\bar{x} = \bar{c}_1\phi_1 + \bar{c}_2\phi_2 + \cdots + \bar{c}_N\phi_N$ be an arbitrary point of X_N. Let

$$g(\bar{c}_1, \bar{c}_2, \ldots, \bar{c}_N) = \|x - \bar{x}\|^2 = \|x - \sum_{i=1}^{N} \bar{c}_i\phi_i\|^2.$$

Then g is a continuous, real-valued function of the N variable $\bar{c}_1, \bar{c}_2, \ldots, \bar{c}_N$. Moreover, since x_N is the *unique* vector in X_N minimizing $\|x - \bar{x}\|^2$ over all $\bar{x}$ in X_N, and since the ϕ_i's are independent,

$$g(c_1, c_2, \ldots, c_N) = \min g(\bar{c}_1, \bar{c}_2, \ldots, \bar{c}_N)$$

over all n-tuples of real numbers $(\bar{c}_1, \bar{c}_2, \ldots, \bar{c}_N)$. Letting $c = (c_1, c_2, \ldots, c_N)$, it follows that $\partial g / \partial c_i(c) = 0$. But $\|x - \bar{x}\|^2 = (x - \bar{x}, x - \bar{x})$. Thus

$$\|x - \bar{x}\|^2 = \|x\|^2 - 2\sum_{j=1}^{N} \bar{c}_j(x, \phi_j) + \sum_{i,j=1}^{N} \bar{c}_i\bar{c}_j(\phi_i, \phi_j).$$

Therefore, for each i, $1 \leqslant i \leqslant N$, we have

$$\frac{\partial g}{\partial c_i}(c) = -2(x, \phi_i) + 2\sum_{j=1}^{N} c_j(\phi_i, \phi_j) = 0.$$

It follows that

$$\sum_{j=1}^{N} (\phi_i, \phi_j)c_j = (\phi_i, x), \qquad 1 \leqslant i \leqslant N.$$

Since the ϕ_i's are independent, this system has a unique solution $(c_1, c_2, \ldots, c_N)'$; thus the Gram matrix (ϕ_i, ϕ_j), $1 \leqslant i,j \leqslant N$, is nonsingular. ■

We have insisted that the functions $\{\phi_1, \phi_2, \ldots, \phi_N\}$ are independent. The case of these functions being dependent is equally interesting and should be examined by the reader, keeping in mind that there exists a unique least squares fit x_N to x from X_N. Let us consider a very simple example.

EXAMPLE A LEAST SQUARES SOLUTION TO AN OVERDETERMINED LINEAR SYSTEM

Consider the simple overdetermined system of linear equations

$$2x - y = 3$$

$$x + y = 1$$

$$x + 3y = 0$$

$$5x + 0y = 2.$$

Let $\phi_1 = (2,1,1,5)'$, $\phi_2 = (-1,1,3,0)'$, and $z = (3,1,0,2)'$. Note that ϕ_1 and ϕ_2 are simply the column vectors of the coefficient matrix of our linear system. It is clear that ϕ_1 and ϕ_2 are linearly independent and span a two-dimensional subspace X_2 of E^4. We can rewrite our system as $x\phi_1 + y\phi_2 = z$. It is clear that this equation may have no solution, but it does have a least squares solution, $x\phi_1 + y\phi_2$. In particular the least squares fit to z from X_2 solves the minimization problem

$$\|z - (x\phi_1 + y\phi_2)\| = \min_{c_1, c_2 \text{ real}} \|z - (c_1\phi_1 + c_2\phi_2)\|,$$

where $\|(z_1, z_2, z_3, z_4)\| = (z_1^2 + z_2^2 + z_3^2 + z_4^2)^{1/2}$. By our algorithm, x and y satisfy the 2×2 system

$$(\phi_1, \phi_1)x + (\phi_1, \phi_2)y = (z, \phi_1)$$

$$(\phi_2, \phi_1)x + (\phi_2, \phi_2)y = (z, \phi_2)$$

But

$$(\phi_1, \phi_1) = 4 + 1 + 1 + 25 = 31$$

$$(\phi_1, \phi_2) = -2 + 1 + 3 + 0 = 2$$

$$(\phi_2, \phi_1) = 2$$

$$(\phi_2, \phi_2) = 1 + 1 + 9 = 11$$

$$(z, \phi_1) = 6 + 1 + 0 + 10 = 7$$

$$(z, \phi_2) = -3 + 1 + 0 + 0 = -2.$$

Thus x and y solve

$$31x + 2y = 7 \qquad \text{and} \qquad 2x + 11y = -2.$$

What has any of this to do with solving operator equations $Ax=y^2$. As the sequel indicates, these matters have a great deal to do with approximating solutions to such equations.

REFERENCES

Mikhlin, S. G. *Variational Methods in Mathematical Physics*, MacMillan, New York, 1964.

Meinardus, G. *Approximation of Functions*, Springer-Verlag, New York, 1967. Translation by L. Schumaker.

Cheney, E. W. *Introduction to Approximation Theory,* McGraw-Hill, New York, 1966.

7

THE FINITE ELEMENT METHOD

7.1 INTRODUCTION

In this chapter we construct so-called *finite element approximations* to solutions to some very simple linear ordinary and partial differential equations. The term "finite element method" has come to be associated with using piecewise polynomials in one, two, and three dimensions together with the so-called Rayleigh-Ritz method and its more general counterpart, the Bubnov-Galerkin method, to approximate solutions to operator equations. In this chapter we concentrate first on a Rayleigh-Ritz formulation of this problem that enables us to view the Rayleigh-Ritz approximate solution to a differential equation as a least squares fit to the actual solution in terms of an energy norm. This point of view initially requires our differential equation to form a positive definite, symmetric, linear operator equation on an appropriate inner product space X.

The seeds for the idea underlying the method were sown in the previous chapter. In particular, let X be a linear subspace of an inner product space Y and suppose A is a linear operator from X into Y (A need not be defined *on* X). Suppose, moreover, that for a given $y \in X$, the equation

$$Ax = y \tag{1}$$

has a solution $x \in X$. Given an N-dimensional subspace X_N of X, we wish to approximate the solution x to this equation by some function $x_N \in X_N$. That is, we seek a function

$$x_N = c_1\phi_1 + c_2\phi_2 + \cdots + c_N\phi_N$$

belonging to X_N such that $\|x - x_N\|$ is small. If A is a *positive definite, symmetric* operator defined on all of X, we know we can define a *new inner product* $(\ ,\ )_A$ in X by

$$\boxed{(u,v)_A = (Au,v)}$$

for all $u, v \in X$ (see Section 6.3). Even when A is not defined on all of X (and this is often the case), we learn that we can extend (Au, v) on the domain of A to define an inner product

$$\boxed{(u,v)_A = a(u,v)}$$

on all of X. Most frequently this is accomplished by a *formal integration by parts*.

Having extended the inner product $(\ ,\ )_A$ to all of X, we know from our least squares arguments of the previous chapter that there exists a unique least squares approximate (or least squares fit) x_N to x from X_N in the norm defined by

$$\|u\|_A = \sqrt{(u,u)_A}$$

for all $u \in X$. Moreover, x_N is the unique solution to the minimization problem

$$\|x-x_N\|_A=\min_{\bar{x}\in X_N}\|x-\bar{x}\|_A.$$

The vector x_N has a name.

DEFINITION RAYLEIGH–RITZ APPROXIMATE

The unique vector $x_N \in X_N$ solving the minimization problem

$$\|x-x_N\|_A=\min_{\bar{x}\in X_N}\|x-\bar{x}\|_A \tag{2}$$

is called the *Rayleigh–Ritz approximate to x from* X_N.

If $X_N=\operatorname{span}\{\phi_1,\phi_2,\ldots,\phi_N\}$, it follows from our least squares algorithm of Section 6.7 that

$$x_N=c_1\phi_1+c_2\phi_2+\cdots+c_N\phi_N,$$

where $(c_1,c_2,\ldots,c_N)$ solves the linear algebraic system

$$\sum_{j=1}^{N}(\phi_i,\phi_j)_A c_j=(x,\phi_i)_A, \qquad 1\leqslant i\leqslant N.$$

But since x must belong to the domain of A, $(x,\phi_i)_A=(Ax,\phi_i)=(y,\phi_i)$. That is,

THEOREM ALGORITHM FOR THE RAYLEIGH–RITZ METHOD

The Rayleigh–Ritz approximate x_N to the unique solution of the positive definite linear equation $Ax=y$ is given by

$$x_N=c_1\phi_1+c_2\phi_2+\cdots+c_N\phi_N,$$

where $(c_1,c_2,\ldots,c_N)$ solves the linear system

$$\sum_{j=1}^{N}(\phi_i,\phi_j)_A c_j=(y,\phi_i), \qquad 1\leqslant i\leqslant N.$$

Moreover, the *matrix* $(\phi_i,\phi_j)_A$, $1\leqslant i,\ j\leqslant N$, being positive definite and symmetric is *nonsingular* (see Exercises). If X_N is a subspace of *spline functions* (piecewise polynomials), the function x_N is called the finite element approximate to x from X_N, and the method is referred to as the *finite element method*. In particular, looking ahead,

DEFINITION FINITE ELEMENT METHOD

The finite element method is simply the Rayleigh–Ritz or the Galerkin method, where X_N is a subspace of X consisting of spline functions.

The Rayleigh–Ritz method can be viewed as a special case of the Galerkin method, which applies to nonsymmetric operators as well as symmetric operators, as the sequel reveals.

Meanwhile, once one has obtained a Rayleigh–Ritz approximate x_N it is only natural to ask how good an approximate x_N is to x. The answer to this question depends, as we have seen throughout the book, on our choice of the norm $\| \ \|$ in which we measure the error $\|x-x_N\|$. The optimum choice is the Tchebycheff norm $\|x-x_N\|_\infty$. Frequently we must be satisfied with the L_2 norm $\|x-x_N\|_2$ or some Sobolev norm involving $\|x-x_N\|_2$ (see Section 7.1). If A is an ordinary differential operator, the quantity $\|x-x_N\|_\infty$ is fairly easily estimated, as we shall see, in terms of powers of the mesh spacing h of the knots of our splines provided some sort of Gårding inequality applies. In the case of partial differential equations, L_2 estimates for classical equations will follow from Friedrichs and Poincaré types of inequality. With a bit more work and some matrix analysis, Tchebycheff estimates are also available for some partial differential equations, although we do not derive such estimates. The aim of all these inequalities is to prove that there exist constants γ independent of N such that

$$\|x-x_N\| \leqslant \gamma \|x-x_N\|_A, \tag{3}$$

where $\| \ \| = \| \ \|_\infty$ or $\| \ \| = \| \ \|_2$. By using spline functions, we are usually able to estimate $\|x-x_N\|_A$ fairly easily. In particular it is most often the case that there exists a constant β and positive integer k independent of N such that

$$\|x-x_N\|_A \leqslant \beta h^k, \tag{4}$$

where h is the mesh spacing associated with our splines X_N. The closeness of x to x_N in the norm $\| \ \|$ then follows through (3) combined with (4). In particular, we find that

$$\|x-x_N\| \leqslant \gamma \|x-x_N\|_A \leqslant \gamma\beta h^k.$$

Moreover, by using splines we can usually assure that the matrix $(\phi_i, \phi_j)_A$ *will be banded and sparse*, a highly desirable computational feature.

Our goal in this chapter is to study classes of differential operators (both ordinary and partial) to which we can apply the Rayleigh–Ritz–Galerkin methods, hence the finite element methods. Appropriate choices of X and of X_N enable us to estimate either $\|x-x_N\|_\infty$ or $\|x-x_N\|_2$, where x_N is our approximate solution. Since appropriate choices of X_N are usually N-dimensional subspaces of spline functions, we focus our attention on the finite element method.

EXERCISES

1. Let A be a positive definite, symmetric linear operator on X into Y where X is a linear subspace of the inner product space Y. Let $\{\phi_1,\phi_2,\ldots,\phi_N\}$ span an N-dimensional subspace of X. Let A_N be the matrix $(A\phi_j,\phi_j)$, $1\leq i, j\leq N$. Prove that A_N is positive definite, symmetric, and nonsingular.

7.2 A SIMPLE APPLICATION

To illustrate the method by a simple example, we confine our attention in the next three sections to the ordinary differential equation

$$Ax(t)=-x''(t)+\sigma(t)x(t)=f(t), \qquad 0\leq t\leq 1.$$

subject to the *homogeneous boundary conditions* $x(0)=x(1)=0$, where $0 \neq\sigma(t)\geq 0$ on $[0,1]$. Note that there is no loss of generality assuming zero boundary values. In particular, suppose $\bar{x}(t)$ is a solution to $A\bar{x}(t)=\bar{f}(t)$ subject to the boundary values $\bar{x}(0)=\alpha$ and $\bar{x}(1)=\beta$, where $\bar{f}(t)$ is given. Let $u(t)$ be any $C^2[0,1]$ function such that $u(0)=\alpha$ and $u(1)=\beta$, and let $Au(t)=\tilde{f}(t)$. Then $\bar{x}(0)-u(0)=0=\bar{x}(1)-u(1)=0$. If $x(t)$ is a solution to the homogeneous boundary value problem, $Ax(t)=A(\bar{x}-u)(t)=\bar{f}(t)-\tilde{f}(t)$. Then $\bar{x}(t)=x(t)+u(t)$ solves $A\bar{x}=\bar{f}$ with $\bar{x}(a)=\alpha$ and $\bar{x}(b)=\beta$.

If both $\sigma(t)$ and $f(t)$ are continuous on $[0,1]$, the theory of ordinary differential equations tells us that there is a unique solution $x(t)$ to this problem and $x(t)$ belongs to $C^2[0,1]$. Initially we let

$$X=\{v(t)\in C^2[0,1]:v(0)=v(1)=0\}.$$

Then X is a linear subspace of $C^2[0,1]$ containing the solution $x(t)$ to our problem (the reader should check that X is indeed a linear space). *In section 7.4 we see that we can lower the smoothness restrictions on X and use much cruder (thus computationally nicer) approximating functions than those we deal with in the first three sections.* For the time being, however, we keep X a subspace of $C^2[0,1]$ as defined. It follows simply that A is a linear operator from Y onto the linear space $Y=C[0,1]$. Note that X is a subspace of $C^2[0,1]$, which is in turn a subspace of Y. Suppose we define an inner product in Y by

$$(f,g)=\int_0^1 f(t)g(t)\,dt.$$

Since $\sigma(t)\geqslant 0$ in $[0,1]$ and $\sigma(t)\neq 0$, we can prove that A is a positive definite linear operator from X into Y. In particular, if $x\in X$

$$(Ax,x)=\int_0^1[-x''(t)+\sigma(t)x(t)]x(t)\,dt.$$

Integrating by parts we find

$$\begin{aligned}-\int_0^1 x''(t)x(t)\,dt&=-x'(t)x(t)|_0^1+\int_0^1[x'(t)]^2\,dt\\&=-[x'(1)x(1)-x'(0)x(0)]+\int_0^1[x'(t)]^2\,dt\\&=\int_0^1[x'(t)]^2\,dt,\end{aligned}$$

since $x\in X$ thus $x(1)=x(0)=0$. It follows from simple continuity arguments that

$$(Ax,x)=\int_0^1\sigma(t)[x(t)]^2\,dt+\int_0^1[x'(t)]^2\,dt>0$$

unless $x(t)\equiv 0$ (see Exercises). Thus A is positive definite. To prove that A is symmetric on X, simply integrate by parts once again. That is, if $x(t)$ and $y(t)$ both belong to X, then

$$\begin{aligned}(Ax,y)&=\int_0^1[-x''(t)+\sigma(t)x(t)]y(t)\,dt\\&=-x'(t)y(t)|_0^1+\int_0^1 x'(t)y'(t)\,dt+\int_0^1\sigma(t)x(t)y(t)\,dt\\&=x(t)y'(t)|_0^1-\int_0^1 x(t)y''(t)\,dt+\int_0^1\sigma(t)x(t)y(t)\,dt\\&=\int_0^1 x(t)[-y''(t)+\sigma(t)y(t)]\,dt\\&=(x,Ay).\end{aligned}$$

Thus A is a positive definite and symmetric linear operator on X into Y. As such, A defines a new inner product $(\ ,\ )_A$ on X (frequently called the energy inner product), which is defined by

$$(x,y)_A=(Ax,y).$$

We have seen that the Rayleigh–Ritz method consists in finding the least squares fit x_N to x in this inner product from an N-dimensional subspace X_N of X. Let $\{\phi_1,\phi_2,\ldots,\phi_N\}$ be a linearly independent subset of X and let $X_N=\text{span}\{\phi_1,\phi_2,\ldots,\phi_N\}$. *Note that each of the functions $\phi_i(t)$ satisfies the boundary values $\phi_i(0)=\phi_i(1)=0$.* (If this were not true, X_N would not be a subspace of X.) The approximate solution to our boundary value problem is a linear combination

$$x_N=c_1\phi_1+c_2\phi_2+\cdots+c_N\phi_N,$$

where $c=(c_1,c_2,\ldots,c_N)'$ solves the linear system

$$\sum_{j=1}^{N}(\phi_i,\phi_j)_A c_j=(y,\phi_i)_A, \qquad 1\leqslant i\leqslant N$$

or

$$A_N c=b,$$

where $b=(b_1,b_2,\ldots,b_N)'$, $b_i=(y,\phi_i)$, and A_N is the matrix $(\phi_i,\phi_j)_A$, $1\leqslant i,j\leqslant N$. Note that the right-hand side b is known, in contrast to the ordinary least squares fit to x from X_N using the nonenergy inner product $(\ ,\ )$, where the coordinates of the right-hand side would be (x,ϕ_i), $1<i<N$.

In this particular case,

$$Ax(t)=-x''(t)+\sigma(t)x(t).$$

Thus after integrating by parts we find

$$(\phi_i,\phi_j)_A=(A\phi_i,\phi_j)=\int_0^1[-\phi_i''(t)+\sigma(t)\phi_i(t)]\phi_j(t)\,dt$$

$$=\int_0^1[\phi_i'(t)\phi_j'(t)+\sigma(t)\phi_i(t)\phi_j(t)]\,dt.$$

To compute x_N, we must first choose a particular N-dimensional subspace of approximating functions. For purposes of illustration we choose a subspace X_N of cubic spline functions with equally spaced knots from a partition $\pi:0=t_0<t_1<\cdots<t_n=1$ of $[0,1]$. Later (Section 7.4) we use

much less complicated subspaces. Letting $S_3(\pi)$ denote the set of a cubic splines with knots at π, we know that $S_3(\pi)$ is a linear space of dimension $n+3$, which is spanned by the linearly independent set of B splines $\{B_{-1}, B_0, B_1, \ldots, B_n, B_{n+1}\}$, where

$$B_i(t)=\begin{cases} (t-t_{i-2})^3 & \text{if } t_{i-2} \leqslant t \leqslant t_{t-1} \\ h^3+3h^2(t-t_{i-1})+3h(t-t_{i-1})^2-3(t-t_{i-1})^3 & \text{if } t_{i-1} \leqslant t \leqslant t_i \\ h^3+3h^2(t_{i+1}-t)+3h(t_{i+1}-t)^2-3(t_{i+1}-t)^3 & \text{if } t_i \leqslant t \leqslant t_{i+1} \\ (t_{i+2}-t)^3 & \text{if } t_{i+1} \leqslant t \leqslant t_{i+2} \\ 0 & \text{elsewhere.} \end{cases}$$

It is helpful to keep the graph of $B_i(t)$ in mind (Figure 7.1). Note that $S_3(\pi)$ is *not* a linear subspace of X because it contains spline functions $s(x)$, which do not vanish at 0 and at 1. Thus we do not want all of $S_3(\pi)$ but only part of it. In particular, let

$$X_N = \{s(t) \in S_3(\pi, 1) : s(0) = s(1) = 0\}.$$

We must find a basis B_N for X_N. Since each $B_i(t)$, $2 \leqslant i \leqslant n-2$ vanishes at $x=0$ and $x=1$, such B_i's belong to X_N. Thus $B_2, B_3, \ldots, B_{n-2}$ serve as part of our basis. None of the functions B_{-1}, B_0, B_1, B_{n-1}, B_n, and B_{n+1} vanish at $x=0$ and $x=1$, so they clearly *do not* belong either to X or to X_N. However, we can use these six functions to find the four additional functions we need to compose a basis for X_N. In particular, we seek constants a, b, c, d, α, β, γ, and δ

$$\tilde{B}_0(t) = aB_{-1}(t) + bB_0(t)$$

$$\tilde{B}_1(t) = cB_0(t) + dB_1(t)$$

$$\tilde{B}_{n-1}(t) = \alpha B_{n-1}(t) + \beta B_n(t)$$

$$\tilde{B}_n(t) = \gamma B_n(t) + \delta B_{n+1}(t),$$

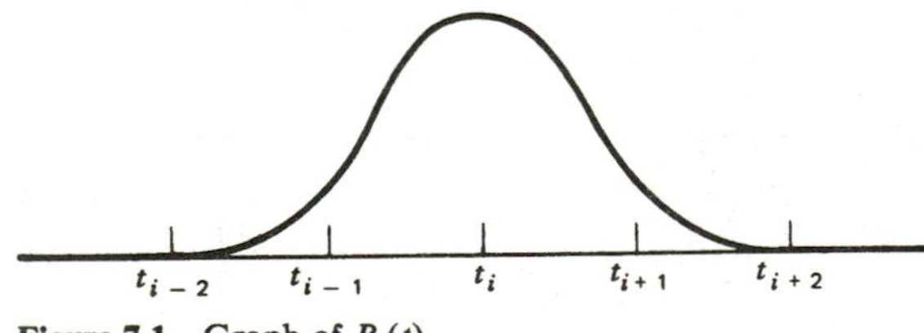

Figure 7.1 Graph of $B_i(t)$.

where $\tilde{B}_0$, $\tilde{B}_1$, $\tilde{B}_{n-1}$, and $\tilde{B}_n$ solve the following interpolation problem:

$$\tilde{B}_0(t_0)=0$$

$$\tilde{B}_0(t_1)=1$$

$$\tilde{B}_1(t_0)=0$$

$$\tilde{B}_1(t_{-1})=1$$

$$\tilde{B}_{n-1}(t_{n+1})=1$$

$$\tilde{B}_{n-1}(t_n)=0,$$

and

$$\tilde{B}_n(t_{n-1})=1$$

$$\tilde{B}_n(t_n)=0.$$

Other choices are also possible (see Exercises), but this choice is easy to compute. Solving these equations we find

$$\tilde{B}_0(t)=B_0(t)-4B_{-1}(t)$$

$$\tilde{B}_1(t)=B_0(t)-4B_1(t)$$

$$\tilde{B}_{n-1}(t)=B_n(t)-4B_{n-1}(t)$$

$$\tilde{B}_n(t)=B_n(t)-4B_{n+1}(t).$$

Letting $\tilde{B}_i(t)=B_i(t)$ for $2\leqslant i\leqslant n-2$, we can prove (see Exercises) that the set $\{\tilde{B}_0,\tilde{B}_1,\ldots,\tilde{B}_n\}$ forms a basis for X_N. The finite element approximate $x_N(t)$ to the unique solution $x(t)$ to our boundary value problem is given by

$$x_N(t)=c_0\tilde{B}_0(t)+c_1\tilde{B}_1(t)+\cdots+c_n\tilde{B}_n(t),$$

where $c=(c_0,c_1,\ldots,c_n)'$ solves the linear system

$$\boxed{\sum_{j=0}^{n}\left(A\tilde{B}_i,\tilde{B}_j\right)c_j=\left(f,\tilde{B}_i\right),}\qquad 0\leqslant i\leqslant n.$$

Since $Ax(t) = -x''(t) + \sigma(t)x(t)$, integration by parts gives

$$\left(A\tilde{B}_i, \tilde{B}_j\right) = \int_0^1 \left[-\tilde{B}_i''(t) + \sigma(t)\tilde{B}_i(t)\right]\tilde{B}_j(t)\,dt$$

$$= \int_0^1 \left[\tilde{B}_i'(t)B_j'(t) + \sigma(t)B_i(t)B_j(t)\right]dt$$

Before carrying out any integrations, we note the important fact that if $2 \leqslant j \leqslant n-2$, then $\tilde{B}_j(t) = 0$ for all $t \geqslant t_{j+2}$ and for all $t \leqslant t_{j-2}$ and the same is true of $\tilde{B}_j'(t)$. Thus

$$\left(A\tilde{B}_i, \tilde{B}_j\right) = 0 \qquad \text{whenever} \quad |i-j| \geqslant 4,$$

when $0 \leqslant i,j \leqslant n$. The result of this simple argument is that the *coefficient matrix* A_N

$$A_N = (a_{ij}^N) = \left(\left(A\tilde{B}_i, \tilde{B}_j\right)\right), \qquad 0 \leqslant i,j \leqslant N$$

is a 7-banded matrix, and the number of integrations involved in obtaining the matrix A_N is nowhere as formidable as it appeared to be at first glance. With complicated σ or f, hand computation of the integrals a_{ij}^N and $(f, \tilde{B}_i)$ with even a modest number of knots is at best formidable, at worst impossible, and, in any event, distasteful. For this reason, one takes recourse to *numerical quadratures* of the same order of accuracy as the error $\|x - x_N\|_\infty$ (see Section 7.4 and 7.5). In this case a two-point Gaussian quadrature rule would be appropriate.

EXERCISES

1. Let $Ax(t) = -x''(t) + \sigma(t)x(t)$, $0 \leqslant t \leqslant 1$, where $\sigma(t) \geqslant 0$ is continuous. Let $X = C^2[0,1]$ and let $X_N = \text{span}\{\phi_1, \phi_2, \ldots, \phi_N\}$ be N-dimensional. Define $(Ax,y) = \int_0^1 Ax(t)y(t)\,dt$. Let $A_N = (a_{ij}^N)$, $1 \leqslant i,\ j \leqslant N$, be the matrix whose terms are $a_{ij}^N = (A\phi_i, \phi_j)$. Prove that A_N is positive definite, symmetric, and nonsingular when $\sigma(t) \not\equiv 0$. What if $\sigma(t) \equiv 0$? Is continuity of $\sigma(t)$ essential? Explain.
2. Prove $X = \{x(t) \in C^2[0,1] : x(0) = x(1) = 0\}$ is a subspace of $C^2[0,1]$. Is $W = \{x(t) \in C^2[0,1] : x(0) = 1 \text{ and } x(1) = 0\}$ a subspace of $C^2[0,1]$? Why?
3. Could we take $X = \{x(t) \in C^1[0,1] : x(0) = x(1) = 0\}$ and still have our method make sense? How about $X = \{x(t) \in C[0,1] : x(0) = x(1) = 0\}$? Explain. What advantage, if any, would there be to such choices of X? (See Section 7.4.)

4. It is not true that every second-order differential equation with two given boundary values has a solution. To see this, consider the equation

$$x''(t)+x(t)=0.$$

Try to find a solution satisfying $x(0)=0$ and $x(\pi)=1$. For what sets of boundary values does this equation have a solution? A unique solution? Can you prove it?

5. Prove that the set $\{\tilde{B}_0, \tilde{B}_1, \ldots, \tilde{B}_n\}$ of cubic splines is linearly independent and does form a basis for $X_N=\{s(t)\in S_3(\pi): s(0)=s(1)=0\}$.
6. What other choices for $\tilde{B}_0$, $\tilde{B}_1$, $\tilde{B}_{n-1}$, and $\tilde{B}_n$ are available? Explain.
7. Compute the matrix A_N given in the "particular example" by programming it on a computer using a quadrature formula.
8. Let $H_3(\pi)$ be the set of piecewise cubic Hermite polynomials with evenly spaced knot $x_i=ih$, $0\leqslant i\leqslant n$ and $X=\{x(t)\in C^2[0,1]: x(0)=x(1)=0\}$. Find a basis for the set $H_N=\{s(t)\in H_3(\pi): s(0)=s(1)=0\}$. Is H_N a subspace of X? Why? What is the dimension of H_N?
9. Let $Ax(t)=-x''(t)+t^3x(t), 0\leqslant t\leqslant 1$. Compute $(A\tilde{B}_0, \tilde{B}_2)$ and $(A\tilde{B}_3, \tilde{B}_3)$. (This laborious problem spells out exactly what is involved in using these methods.)

7.3 AN ELEMENTARY ERROR ANALYSIS

We have seen how to compute the finite element approximate $x_N(t)$ to the solution $x(t)$ of a simple ordinary differential equation. For $x_N(t)$ to qualify as a good approximate to $x(t)$ we would like the error $\|x-x_N\|$ to be small where $\|\ \|$ is some norm on the domain X of our differential operator. The optimum choice of norm is, of course, the Tchebycheff norm $\|\ \|_\infty$. In this section, we prove that $\|x-x_N\|_\infty$ can indeed be made small (it can be made as small as we want for large enough N). In fact, we obtain an upper bound on $\|x-x_n\|_\infty$ in terms of the mesh spacing h of the knots of the spline functions X_N used in obtaining $x_N(t)$. The ideas underlying this error analysis are very simple and are best demonstrated by an uncomplicated problem. For this reason, we continue with the same second-order, ordinary differential equation

$$Ax(t)=-x''(t)+\sigma(t)x(t)=f(t), \qquad 0\leqslant t\leqslant 1,$$

subject to the boundary values $x(0)=x(1)=0$, where $0\not\equiv\sigma(t)\geqslant 0$ is continuous on $[0,1]$.

Let $x(t)$ be the unique solution to this equation and let $x(t)$ be the cubic spline computed in the last section which formed the finite element or Rayleigh–Ritz approximate to $x(t)$ from X_N. We want to estimate the error

$$\|x - x_N\|_\infty = \max_{0 \leqslant t \leqslant 1} |x(t) - x_N(t)|.$$

We know that

$$\|x - x_N\|_A = \min_{\bar{x} \in X_N} \|x - \bar{x}\|_A, \tag{1}$$

where $\|x - x\|_A = [(A(x - \bar{x}), x - \bar{x})]^{1/2}$. Specifically

$$(\|x - \bar{x}\|_A)^2 = \int_0^1 [A(x - \bar{x})(t)](x - \bar{x})(t)\, dt$$

$$= \int_0^1 -(x'' - \bar{x}'')(x - \bar{x})(t)\, dt + \int_0^1 \sigma(t)[x(t) - \bar{x}(t)]^2.$$

Integrating by parts and recalling that x and $\bar{x}$ belong to X so that $x(0) = x(1) = \bar{x}(0) = \bar{x}(1) = 0$, we see that

$$(\|x - \bar{x}\|_A)^2 = \int_0^1 [x'(t) - \bar{x}'(t)]^2\, dt + \int_0^1 \sigma(t)[x(t) - \bar{x}(t)]^2\, dt.$$

Since $0 \not\equiv \sigma(t) \geqslant 0$ and $\sigma(t)$ is continuous, it follows that

$$(\|x - \bar{x}\|_A)^2 \leqslant \int_0^1 [x'(t) - \bar{x}'(t)]^2\, dt + \|\sigma\|_\infty \int_0^1 [x(t) - \bar{x}(t)]^2\, dt.$$

But for any continuous functions $f(t)$ and $g(t)$

$$\|f - g\|_2 = \int_0^1 [f(t) - g(t)]^2\, dt;$$

thus

$$(\|x - \bar{x}\|_A)^2 \leqslant (\|x' - \bar{x}'\|_2)^2 + \gamma(\|x - \bar{x}\|_2)^2, \tag{2}$$

where $\gamma = \|\sigma\|_\infty$.

Since

$$X_N = \{s(t) \in S_3(\pi, 1) : s(0) = s(1) = 0\}$$

is the set of all cubic splines $s(t)$ with knots at π interpolating zero boundary data, $s(0) = s(1) = 0$. Let $\bar{x}_N(t)$ be the unique spline interpolate to the solution $x(t)$ of our differential equation. The accuracy of our approximate $x_N(t)$ depends on the smoothness of the solution $x(t)$. If $f(t)$ is continuous, $x(t)\in C^2[0,1]$. On the other hand, if $f(t)\in C^2[0,1]$, then $x(t)\in C^4[0,1]$. In general if $f(t)\in C^k[0,1]$ for some integer $k>0$, then $x(t)\in C^{k+2}[0,1]$. Let us assume for the moment that $f(t)\in C^2[0,1]$, so that $x(t)\in C^4[0,1]$. Our optimal error estimates for spline approximates (see Chapter 4) then apply to $x(t)-\bar{x}_N(t)$. In particular there exist constants M_0 and M_2 independent of h but dependent on x such that

$$\|x-\bar{x}_N\|_2 \leqslant \sqrt{M_0}\, h^4 \qquad \text{and} \qquad \|x'-\bar{x}_N'\|_2 \leqslant \sqrt{M_1}\, h^3, \tag{3}$$

where $h=\max(t_{i+1}-t_i)$ is the mesh spacing of our partition π. Letting $\bar{x}(t)=\bar{x}_N(t)$ in inequality (2) and combining the result with (3), we find

$$(\|x-\bar{x}_N\|_A)^2 \leqslant M_1 h^6 + \gamma M_0 h^8.$$

The constants M_1 and M_0, given in Chapter 4, depend on $x^{(4)}(t)$. Thus we have

$$\|x-\bar{x}_N\|_A \leqslant Kh^3,$$

where $K=(M_1+\gamma M_0 h^2)^{1/2}$. But the finite element approximate $x_N(t)$ to $x(t)$ minimizes $\|x-\bar{x}\|_A$ over all $\bar{x}\in X_N$. Thus we arrive at *the first crucial inequality*

$$\|x-x_N\|_A \leqslant \|x-\bar{x}_N\|_A \leqslant Kh^3. \tag{4}$$

Thus for small h, $x_N(t)$ is an excellent approximate to $x(t)$ in the energy norm.

But we want to estimate $\|x-x_N\|_\infty$. Obtaining this estimate is equally simple, requiring no more that a bit of calculus. In particular, suppose $g(t)\in C^1[0,1]$, where $g(0)=0$. Then

$$\int_0^t 2g(t)g'(t)\,dt=[g(t)]^2, \qquad 0\leqslant t\leqslant 1.$$

Thus invoking Schwarz's inequality and the continuity of $g(t)$ we have

$$[g(t)]^2=2\left[\int_0^1 g(t)g'(t)\,dt\right] \leqslant 2\left(\int_0^1 [g(t)]^2\right)^{1/2}\left(\int_0^1 [g'(t)]^2\,dt\right)^{1/2}.$$

But $[g(t)]^2 \leqslant (\|g\|_\infty)^2$; thus

$$[g(t)]^2 \leqslant 2\|g\|_\infty\left(\int_0^1 [g'(t)]^2\,dt\right)^{1/2} = 2\|g\|_\infty\|g'\|_2$$

for all t in $[0,1]$. Taking the maximum of the left-hand side over the entire interval $[0,1]$, we have

$$\|g\|_\infty^2 \leqslant 2\|g\|_\infty\|g'\|_2.$$

For $g(t)$ not identically zero, $\|g\|_\infty > 0$, and we can divide to obtain

$$\|g\|_\infty \leqslant 2\|g'\|_2. \tag{5}$$

Now $x(t)-x_N(t)=0$ when $t=0$. If $x(t)-x_N(t)\equiv 0$, there is nothing to estimate because $x(t)\equiv x_N(t)$. If not, (5) applies and we obtain the *second crucial inequality*

$$\|x-x_N\|_\infty \leqslant 2\|x'-x_N'\|_2. \tag{6}$$

But

$$(\|x-x_N\|_A)^2 = \int_0^1 [x'(t)-x_N(t)]^2\,dt + \int_0^1 \sigma(t)[x(t)-x_N(t)]^2\,dt$$

$$\geqslant \int_0^1 [x'(t)-x_N'(t)]^2\,dt;$$

so that

$$\|x'-x_N'\|_2 \leqslant \|x-x_N\|_A. \tag{7}$$

Combining (6), (7), and (4), we have

$$\boxed{\|x-x_N\|_\infty \leqslant 2Kh^3.} \tag{8}$$

Actually we have proved a theorem.

THEOREM (ERROR ESTIMATE)

Let $x(t)$ be the unique solution to the differential equation

$$-x''(t)+\sigma(t)x(t)=f(t), \qquad 0\leqslant t\leqslant 1$$

satisfying boundary conditions $x(0)=x(1)=0$, where $0\not\equiv\sigma(t)\geqslant 0$ is continuous in $[0,1]$. Let $x_N(t)$ be the cubic spline Rayleigh–Ritz approximate to $x(t)$ satisfying $x_N(0)=x_N(1)=0$. If $f\in C^2[0,1]$, then there exists a constant K independent of N such that

$$\boxed{\|x-x_N\|_\infty \leqslant Kh^3,}$$

where h is the mesh spacing between the knots of $x_N(t)$.

If $f(t)$ is not so smooth, we cannot anticipate such rapid convergence. This is because with less smoothness on $f(t)$, we are guaranteed less smoothness on $x(t)$; thus the inequalities (3) must be modified. For example, if f is in $C^1[0,1]$, then $x(t)\in C^3[0,1]$. We leave this as an exercise for the student to prove (after a brief review of Chapter 4).

EXTENDED ERROR ESTIMATE

Let $x(t)$ be the unique solution to

$$-x''(t)+\sigma(t)x(t)=f(t), \qquad 0\leqslant t\leqslant 1,$$

satisfying the boundary conditions $x(0)=x(1)=0$. Let $x_N(t)$ be the cubic spline Rayleigh–Ritz approximate to $x(t)$ satisfying $x_N(0)=x_N(1)=0$. If $f\in C^j[0,1]$, $0\leqslant j\leqslant 2$, there exists a constant K_j independent of N such that

$$\|x-x_N\|_\infty \leqslant K_j h^{j+1},$$

where h is the mesh spacing between the knots of π.

These error estimates can frequently be improved if the operator A is *coercive* (see Section 7.9). For example, one can prove that

$$\|x-x_N\|_\infty = 0(h^4),$$

when $x\in C^4[a,b]$ and the mesh spacing is even.

We are not, of course, obligated to choose cubic spline functions for X_N. We may use higher degree splines of piecewise cubic or quintic Hermites, or even piecewise linear Lagrange polynomials (see next section). In general our piecewise polynomial subspace X_N should be made up of functions that are at least $C[0,1]$ functions; a different choice of X_N modifies both the inequalities (3) and our final estimate (8). The Exercises illustrate these matters. The underlying simple idea, however, is always the same chain of "crucial inequalities," and this generalizes to higher order ordinary operators and to partial differential operators.

EXERCISES

1. Compute the finite element approximate to

$$Au(t) = -u''(t) + e^{t^2}u(t) = t + 2e^t$$
$$u(0) = u(1) = 0$$

from an N-dimensional subspace X_N consisting of piecewise polynomials with 5 knots, 10 knots, 15 knots, where
(a) X_N = cubic splines
(b) X_N = piecewise cubic Hermite polynomials
(c) X_N = piecewise linear Lagrange polynomials.
Compare the size of the linear systems, the complexity of inverting the system, and the numerical accuracy of each method.

2. Can you use finite elements from a basis of piecewise cubic Hermite polynomials to approximate the solution to the boundary value problem

$$-u''(t) + \sigma(t)u(t) = f(t)$$
$$u(0) = u(1) = 0,$$

where $0 \neq \sigma(t) \geqslant 0$? Explain. Can you use piecewise linear Lagrange polynomials? Estimate the error in both cases.

3. Experiment with finite elements to approximate the solution of the boundary value problem

$$-u''(t) + \frac{1}{t}u'(t) + u(t) = -\frac{t}{2} + \frac{t^3}{6}.$$
$$u(0) = 0,\ u(1) = -\frac{1}{3}.$$

Are spline functions the best choice of approximates? What theoretical difficulties, if any, are encountered? Explain.

4. Is it always necessary to have the functions in the linear spaces X and X_N satisfy the boundary values of the ordinary differential equation whose solution you are trying to approximate, or is some more general scheme available? Explain.

7.4 LOWERING THE SMOOTHNESS REQUIREMENTS—CHOICE OF LINEAR SPACE

In the preceding two sections we computed the Rayleigh–Ritz or finite element approximate $x_N(t)$ to the solution $x(t)$ to the boundary value problem

$$\begin{cases} Ax(t) = -x''(t) + \sigma(t)x(t) = f(t) \\ \qquad x(0) = x(1) = 0 \end{cases}$$

from a subspace X_N of cubic spline functions with knots at the points t_i, $0 \leqslant i \leqslant n$, of a partition $\pi: 0 = t_0 < t_i < \cdots < t_n = 1$ of the interval $[0, 1]$. The function $x_N(t)$ belonged to the same smoothness class $C^2[0,1]$ as the solution $x(t)$ to our equation. It is only natural to ask whether this much smoothness is needed. Might we, for example, have looked for $x_N(t)$ among the set $S_N = X_N$, (S for step) of *step functions* (piecewise constant functions illustrated in Figure 7.2) with knots at the t_i's. Or might the solution have appeared among the set $H_N = X_N$ (H for hat) of piecewise linear functions (splines of degree 1) with knots again at the t_i's (see Figure 7.3)? In other words, we are concerned with the problem of *choice of linear space* in which to formulate the finite element method.

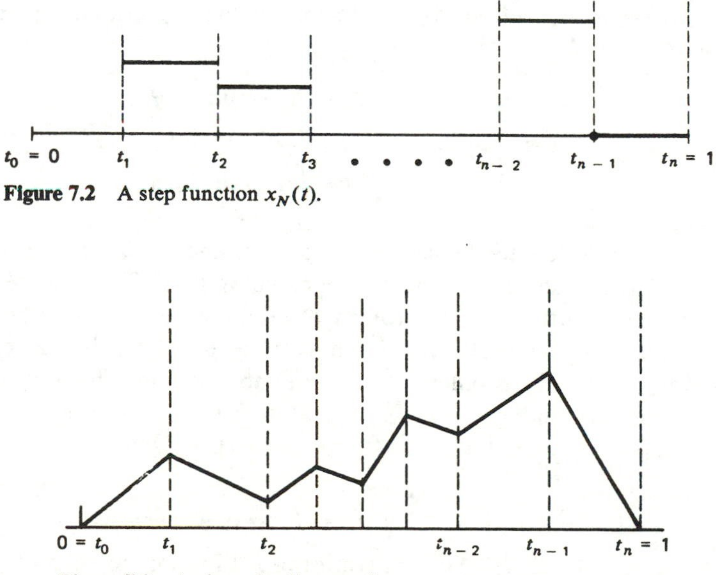

Figure 7.2 A step function $x_N(t)$.

Figure 7.3 A piecewise linear spline.

To answer this question we must look a bit more closely at the energy inner product

$$(u,v)_A=\int_0^1 v(t)Au(t)\,dt=(Au,v).$$

The right-hand side of this equation makes sense if and only if $Au(t)$ is defined at almost all t in $[0,1]$ and if the integral on the right-hand side exists in, at least, the Lebesgue sense. Until now, we have insisted that both $u(t)$ and $v(t)$ belong to the set $X=\{z(t)\in C^2[0,1]: z(0)=z(1)=0\}$. Note that in this case

$$(u,v)_A=\int_0^1 v(t)[-u''(t)+\sigma(t)u(t)]\,dt.$$

Integrating the right-hand side by parts and keeping in mind the boundary conditions $u(0)=u(1)=v(0)=v(1)=0$, one easily verifies that

$$(u,v)_A=\int_0^1 [u'(t)v'(t)+\sigma(t)u(t)v(t)]\,dt. \tag{1}$$

The point we want to make is simply that the right-hand side of (1) makes sense just so long as the Lebesgue integrals of $u'(t)v'(t)$ and of $u(t)v(t)$ are defined. For this reason, it seems that our method should apply to any linear space X to which the solution $x(t)$ belongs and on which the bilinear form $a(u,v)=(u,v)_A$ defined by (1) forms an inner product. One such X we can choose is $X=\hat{X}$, where

$$\hat{X}=\left\{ v(t): \quad \begin{array}{l} v(0)=v(1)=0, \\ v'(t)\in L_2[0,1], \text{ and} \\ v(t)=\int_0^t v'(t)\,dt, \end{array} \right\}$$

where integration denotes Lebesgue integration and $L_2[0,1]$ is the set of all functions that are Lebesgue square integrable on $[0,1]$. This choice of X is one favored by mathematicians for its theoretic ease. On the other hand, most engineers will counter that one hardly needs so much mathematical machinery to work the problem in a very simple setting. This is quite true, provided one can endure some fussing at interior knots. For exampls, if $\pi: 0=t_0<t_1<\cdots<t_n=1$ is a partition of $[0,1]$, we could take $X=X(\pi)$ where

$$X(\pi)=\left\{ v(t)\in C[0,1]: \quad \begin{array}{l} v(0)=v(1)=0,\ v'(t) \text{ exists and is} \\ \text{a continuous function on each of} \\ \text{the intervals } [t_i,t_{i+1}]. \end{array} \right\}$$

Note that if $v(t) \in X(\pi)$, then $v'(t_i+)$ and $v'(t_i-)$ both exist but may not be equal at interior knots t_i. Thus $v'(t)$ is piecewise continuous and possibly *undefined* at the interior knots $t_1, t_2, \ldots, t_{n-1}$. For example (see Exercises), the set X_N of piecewise polygonal lines satisfying zero boundary data, $X_N = \{v(t) \in H_N : v(0) = v(1) = 0\}$, is a subspace of $X(\pi)$ (see Figure 7.3). Even though $v'(t)$ is not well defined for arbitrary members of $X(\pi)$, it is still true that $\int_0^t v'(t)\,dt$ is defined and that $v(t) = \int_0^t v'(t)\,dt$. Common sense dictates the definition of $\int_0^t v'(t)\,dt$. In particular it is defined interval by interval. That is, if $t \in [t_{j-1}, t_j]$ we define

$$\int_0^t v'(t)\,dt = \sum_{i=1}^{j-1} \int_{t_{i-1}}^{t_i} v'(t)\,dt + \int_{t_i}^{t} v'(t)\,dt.$$

By way of strongly endorsing the mathematicians' point of view, we note that for all choices of partition $\pi, X(\pi) \subset \hat{X}$, and it is not necessary to define exactly what one means by $\int_0^t v'(t)\,dt$ for each new change of π. Thus the mathematicians' choice of $X = \hat{X}$ takes all the drudgery out of the analysis and enables us to use even more general sets of approximating functions than functions from $X(\pi)$. By way of endorsing the engineers' viewpoint we note that in practice spaces like $X(\pi)$ are all we ever use anyway, and in integrating $v'(t)$ or a product of such functions (which is what one does in the finite element method), the integral used is exactly the sum of integrals given previously. Thus the mathematicians' point of view takes the drudgery out of the analysis, while the more practical point of view removes the obscurity for those who are unfamiliar with Lebesgue integration.

In any event, for either choice of X, the positive definite bilinear form $(u,v)_A$ given by (1) still defines an (energy) inner product on X with corresponding (energy) norm $\|u\|_A = \sqrt{(u,u)_A}$. If $X_N = \text{span}\{\phi_1, \phi_2, \ldots, \phi_N\}$ is an N-dimensional subspace of X, the Rayleigh-Ritz or energy norm least squares approximate $x_N(t)$ to the unique solution $x(t)$ of our differential equation is still given by

$$x_N(t) = c_1\phi_1(t) + c_2\phi_2(t) + \cdots + c_N\phi_N(t),$$

where $c = (c_1, c_2, \ldots, c_N)'$ solves the same linear system

$$\sum_{j=1}^{N} (\phi_i, \phi_j)_A c_j = (x, \phi_j)_A = (f, \phi_i), \qquad 1 \leq i \leq N.$$

The only possible computational difference from our arguments in Sections 7.1 and 7.2 is that this time we must take

$$a(\phi_i,\phi_j)=(\phi_i,\phi_j)_A=\int_0^1[\phi_i'(t)\phi_j'(t)+\sigma(t)\phi_i(t)\phi_j(t)]\,dt$$

instead of $(\phi_i,\phi_j)_A=(A\phi_i,\phi_j)$. This, of course, is because $A\phi_i(t)$ need no longer be defined at all points of $[a,b]$, since we have made no guarantees about $\phi_i''(t)$ in our new choice of X. However, $x(t)$, our solution, still belongs to $C^2[0,1]$. Thus it is still true that $(x,\phi_i)_A=(Ax,\phi_i)=(f,\phi_i)=\int_0^1 f(t)\phi_i(t)\,dt$. If X_N is a subspace of piecewise polynomials, $x_N(t)$ is still called a finite element fit to $x(t)$, and our method is still referred to as *the finite element method*.

We are now able to answer the initial queries of this section. That is, can we choose $X_N=S_N$ (step functions) or $X_N=H_N$ (piecewise linear functions)? The answer to the first question is left as an exercise for the reader. The answer to the second question is "indeed yes," so long as our functions satisfy zero boundary data. In particular let

$$X_N=\{v(t)\in H_N : v(0)=v(1)=0\}.$$

Then X_N is the set of piecewise Lagrange polynomials of degree one (see Figure 7.4) interpolating zero boundary data, and X_N is a subspace of X. If $x_N(t)\in X_N$, then $x_N'(t)\in S_N$ is a step function (piecewise constant) with discontinuities (jumps) at the interior knots $t_1,t_2,\ldots,t_{n-1}$ (i.e., $x_N'(t_i)$ is undefined for $1\leqslant i\leqslant n-1$). Let $x_N(t)$ be the finite element approximate to

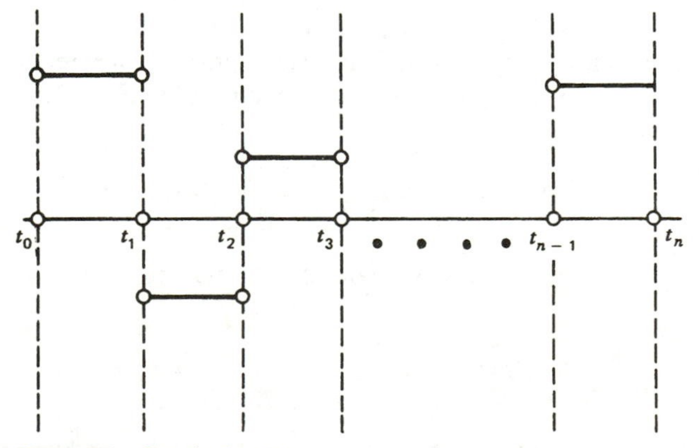

Figure 7.4 Graph of $x_N'(t)$.

$x(t)$ from X_N. Then we have

$$\|x-x_N\|_A^2=\int_0^1\left\{[x'(t)-x_N'(t)]^2+\sigma(t)[x(t)-x_N(t)]^2\right\}dt.$$

An error analysis similar to that of Section 7.3 now applies. In particular, let $\bar{x}_N(t)$ be the piecewise Lagrange polynomial of degree 1 interpolating $u(t)$ at the knots $t_0<t_1<\cdots<t_n$. Since $x(t)\in C^2[0,1]$, our error analysis of Chapter 2 gives us

$$\|x-\bar{x}_N\|_\infty\leqslant M_0h^2$$

and

$$\|x'-\bar{x}_N'\|_\infty\leqslant M_1h,$$

where h is our mesh spacing, $M_0=\|x''\|_\infty/4$, and $M_1=\|x''\|_\infty$. It follows that

$$\|x-\bar{x}_N\|_A\leqslant Kh,$$

where $K=(1+\|\sigma\|_\infty)^{1/2}$. But

$$\|x-x_N\|_A=\min_{\bar{x}\in X_N}\|x-\bar{x}\|_A;$$

thus $\|x-x_N\|_A\leqslant\|x-\bar{x}_N\|_A$. Therefore,

$$\|x-x_N\|_A\leqslant Kh.$$

To estimate $\|x-x_N\|_\infty$, simply note that inequality (6) of the previous section

$$\|x-x_N\|_\infty\leqslant 2\|x-x_N'\|_2\leqslant 2\|x-x_N'\|_A$$

still applies (see Exercises). Thus

$$\boxed{\|x-x_N\|_\infty\leqslant 2Kh.}$$

Using coerciveness (see Section 7.9), one can prove

$$\|x-x_N\|_2=O(h^2)$$

so that the order of convergence is the order of best approximation. In any event, we have proved

ERROR ESTIMATE

Let $x(t)$ be the unique solution to the differential equation

$$-x''(t)+\sigma(t)x(t)=f(t), \qquad 0\leqslant t\leqslant 1$$

satisfying the boundary conditions $x(0)=x(1)=0$, where $0\not\equiv\sigma(t)\geqslant 0$ is continuous on $[0,1]$. Let $x_N(t)$ be the piecewise linear finite element approximate to $x(t)$ satisfying $x(0)=x(1)=0$. If $f\in C[0,1]$, there exists a constant $K>0$ independent of N such that

$$\boxed{\|x-x_N\|_\infty \leqslant Kh,}$$

where h is the mesh spacing between the knots of $x_N(t)$.

To compute $x_N(t)$ we first note that X_N is spanned by the "hat functions" $\phi_j(t)$, $1\leqslant j\leqslant n-1$, shown in Figure 7.5 and given by

$$\phi_j(t)=\begin{cases}\dfrac{(t_{j+1}-t)}{t_{j+1}-t_j}, & \text{when} \quad t_j\leqslant t\leqslant t_{j+1}\\[2ex] \dfrac{t-t_{j-1}}{t_j-t_{j-1}}, & \text{when} \quad t_{j-1}\leqslant t\leqslant t_j\\[2ex] 0, & \text{elsewhere.}\end{cases}$$

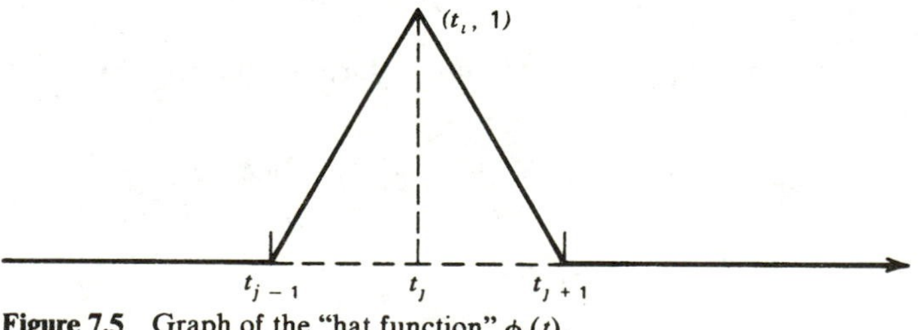

Figure 7.5 Graph of the "hat function" $\phi_j(t)$.

Then

$$x_N(t)=\sum_{j=1}^{n-1} c_j\phi_j(t),$$

where $c=(c_i,c_2,\ldots,c_{n-1})'$ solves the linear system

$$\sum_{j=1}^{n-1}(\phi_i,\phi_j)_A c_j=(f,\phi_i), \qquad 1\leqslant i\leqslant n-1.$$

But

$$(\phi_i,\phi_j)_A=\int_0^1[\phi_i'(t)\phi_j'(t)+\sigma(t)\phi_i(t)\phi_j(t)]\,dt$$

and

$$(f,\phi_i)=\int_0^1 f(t)\phi_i(t)\,dt.$$

Since

$$\phi_j'(t)=\begin{cases}\dfrac{-1}{t_{j+1}-t_j} & \text{on } (t_j,t_{j+1})\\[2mm] \dfrac{1}{t_j-t_{j-1}} & \text{on } (t_{j-1},t_j)\\[2mm] 0 \quad \text{when } t<t_{i-1} \text{ and } t>t_{i+1}, & \end{cases}$$

we have

$$(\phi_j,\phi_j)_A=0,\qquad \text{when}\quad i\leqslant j-2\qquad \text{or}\qquad i\geqslant j+2.$$

Thus the coefficient matrix of our linear system (1) is a *three-banded* $(n-1)\times(n-1)$ square matrix:

$$\begin{bmatrix} a_{11} & a_{12} & 0 & 0 & 0 & \cdots & \cdots \\ a_{21} & a_{22} & a_{23} & 0 & 0 & \cdots & \cdots \\ 0 & a_{32} & a_{33} & a_{34} & 0 & \cdots & \cdots \\ & \vdots & & \ddots & \ddots & \ddots & a_{n-2,n-1} \\ 0 & 0 & \cdots & & 0 & a_{n-1,n-2} & a_{n-1,n-1} \end{bmatrix}.$$

The nonzero terms of the ith row of this matrix are $a_{i,i-1}, a_{ii}, a_{i,i+1}$ where

$$a_{i,i-1}=\int_{t_{i-1}}^{t_i}[\phi_{i-1}'(t)\phi_i'(t)+\sigma(t)\phi_{i-1}(t)\phi_i(t)]\,dt$$

$$=\frac{1}{(t_i-t_{i-1})^2}\int_{t_{i-1}}^{t_i}[-1+\sigma(t)(t-t_{i-1})(t_i-t)]\,dt$$

$$a_{ii}=\int_{t_{i-1}}^{t_i}\left([\phi_i'(t)]^2+\sigma(t)[\phi_i(t)]^2\right)dt+\int_{t_i}^{t_{i+1}}\left([\phi_i'(t)]^2+\sigma(t)[\phi_i(t)]^2\right)dt$$

$$=\frac{1}{(t_i-t_{i-1})^2}\int_{t_{i-1}}^{t_i}\left[1+\sigma(t)(t-t_{i-1})^2\right]dt$$

$$+\frac{1}{(t_{i+1}-t_i)^2}\int_{t_i}^{t_{i+1}}\left[1+\sigma(t)(t_{i+1}-t)^2\right]dt$$

and

$$a_{i,i+1}=\int_{t_i}^{t_{i+1}}[\phi_{i+1}'(t)\phi_i'(t)+\sigma(t)\phi_{i+1}(t)\phi_i(t)]\,dt$$

$$=\frac{1}{(t_{i+1}-t_i)^2}\int_{t_i}^{t_{i+1}}[-1+\sigma(t)(t_{i+1}-t)(t-t_i)]\,dt.$$

For purposes of computation it is simplest to put these integrals on a machine, using a quadrature formula such as the trapezoidal rule. With this choice of quadrature rule one can prove the same order accuracy for the error

$$\|x-\bar{x}_N\|_\infty,$$

where $\bar{a}_N=\sum_{j=1}^{n}\bar{c}_j\phi_j$, and $c=(\bar{c}_1,\bar{c}_2,\ldots,\bar{c}_{n-1})$ is the solution to the linear system obtained replacing each $(\ ,\ )_A$ by such a numerical integration.

We leave is as an exercise to the reader to prove the availability of a variety of other choices of X_N (e.g., piecewise cubic Hermites). Moreover, an analogous error analysis obtains where the power of h (h the mesh spacing) depends both on the smoothness of the right-hand side and on the choice of X_N. To give a measure of the speed of convergence, we have computed the piecewise cubic Hermite approximate $x_N(t)$ to the solution $x(t)$ to the boundary value problem

$$-x''(t)+a^2x(t)=2ab\cos at$$

$$x(0)=x(1)=0,$$

where $a=4\pi$ and $b=5$. Using an even mesh spacing $h=\frac{1}{8}$ on the CDC 6, we obtained the solution to eight-decimal place accuracy using machine quadratures and with a fairly crude program that is applicable to the same equation with a^2 replaced by $\sigma(t)$.

EXERCISES

1. Let $X_N=\{v(t)\in H_N : v(0)=v(1)=0\}$. Prove that X_N is a subspace of $X(\pi)$. Of $\hat{X}$.
2. Let $0\not\equiv\sigma(t)\geqslant 0$. Does $(u,v)_{\hat{A}}=\int_0^1[u'(t)v'(t)+\sigma(t)u(t)v(t)]\,dt$ define an inner product on $X(\pi)$? On $\hat{X}$? Prove your answer.
3. Let $X_N=\{v(t)\in H_N : v(0)=v(1)=0\}$ and let $X=X(\pi)$. Let $x(t)\in X$ and $x_N(t)\in X_N$. Prove

$$[x(t)-x_N(t)]^2=\int_0^t 2[x(t)-x_N(t)][x'(t)-x_N'(t)]\,dt.$$

   What has this to do with the error analysis of this section?
4. Can one use the space $X_N=\{v(t)\in S_N : v(0)=v(1)=0\}$ of step functions to approximate $x(t)$, the unique solution to the differential equation of this section? Explain.

7.5 SOME PRACTICAL CONSIDERATIONS

In the previous sections we saw that we could use piecewise linear Lagrange polynomials, cubic splines, and cubic Hermites, among many other choices of approximating functions, to approximate the solution u to the boundary value problem

$$\begin{cases} Lu=-u''+\sigma u=f \\ u(0)=u(1)=0, \qquad \sigma>0. \end{cases} \tag{1}$$

With all these choices at hand, which X_N do we choose? Initially the answer seems obvious: use the subspace that guarantees the smallest error. However, this may not be the best answer if computational complexity is considered. To illustrate, we compare computation and error analysis for the finite element approximates u_N and $\hat{u}_N$ to (1) from subspaces $X_N=\text{span}\{\phi_i\}_{i=0}^{i=n}$ and $\hat{X}_N=\text{span}\{\hat{\phi}_i\}_{i=1}^{i=n-1}$ of cubic B splines and cardinal piecewise linear Lagrange polynomials, respectively, satisfying zero boundary data. Let h be the mesh of our partition π_N of $[0,1]$. In this comparison we consider the following properties.

Error Estimates

Splines guarantee a better fit, in addition to being a smoother approximate. In particular

$$\|u-u_N\|_\infty=o(h^3)$$
$$\|u-\hat{u}_N\|_\infty=o(h).$$

In fact under strong coercivity (see Section 7.9), we can prove $\|u-u_N\|_2 = o(h^4)$ while $\|u-\hat{u}_N\|_2 = o(h^2)$.

Size of the Linear System

Splines

System (2) is $n+1 \times n+1$:

$$u_N = b_0\phi_0 + b_1\phi_1 + \cdots + b_n\phi_n$$

$$\sum_{j=0}^{n} (\phi_i, \phi_j)_A b_j = \sum_{j=0}^{n} [a(\phi_i, \phi_j)] b_j = (f, \phi_i), \qquad 0 \leqslant i \leqslant n. \tag{2}$$

On the other hand, we have piecewise linear Lagrange polynomials.

Piecewise Linear Lagrange Polynomials

System (3) is $n-1 \times n-1$.

$$u_N = c_1\hat{\phi}_1 + c_2\hat{\phi}_2 + \cdots + c_{n-1}\hat{\phi}_{n-1}$$

$$\sum_{j=1}^{n-1} \left[a\left(\hat{\phi}_i, \hat{\phi}_j\right)\right] c_j = \sum_{j=1}^{n-1} \left(\hat{\phi}_i, \hat{\phi}_j\right)_A c_j = \left(f, \hat{\phi}_i\right), \qquad 1 \leqslant i \leqslant n-1. \tag{3}$$

Thus the sizes of the two systems just described are comparable, neither one being preferable to the other by virtue of size. These systems should be compared with the system resulting from using piecewise cubic Hermites (see Exercises).

Formation of the Matrix Involved and Numerical Inversion of the Resultant System

Let

$$a_{ij} = a(\phi_i, \phi_j) = \int_0^1 [\phi_i'\phi_j' + \sigma\phi_i\phi_j] \tag{2}$$

and

$$\hat{a}_{ij} = a\left(\hat{\phi}_i, \hat{\phi}_j\right) = \int_0^1 \left[\hat{\phi}_i'\hat{\phi}_j' + \sigma\hat{\phi}_i\hat{\phi}_j\right]. \tag{3}$$

The matrix $(\hat{a}_{ij})$ is much easier to form than the matrix (a_{ij}) because the functions $\hat{\phi}_i$ are simpler and have *smaller compact support*. Moreover, the matrix $(\hat{a}_{ij})$ *is three-banded*, since $\hat{\phi}_i\hat{\phi}_j = 0$ whenever $|i-j| \geqslant 2$. On the other hand, the matrix (a_{ij}) *is seven-banded*, since $\phi_i\phi_j = 0$ whenever $|i-j| \geqslant 4$. It follows that the system (3) is much more sparse than (2), hence is easier to invert. The seven-banded width of (a_{ij}) is the price to be paid for the more rapid convergence. Again, it is interesting to compare (a_{ij}) and $(\hat{a}_{ij})$ to the piecewise cubic Hermite case (see Exercises). Note too that if $\sigma(t)$ is at all complicated, one must resort to a *numerical quadrature* to obtain either of the matrices (a_{ij}) and $(\hat{a}_{ij})$. We have more to say about this later.

Reducing the Size of the System

Since splines promise increased speed of convergence, we can certainly use a much smaller system and still obtain better accuracy using splines than piecewise linear Lagranges. There is, however, a point at which the accuracy of a spline approximate is surpassed by the piecewise linear approximates (see Exercises). When dealing with very complex and large-scale finite element approximates (i.e., n very large), these considerations become quite important. Their significance increases in the two-variable case, as we shall see, since the size of the systems involved is roughly order $n^2 \times n^2$.

EXERCISES

1. Set up the finite element approximating scheme to problem (1) using a basis of piecewise cubic Hermite polynomials. Give the basis $\{\phi_i\}_{i=1}^{I=2N}$ specifically and compare the numerical efficiency of this approximation with finite elements using piecewise linear Lagranges and cubic splines. In particular, compare properties 1, 2, 3, and 4 in all three cases.
2. Repeat problem (1) using quintic splines.
3. Let n be fixed and let

$$u_N = c_1\hat{\phi}_1 + c_2\hat{\phi}_2 + \cdots + c_{n-1}\hat{\phi}_{n-1}$$

be the finite element approximate to the solution u of (1) from $\hat{X}_N$ and let

$$u_M = c_0\phi_0 + c_1\phi_1 + \cdots + c_m\phi_m$$

be the cubic spline approximate to u from span $\{\phi_i\}_{i=0}^{i=m}$. For what m's, if any, is the cubic spline a worse fit than the piecewise linear Lagrange? Explain.

7.6 APPLICATION TO THE DIRICHLET PROBLEM

So far we have focused on solving a simple, second-order, ordinary differential equation. The most common application of the finite element method is to the numerical solution of partial differential equations, and one of the simplest of these is the *Dirichlet problem*. Suppose D is a closed region of the plane bounded by a simple closed curve C (see Figure 7.6). Given functions f and g, we seek a function $u(x,y)$ such that

$$\Delta u = \frac{\partial^2 u}{\partial x^2} + \frac{\partial^2 u}{\partial y^2} = f$$

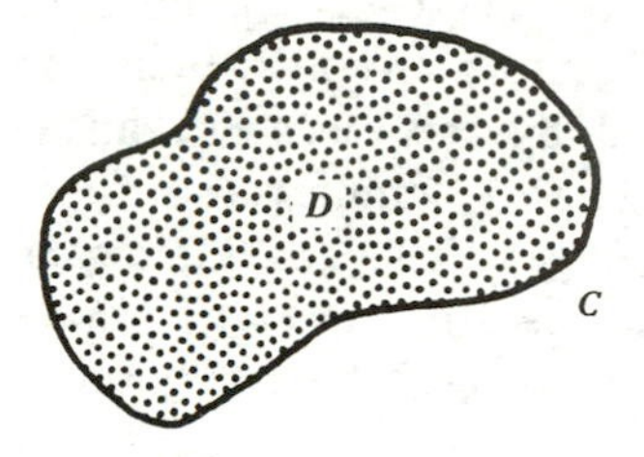

$u(x, y) = g(x, y)$ on C

Figure 7.6 Domain D of the Dirichlet operator.

on the region D, and $u(x,y) = g(x,y)$ at points (x,y) on the boundary C of D. Suppose f is continuous on D, and there exists a unique solution $u(x,y)$ to this boundary value problem which belongs to $C^{2,2}[D]$. Theoretically, we can assume $g(x,y) \equiv 0$ on C. In particular if $g \not\equiv 0$ on C and $\Delta u = f$, we can extend g to all of D and let $\bar{u} = u - g$. Then $\bar{u} \equiv 0$ on C and $\Delta \bar{u} = f - \Delta g$. Thus to find u, we can first solve for $\bar{u}$ and then let $u = \bar{u} + g$. Computationally, the difficulty with this logic is that it is seldom easy to extend g to all of D. Be that as it may, we begin by restricting ourselves to the homogeneous problem

$$-\Delta u = f$$

$$u \equiv 0 \qquad \text{on } C.$$

To solve this problem by the Rayleigh–Ritz or finite element method, we need an inner product space Y in which to imbed the problem. The natural choice is $Y = L_2[D]$, where $L_2[D]$ is the set of all functions u and v which are square integrable on D together with the inner product

$$(u,v) = \int_D\int uv \, dx \, dy.$$

Presently we learn that the operator $A = -\Delta$ is positive definite and symmetric on a subspace of Y and that as such A dictates an energy inner product $(\ ,\ )_A$ on a somewhat larger subspace X of Y in the usual way. For the moment, however, let us proceed formally. We shall need *Green's identities in the plane*. To this end we recall some elementary two-variable calculus.

Suppose $M(x,y)$ and $N(x,y)$ are two real-valued functions for which $\partial M/\partial x$ and $\partial N/\partial y$ are integrable on D with

$$M(x_2,y)-M(x_1,y)=\int_{x_1}^{x_2}\frac{\partial M}{\partial t}(t,y)\,\partial t$$

and

$$N(x,y_2)-N(x,y_1)=\int_{y_1}^{y_2}\frac{\partial N}{\partial s}(x,s)\,\partial s,$$

where (x,y) is any interior point of D and (x_1,y), (x_2,y), (x,y), and (x,y_2) are boundary points of D as illustrated in Figure 7.7. It follows that x_1 and x_2 depend on y and y_2 depend on x. Recalling the definition of a line integral, we obtain Green's identities in the planes

$$\iint_D \frac{\partial M}{\partial x}\,dx\,dy=\int_C M(x,y)\,dy$$

and

$$\iint_D \frac{\partial N}{\partial y}\,dx\,dy=-\int_C N(x,y)\,dx,$$

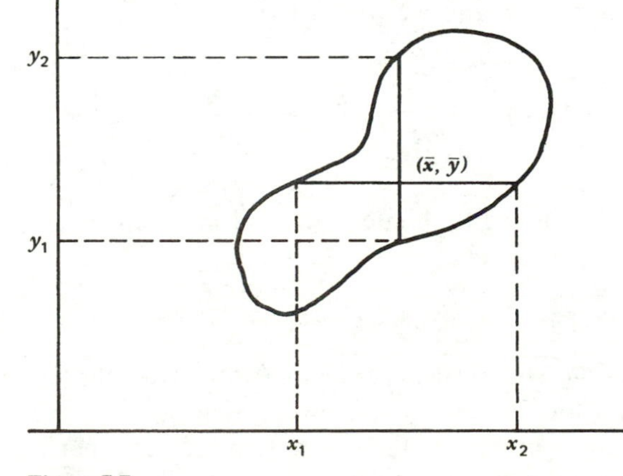

Figure 7.7

where the right-hand integrals are line integrals along the curve C. Suppose we let

$$Au = -\Delta u$$

and integrate formally by parts, using Green's identities. Then if u were in the domain of A, we would have

$$(Au, v) = -\int_D\int \left(\frac{\partial^2 u}{\partial x^2} + \frac{\partial^2 u}{\partial y^2}\right) v\,dx\,dy.$$

But

$$\int_D\int \frac{\partial^2 u}{\partial x^2} v\,dx\,dy = -\int_D\int \left\{\frac{\partial}{\partial x}\left[v\frac{\partial u}{\partial x}\right] - \left(\frac{\partial u}{\partial x}\frac{\partial v}{\partial x}\right)\right\} dx\,dy$$

$$= -\int_C v\frac{\partial u}{\partial x}\,dy + \int_D\int \frac{\partial u}{\partial x}\frac{\partial v}{\partial x}\,dx\,dy$$

and

$$-\int_D\int \frac{\partial^2 u}{\partial y^2} v\,dx\,dy = -\int_D\int \left\{\frac{\partial}{\partial y}\left[v\frac{\partial u}{\partial y}\right] - \left(\frac{\partial u}{\partial y}\frac{\partial v}{\partial y}\right)\right\} dx\,dy$$

$$= \int_C v\frac{\partial u}{\partial y}\,dx + \int_D\int \frac{\partial u}{\partial y}\frac{\partial v}{\partial y}\,dx\,dy.$$

If v vanished on C, the boundary of D, the line integrals on the last two equations would vanish and we would have

$$(Au, v) = \int_D\int \left\{\frac{\partial u}{\partial x}\frac{\partial v}{\partial x} + \frac{\partial u}{\partial y}\frac{\partial v}{\partial y}\right\} dx\,dy. \tag{1}$$

If both u and v belonged to the domain of A and vanished on C, we would find that

$$(Au, v) = (u, Av)$$

and the expression (Au, u) was positive definite. Thus there is a subspace of functions from the domain of A on which A is a positive definite, symmetric linear operator. Rather than seek out this particular subspace, however, we simply borrow the right-hand side of (1) and seek a subspace

X of Y on which the expression

$$\boxed{(u,v)_A = \int_D\int \left\{ \frac{\partial u}{\partial x}\frac{\partial v}{\partial x} + \frac{\partial u}{\partial y}\frac{\partial v}{\partial y} \right\} dx\,dy} \tag{2}$$

forms an inner product. We saw in Section 7.4 that there is a great advantage to keeping the smoothness requirements in X as low as possible, to enlarge our admissible set of approximating functions. This is even more desirable in the case of several variables, smooth approximates being either unavailable or too complicated computationally. For example, one popular choice of approximating functions X_N is the set of piecewise planes on a triangular grid (see Chapter 5 on the method of plates and Figure 7.8). Such approximating functions are not differentiable across the edges of triangles, and our choice of X must be generous enough to accommodate such a selection of X_N and to admit the validity of Green's identities.

The best choice is the Sobolev space $W_2^1[D]$ (see Section 7.12) where generalized Friedrichs and Poincaré inequalities obtain via the embedding theorem. However, to avoid introducing the functional analysis necessary to solve the problem in this more catholic setting, we proceed more modestly and let

$$X = \left\{ v(x,y): \begin{array}{l} v(x,y)=0 \text{ on } C; \\ v(x,y) \text{ is absolutely continuous in} \\ \text{both } x \text{ and } y \text{ on } D; \text{ and} \\ \partial v/\partial x \text{ and } \partial v/\partial y \text{ belong to } L_2[D]. \end{array} \right\}$$

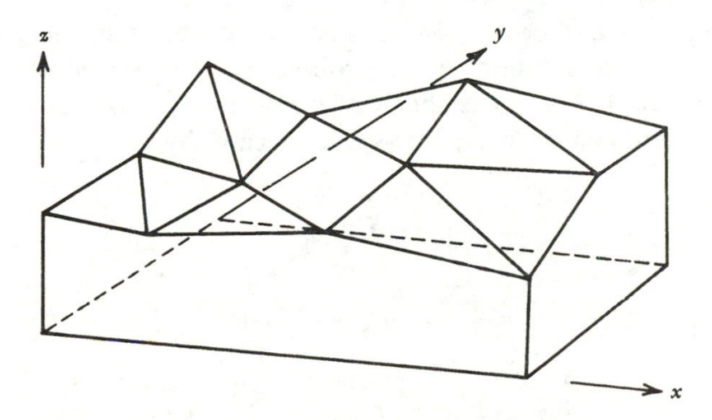

Figure 7.8 Approximate from the method of plates.

Those unfamiliar with the Lebesgue theory can take X to be the set of all functions $v(x,y)$ defined on D and possessing the following properties:

1. $v(x,y)=0$ on C

2. $v(\bar{x},y)=\int_{y_1}^{y}\frac{\partial v}{\partial y}(\bar{x},s)\partial s, \qquad (\bar{x},y_1)\in C$

3. $v(x,\bar{y})=\int_{x_1}^{x}\frac{\partial v}{\partial x}(t,\bar{y})\partial t, \qquad (x_1,\bar{y})\in C$

4. $\int_D\int\left(\frac{\partial v}{\partial x}\right)^2$ and $\int_D\int\left(\frac{\partial v}{\partial y}\right)^2$ exist and are finite.

Note that condition 1 implies $v(\bar{x},y_1)=v(x_1,\bar{y})=0$, whereas conditions 2 and 3 are tantamount to the absolute continuity requirement and necessary to the proof of Green's identities. Condition 4 is a technical one guaranteeing the existence of the double integrals $(\partial u/\partial x)(\partial v/\partial x)$ and $(\partial u/\partial y)(\partial v/\partial y)$ for all u and v in X. It is clear that in either case X *is a linear space*. Moreover, under these restrictions on X we can prove that

$$\boxed{(u,v)_A=\int_D\int\left\{\frac{\partial u}{\partial x}\frac{\partial v}{\partial x}+\frac{\partial u}{\partial y}\frac{\partial v}{\partial y}\right\}dx\,dy}$$

exists and defines an (energy) *inner product* on X (see Exercises) and that the piecewise planes of Figure 7.8 belong to this X.

Letting $X_N=\operatorname{span}\{\phi_1,\phi_2,\dots,\phi_N\}$ be an N-dimensional subspace of X and u be the unique solution to our problem, the closest point u_N to u in the energy norm $\|u-u_N\|_A$ is the Rayleigh–Ritz approximate to u from X_N. This is given, just as in the one-variable case, by

$$u_N=\sum_{j=1}^{N}c_j\phi_j,$$

where $c=(c_1,c_2,\dots,c_N)'$ solves the linear system

$$\sum_{j=1}^{N}(\phi_i,\phi_j)_A c_j=(f,\phi_i), \qquad 1\leqslant i\leqslant N$$

and

$$(\phi_i,\phi_j)_A = \int_D\int \left\{ \frac{\partial\phi_i}{\partial x}\frac{\partial\phi_j}{\partial x} + \frac{\partial\phi_i}{\partial y}\frac{\partial\phi_j}{\partial y} \right\} dx\,dy.$$

If X_N is a subspace of piecewise polynomials in two variables, u_N is said to be a finite element approximate to u and the (Rayleigh–Ritz) method is referred to as the finite element method.

Ideally, we would like to estimate the error $\|u-u_N\|_\infty = \max|u(x,y)-u_N(x,y)|$ on D. This error, although available, is not as easily estimated as the error in the L_2 norm

$$\|u-u_N\|_2 = \left(\int_D\int [u(x,y)-u_N(x,y)]^2 dx\,dy\right)^{1/2}.$$

The size of $\|u-u_N\|_2$ depends on our choice of X_N, on the mesh spacing of our partition of D associated with X_N, and on f. The differential inequality strategic to obtaining this estimate is a special case of Friedrichs's inequality.

FRIEDRICHS' FIRST INEQUALITY

Let $[a,b]\times[c,d]$ be a rectangle containing D. For all $v\in X$

$$\|v\|_2 \leqslant \sqrt{b-a}\,\|v\|_A,$$

where $K=\min\{b-a,c-d\}$.

Proof. Let $R=[a,b]\times[c,d]$, a rectangle containing D (see Figure 7.9). Note that if $v\in X$, $(x,y)\in D$, and $(x,y_1)\in C$, we have

$$v(x,y) = \int_{y_1}^{y} \frac{\partial v}{\partial y}(x,t)\,\partial t.$$

Invoking the Schwarz inequality we have

$$[v(x,y)]^2 \leqslant (d-c)\int_{y_1}^{y}\left[\frac{\partial v}{\partial y}(x,t)\right]^2 dt.$$

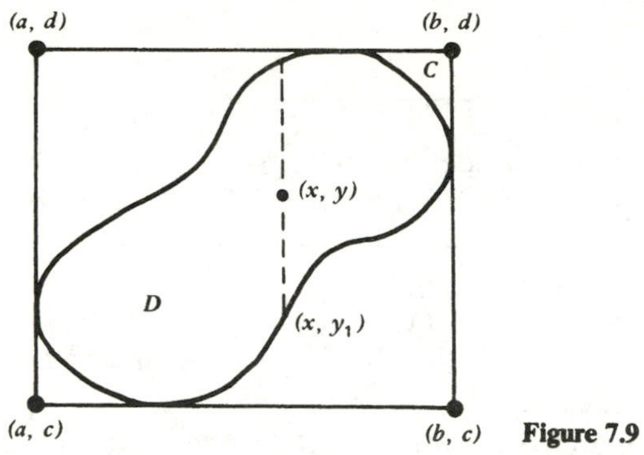

Figure 7.9

Define $\bar{v}(x,y)$ on D by

$$\bar{v}(x,y)=\begin{cases} v(x,y) & \text{if } (x,y)\in D \\ 0 & \text{if } (x,y)\in R \text{ but not to } D. \end{cases}$$

Then we write

$$[\bar{v}(x,y)]^2 \leqslant (d-c)\int_c^d \left[\left(\frac{\partial v}{\partial x}\right)^2+\left(\frac{\partial v}{\partial y}\right)^2\right] dy.$$

Integrating this inequality over the rectangle $[a,b]\times[c,d]$, we find

$$\int_c^d\int_a^b [\bar{v}(x,y)]^2 dx\,dy \leqslant (d-c)^2 \int_c^d\int_a^b \left[\left(\frac{\partial \bar{v}}{\partial x}\right)^2+\left(\frac{\partial \bar{v}}{\partial y}\right)^2\right] dx\,dy.$$

But

$$\int_c^d\int_a^b [\bar{v}(x,y)]^2 dx\,dy = \int_D\int [v(x,y)]^2 dx\,dy$$

and

$$\int_a^b\int_c^d \left[\left(\frac{\partial \bar{v}}{\partial x}\right)^2+\left(\frac{\partial \bar{v}}{\partial y}\right)^2\right] dx\,dy = \int_D\int \left[\left(\frac{\partial v}{\partial x}\right)^2+\left(\frac{\partial v}{\partial y}\right)^2\right] dx\,dy.$$

It follows that

$$\|v\|_2 \leqslant (d-c)\|v\|_A$$

as was to be proved. ■

To see how finite element analysis applies to this problem, we choose a particular X_N and D. For simplicity, suppose D is a region of the plane with piecewise polygonal boundary C composed of a finite number of line segments. Let $\tau = \{T_1, T_2, \ldots, T_N\}$ be a triangulation of D (see Figure 7.10). Let

$$X_N = \{v(x,y) \in X : v(x,y) = a_i x + b_i y + c_i \qquad \text{on triangle } T_i\}.$$

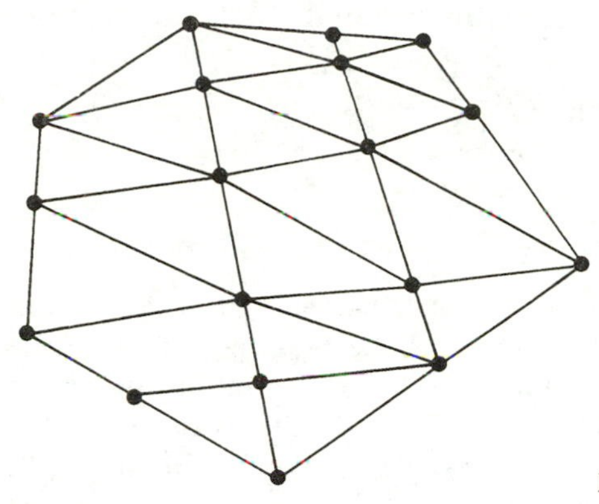

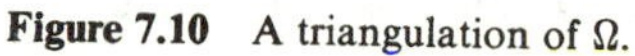

Figure 7.10 A triangulation of Ω.

It is clear that X_N is contained in X and is simply the set of continuous piecewise planes that vanish on C (see the *method of plates* in Chapter 5). Let $u_N(x,y)$ be the finite element (or Rayleigh–Ritz) approximate to the solution $u(x,y)$ from X_N. Then

$$\|u - u_N\|_A = \min_{\bar{u} \in X_N} \|u - \bar{u}\|_A.$$

Let $\bar{u}_N(x,y)$ be the unique member of X_N that interpolates $u(x,y)$ at the vertices of the triangles T_i of T. Since $u \in C^{2,2}[D]$, it follows from the error estimates of Chapter 5 that there are constants M_0 and M_1 such that

$$\|u - \bar{u}_N\|_\infty \leqslant M_0 \frac{h^2}{\sin\theta},$$

and both

$$\left\| \frac{\partial(u - \bar{u}_N)}{\partial x} \right\|_\infty \leqslant M_1 \frac{h}{\sin\theta}$$

and

$$\left\| \frac{\partial(u - \bar{u}_N)}{\partial y} \right\|_\infty \leqslant M_1 \frac{h}{\sin\theta},$$

where h is the mesh of τ (i.e., the length of the longest side among the edges of all triangles of τ) and θ is the smallest angle among the angles of triangles of τ. Thus

$$\|u-\bar{u}_N\|_A \leqslant \beta\frac{h}{\sin\theta},$$

where $\beta=(b-a)(d-c)M$ and $M=\max\{M_1, M_0(h/\sin\theta)\}$. But $\|u-u_N\|_A < \|u-\hat{u}_N\|_A$. Invoking Friedrichs's inequality, we find

$$\|u-u_N\|_2 < K\beta\frac{h}{\sin\theta}.$$

That is, we have proved

THEOREM AN ERROR ESTIMATE

Let $u(x,y)$ be the unique solution to $\Delta u=f$ on D belonging to $C^{2,2}[D]$ and subject to the boundary condition $u=0$ on C. Let $u_N(x,y)$ be the finite element approximate to $u(x,y)$ from the space X_N of continuous piecewise planes on a triangulation $\tau=\{T_1, T_2, \ldots, T_N\}$ of D that vanish on the boundary C of D. If h is the length of the triangle edge of maximum length among the triangles of τ and θ is the smallest angle among triangles of τ, there exists a constant γ independent of τ and of h such that

$$\boxed{\|u-u_N\|_2 \leqslant \gamma\frac{h}{\sin\theta}.}$$

We see in Section 7.9 that $o(h^2)$ convergence rates are available. That is, we can prove

$$\|u-u_N\|_2 \leqslant \gamma_1\frac{h^2}{\sin\theta}.$$

The computation of u_N is itself an interesting problem. Since *the linear system is computed triangle by triangle*, the ordering of the vertices of the triangles of τ is very important. The graph of a typical shape function $\phi_i(p)$ appears in Figure 7.11. Since $\phi_i(p) \in X_N$, $\phi_i(p) \equiv 0$ on C. Moreover this choice of *ϕ_i's guarantees the bandedness of the $n \times n$ matrix $A_N=(a_{ij})$, where* $a_{ij}=(\phi_i, \phi_j)_A$, $1<i,j<n$. That is, $\phi_i\phi_j \neq 0$ only when p_i and p_j are shared vertices of some triangle of τ. The number of nonzero bands of the matrix A_N depends on the ordering of the vertices p_i, hence on the ordering of the functions ϕ_i. This ordering is very important when large-scale computa-

tions are performed, since it can considerable simplify the structure of the matrix A_N.

The reader can easily see that the piecewise plates are not our only choices of X_N. We could have selected the six-point scheme of Chapter 4, using vertices and midpoints of triangles of T, as in Figure 7.12, to obtain order h^2 convergence:

$$\boxed{\|u-u_N\|_2 \leqslant K\frac{h^2}{\sin\theta}}$$

if $u\in C^{3,3}[D]$. If we had used a rectangular grid and $u\in C^{k+1,k+1}[D]$, we could have chosen a subspace of tensor products, splines, or of piecewise

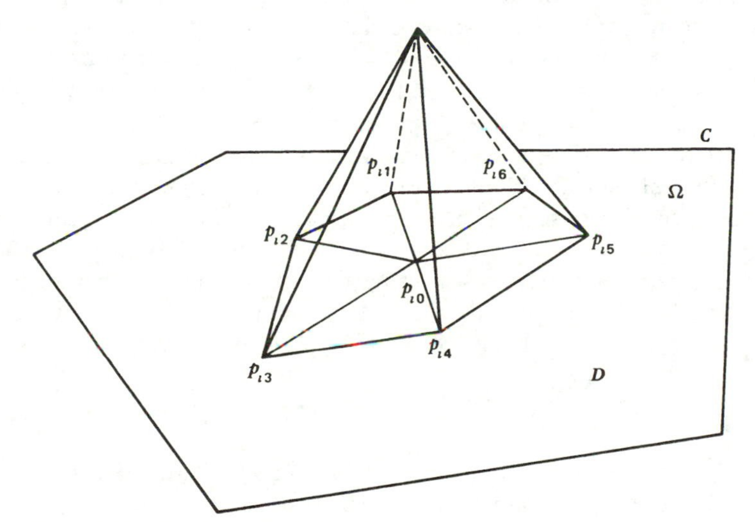

Figure 7.11 Graph of a basis function $\phi_i(x,y)$.

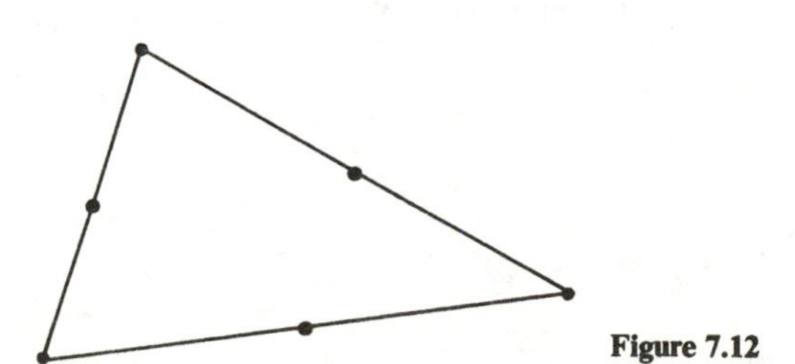

Figure 7.12

polynomials of sufficiently high degree between knots to guarantee

$$\boxed{\|u-u_N\|_2 \leqslant Mh^k}$$

(see Exercises). In all cases, an appropriate choice of basis functions guarantees the bandedness of the matrix $A=(a_{ij})$, where $a_{ij}=(\phi_i,\phi_j)_A$. The number of interior knots dictates the size of the matrix A, and a particular triangularization versus a subdivision into a grid of rectangles or parallelograms together with ordering of vertices determines the *bandwidth of* A.

As an example with a rectangular grid π_n (see Figure 7.13), we could obtain an order h accuracy scheme by simply choosing a basis $\{\phi_1,\phi_2,\ldots,\phi_n\}$ of piecewise bilinear Lagrange polynomials solving $\phi_i(p_j) = \delta_{ij}$, $1 \leqslant i,j \leqslant n$ at the interior knots $\{p_1,p_2,\ldots,p_n\}$ of the grid π_n. Each ϕ_i is a tensor product and vanishes on the boundary of D. The graph of such a $\phi_i(x,y)$ corresponding to an interior knot p_i of our grid appears in Figure 7.14. For computation of such ϕ_i's and a rigorous error analysis, see Chapter 5. The matrix $A_n=(\phi_i,\phi_j)_A$ associated with the finite element method is once again a sparse, banded matrix. That is, $\phi_i\phi_j=0$ unless p_i and p_j are the shared vertices of some rectangle of our grid arising from π. The number of nonzero bands once again depends on our ordering of the vertices p_i (see Exercises). The error estimates of Chapter 5 guarantee at least order h convergence, and coercivity (see Section 7.9) guarantees order h^2 convergence. It is easy to see, using the approaches of Chapter 5, how this method and the method over a triangulation of D can be used to obtain order $h^3,h^4,\ldots,$ schemes, provided $u(x,y)$ is sufficiently smooth.

EXERCISES

Let $D=[0,1]^2$. Let $\pi_N=\{(x_i,y_j):0\leqslant i,j\leqslant n\}$ be the knots of a rectangular grid on D. Let τ_N be a triangulation of D resulting from π_N. Let X_N be the set of piecewise bilinear Lagrange polynomials with knots at π_N that vanish on the boundary of D, and let $\tilde{X}_N$ be the set of piecewise linear Lagrange polynomials over the triangulation τ_N that also vanish on the boundary of D. Consider the Dirichlet problem

$$\begin{cases} \Delta u = f \\ u=0 \text{ on the boundary of } D. \end{cases}$$

1. Find the dimension of X_N and of $\tilde{X}_N$.

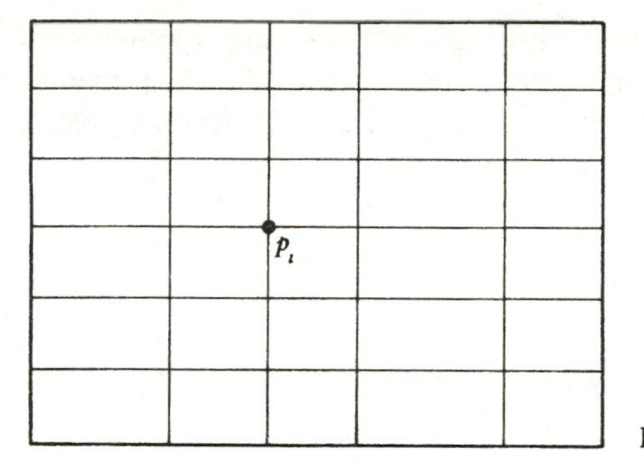

Figure 7.13

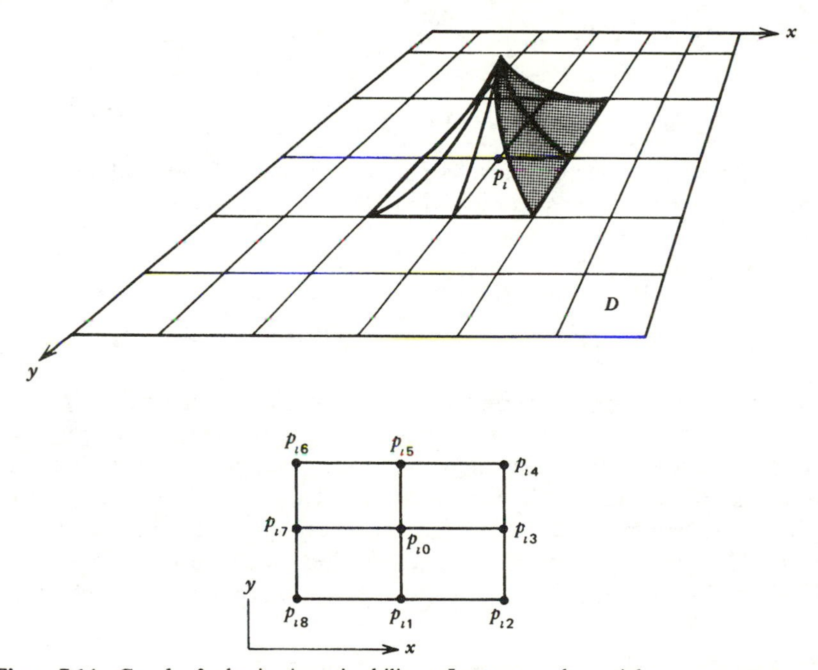

Figure 7.14 Graph of a basis piecewise bilinear Lagrange polynomial on a rectangular grid.

2. Let A_N and $\tilde{A}_N$ be the matrices arising from approximating the solution to the equation using finite elements from X_N and from $\tilde{X}_N$. What is the best way of ordering your basis $\{\phi_i\}$ and $\{\tilde{\phi}_i\}$ for X_N and $\tilde{X}_N$, respectively, to minimize the bandwidths of A_N and $\tilde{A}_N$? Explain. What is the resultant bandwidth of each of these matrices? What is the size of the two matrices? Compare the efficiency of the two methods.

3. Program your computer to obtain the finite element approximates to the solution of the foregoing Dirichlet problem using finite elements from X_N and $\tilde{X}_N$. Let $n=5$, 10, 15 and compute u_N (the finite element approximate) when $f=f_1, f_2$, and f_3, where

$$f_1(x,y)=1$$

$$f_2(x,y)=\begin{cases} 1 & \text{when} \quad x \leqslant y \\ 0 & \text{when} \quad x>y \end{cases}$$

$$f_3(x,y)=\frac{1}{x-y}.$$

Discuss your results.

7.7 THE MIXED BOUNDARY VALUE PROBLEM

The arguments of the previous section generalize to many other partial differential equations. In this section we look at the *mixed boundary value problem*.

$$\begin{aligned} Au=-\Delta u=f \qquad & \text{on } D \\ \sigma u+\frac{\partial u}{\partial n}=0 \qquad & \text{on the boundary } C \text{ of } D, \end{aligned}$$

where D is a closed region bounded by a simple closed curve C; σ, f, and g are given functions, $\sigma>0$, on the boundary; and $\partial u/\partial n$ denotes the outward directed normal derivative of u along C. Suppose this problem has a unique solution $u \in C^{2,2}[D]$ for continuous f and σ. To solve the problem by the finite element method, it is again necessary to require the approximating functions to satisfy the boundary data and to use Green's identities. Following the arguments of Section 7.5, an appropriate choice of X is

$$X=\left\{ u(x,y) \in C[D]: \begin{array}{l} \left(\dfrac{\partial u}{\partial x}\right)^2 \text{ and } \left(\dfrac{\partial u}{\partial y}\right)^2 \text{ are integrable on } D, \\ u(x,y) \text{ is absolutely continuous} \\ \text{in both } x \text{ and } y \text{ on } D, \\ \sigma u+\dfrac{\partial u}{\partial n}=0 \quad \text{on } C. \end{array} \right\}$$

These conditions are equivalent to conditions 2, 3, and 4 of the previous section, where u no longer vanishes on the boundary. If $u(x,y)$ and $v(x,y)$

both satisfy the boundary conditions $\partial u/\partial n+\sigma u=\partial y/\partial n+\sigma v=0$, we have

$$(-\Delta u,v)=\iint_D\left(\frac{\partial u}{\partial x}\frac{\partial v}{\partial x}+\frac{\partial u}{\partial y}\frac{\partial v}{\partial y}\right)-\int_C v\frac{du}{dn},$$

after integrating formally by parts and invoking Green's formulas. But $\sigma u=-\partial u/\partial n$ by assumption. Thus

$$(-\Delta u,v)=\iint_D\left(\frac{\partial u}{\partial x}\frac{\partial v}{\partial x}+\frac{\partial u}{\partial y}\frac{\partial v}{\partial y}\right)+\int_C \sigma uv,$$

where the integral on the far right is a line integral along C. Moreover, it follows by the same formal reasoning that

$$(-\Delta u,v)=(u,-\Delta v).$$

This line of argument leads us to define an *inner product* $(\ ,\ )_A$ *on* X by

$$\boxed{(u,v)_A=\iint_D \nabla u\cdot\nabla v+\int_C \sigma uv,}$$

where

$$\nabla u\cdot\nabla v=\frac{\partial u}{\partial x}\frac{\partial v}{\partial x}+\frac{\partial u}{\partial y}\frac{\partial v}{\partial y}=\operatorname{grad} u\cdot\operatorname{grad} v.$$

The simple proof that $(\ ,\ )_A$ is indeed an inner product on X is left as an exercise for the student. Letting $X_N=\operatorname{span}\{\phi_1,\phi_2,\ldots,\phi_N\}$ be an N-dimensional subspace of X, the finite element approximate u_N to the unique solution u to our mixed boundary value problem is given by

$$u_N=c_1\phi_1+c_2\phi_2+\cdots+c_N\phi_N,$$

where as usual $c=(c_1,c_2,\ldots,c_N)'$ satisfies the linear system

$$\sum_{j=1}^{N}(\phi_i,\phi_j)_A c_j=(f,\phi_i)$$

with

$$(f,\phi_i)=\iint_D f\phi_i$$

and

$$(\phi_i,\phi_j)_A=\iint_D \nabla\phi_i\cdot\nabla\phi_j+\int_C \sigma\phi_i\phi_j.$$

Thus in computing u_N one must evaluate the line integrals $\int_C \sigma\phi_i\phi_j$ along C as well as the double integrals $\iint_D \nabla\phi_i \cdot \nabla\phi_j$.

To estimate the *error* $\|u-u_N\|_2$, we need another inequality due to Friedrich [20].

FRIEDRICHS'S SECOND INEQUALITY

For each continuous $\sigma>0$ on C and each $u\in X$,

$$\iint_D u^2 \leqslant \gamma\left\{\iint_D \nabla u\cdot\nabla u + \int_C u^2\right\},$$

where γ is a constant independent of u but dependent on σ.

Proof. Let $u\in X$, and for simplicity suppose D can be properly enclosed in a rectangle $R=[0,a]\times[0,b]$. Define $u(x,y)=0$ for all (x,y) in R but not in D and let $u=fv$, where f is a function to be defined presently. Noting that

$$\begin{aligned}\nabla u\cdot\nabla u &= \left(\frac{\partial u}{\partial x}\right)^2+\left(\frac{\partial u}{\partial y}\right)^2\\ &= f^2[\nabla v\cdot\nabla v]-v^2f\Delta f+\frac{\partial}{\partial x}\left[v^2f\frac{\partial f}{\partial x}\right]+\frac{\partial}{\partial y}\left[v^2f\frac{\partial f}{\partial y}\right],\end{aligned}$$

we see that by omitting the first term on the right we have

$$\nabla u\cdot\nabla u \geqslant -v^2f\Delta f+\frac{\partial}{\partial x}\left[v^2f\frac{\partial f}{\partial x}\right]+\frac{\partial}{\partial y}\left[v^2f\frac{\partial f}{\partial y}\right].$$

Integrating over D and formally invoking Green's identities, we find

$$\iint_D \nabla u\cdot\nabla u \geqslant -\iint_D v^2f\Delta f+\int_C v^2f\frac{\partial f}{\partial n}\,ds.$$

Thus

$$-\iint_D v^2f\Delta f \leqslant \iint_D \nabla u\cdot\nabla u-\int_C v^2f\frac{\partial f}{\partial n}.$$

It follows that

$$-\iint_D v^2 f \Delta f \leqslant \iint_D \nabla u \cdot \nabla u + \left| \int_C v^2 f \frac{\partial f}{\partial n} \right|.$$

Let $f = \sin(\pi x/a)\sin(\pi y/b)$. Then $\Delta f = -\pi^2[1/a^2 + 1/b^2]f$, and f vanishes nowhere on D. Substituting into the last inequality, the integral on the left becomes

$$\pi^2\left[\frac{1}{a^2} + \frac{1}{b^2}\right]\iint_D u^2.$$

On the other hand, the integral on the right can be estimated as

$$\left|\int_C v^2 f \frac{\partial f}{\partial n}\right| = \left|\int_C u^2 \frac{1}{f}\frac{\partial f}{\partial n}\right| \leqslant \int_C u^2 \frac{1}{f}\left|\frac{\partial f}{\partial n}\right|.$$

Along C the function $(1/f)|\partial f/\partial n|$ is obviously bounded. Let $0 \leqslant (1/f)|\partial f/\partial n| \leqslant d$ on C. Then

$$\left|\int_C v^2 f \frac{\partial f}{\partial n}\right| \leqslant d\int_C u^2.$$

Letting γ be the larger of the numbers $\pi^{-2}[1/a^2 + 1/b^2]^{-1}$ and $d\pi^{-2}[1/a^2 + 1/b^2]^{-1}$, Friedrichs's inequality now follows. ∎

To obtain error estimates and, consequently, the convergence of the Rayleigh-Ritz method in the case of the mixed boundary value problem, we need the following immediate consequence of Friedrichs's inequality.

COROLLARY

For all $u \in X$,

$$\boxed{\|u\|_2 \leqslant \beta \|u\|_A,}$$

where β is a constant independent of u.

Proof. Recall that $[\|u\|_2]^2 = \iint_D u^2$. From Friedrichs's second inequality

$$\frac{1}{\gamma}[\|u\|_2]^2 \leqslant \iint_D \nabla u \cdot \nabla u + \int_C u^2.$$

Let $\sigma_0 = \min \sigma$ on C. Since $\sigma > 0$ on C, $\sigma_0 > 0$. Thus

$$\sigma_0 \int_C u^2 \leqslant \int_C \sigma u^2.$$

Letting $\sigma_1 = \min(\sigma_0, 1)$ we have

$$\sigma_1 \left[\iint_D \nabla u \cdot \nabla u + \int_C u^2 \right] \leqslant (u,u)_A = [\|u\|_A]^2.$$

But invoking Friedrichs's first inequality we find

$$\|u\|_2 \leqslant \beta \|u\|_A,$$

where $\beta = (\gamma/\sigma_1)^{1/2}$. ■

This corollary now leads us in the usual way to error estimates for $\|u - u_N\|_2$. For example, if X_N is the set of piecewise planes corresponding to some triangulation π of D with mesh spacing h, it follows easily (see Exercises) that for a polygonally bounded domain D,

$$\|u - u_N\|_A = o(h)$$

and

$$\boxed{\|u - u_N\|_2 = o(h).}$$

Using a six-point scheme corresponding to a triangulation of D together with midpoints of edges, it follows as in the previous section that

$$\boxed{\|u - u_N\|_2 = o(h^2)}$$

if $u \in C^{3,3}[D]$.

Just as in the case of ordinary differential equations, if f is of smoothness class $C^{k,k}[D]$ and u belongs to $C^{k+2,k+2}[D]$, with an appropriate choice of X_N, we can anticipate order h^k convergence of u_N to u in the L_2

norm. Also, by employing classical Péron-Frobenius type arguments used in finite difference methods, one can prove order h^k convergence in the Tchebycheff norm $\|u-u_N\|_\infty$. Such arguments are frequently unavailable for more complicated second order operators. This phenomena in fact, accounts in part for the preference of finite element methods over finite difference methods. Namely, convergence is much more easily handled using finite elements, as are more complicated domains.

7.8 THE NEUMANN PROBLEM

In this section we look briefly at another well-known partial differential equation, namely, the *Neumann problem*. We consider

$$\begin{cases} -\Delta u = f & \text{on } D \\ \dfrac{du}{dn} = g & \text{on the boundary } C \text{ of } D, \end{cases}$$

where D is, again, a closed region of the plane bounded by a simple closed curve C, and f and g are given functions. This problem is considerably more complicated than either the Dirichlet problem or the mixed boundary value problem. This is partly because the problem does *not* have a unique solution even on the nicest of domains D. In particular, note that if $u_0(x,y)$ is a solution of the problem, $u_0+\gamma$ is also a solution where γ is any constant. Thus the problem has infinitely many solutions. A standard means of overcoming this difficulty is to require

$$\iint_D f = 0$$

and to seek a particular solution u to the problem subject to the constraint

$$\iint_D u = 0.$$

Under these two conditions, one can prove that the problem has a unique solution on sufficiently regular domains D.

With this in mind one can view the problem as that of finding a solution u or an approximate solution u_N subject to the constraint $\iint_D u = 0$. Integrating formally by parts, we find

$$(-\Delta u, v) = \iint_D \nabla u \cdot \nabla v$$

if $\partial u/\partial n=\partial v/\partial n=0$ on C. Following Section 7.4 we can prove that if

$$X=\left\{u(x,y)\in C[D]:\ \begin{array}{l}\iint_D u=0,\\ \left(\frac{\partial u}{\partial x}\right)^2 \text{ and } \left(\frac{\partial u}{\partial y}\right)^2 \text{ are integrable on } D,\\ u(x,y) \text{ is absolutely continuous}\\ \text{in each variable on } D,\\ \frac{du}{dn}=0 \quad \text{on } C,\end{array}\right\}$$

then

$$(u,v)_A=\iint_D \nabla u\cdot\nabla v$$

defines an inner product on X. The absolute continuity requirement, as we have seen, reduces to forcing

$$u(x,y_2)-u(x,y_1)=\int_{y_1}^{y_2}\frac{\partial u}{\partial y}(x,s)\partial s$$

and

$$u(x_2,y)-u(x_1,y)=\int_{x_1}^{x_2}\frac{\partial u}{\partial x}(t,y)\partial t$$

for all (x,y_i) and (x_i,y), $i=1,2$, in D. In this case we must choose our approximating function $u_N=c_1\phi_1+c_2\phi_2+\cdots+c_N\phi_N$ from a subspace X_N of X whose members $\bar{u}$ would automatically be subject to the orthogonality constraint $\iint_D\bar{u}=0$ (i.e., $\bar{u}$ is orthogonal to 1).

As a specific example, suppose D is a rectangle $[a,b]\times[c,d]$ and let $\pi_x: a=x_0<x_1<\cdots<x_n=b$ and $\pi_y: c=y_0<y_1<\cdots<y_m=b$ be partitions of $[a,b]$ and $[c,d]$, respectively. For simplicity, assume $x_i=x_0+ih_1$, $1\leqslant i\leqslant n$, and $y_j=y_0+jh_2$, $1\leqslant j\leqslant m$; this means that the mesh spacing is even in both the x and y directions. Let $h=\max\{h_1,h_2\}$ and let $\tilde{\psi}_i(x)$ be the piecewise Lagrange polynomial of degree 1 with knots at points of π_x shown in Figure 7.15. It is clear that since

$$\int_a^b\tilde{\psi}_i(x)dx=0 \qquad \text{when} \quad 1\leqslant i\leqslant n-3,$$

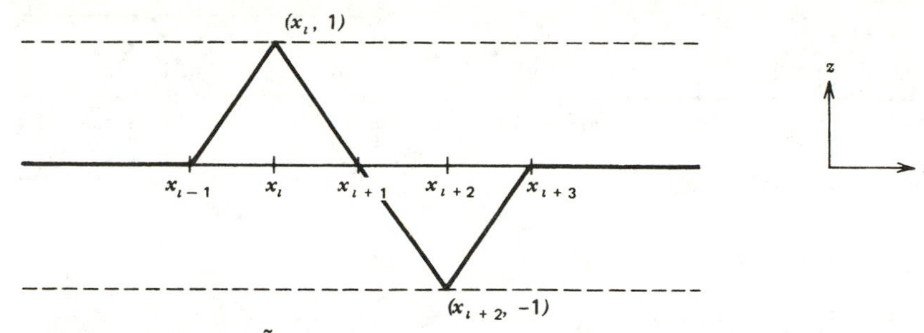

Figure 7.15 Graph of $\tilde{\psi}_i(x)$.

these particular $\tilde{\psi}_i$'s are orthogonal to 1. To obtain one-dimensional splines orthogonal to 1 corresponding to the other knots x_0, x_{n-2}, x_{n-1}, and x_n, let

$$\psi_0(x) = \tilde{\psi}_0(x) + \tilde{\psi}_n(x)$$

$$\psi_{n-2}(x) = \tilde{\psi}_{n-2}(x) + \tilde{\psi}_{-2}(x)$$

$$\psi_{n-1}(x) = \tilde{\psi}_{n-1}(x) + -\tilde{\psi}_{-1}(x)$$

and

$$\psi_n(x) = \tilde{\psi}_n(x) + \tilde{\psi}_{-2}(x),$$

whose graphs appear in Figure 7.16. Here we have introduced additional knots $x_{-1} = x_0 - h$, and $x_{-2} = x_0 - 2h$. Moreover, it is clear from the graphs that with even mesh spacing, if we let

$$\psi_i(x) = \tilde{\psi}_i(x), \qquad 1 \leqslant i \leqslant n-3,$$

each of the functions $\psi_i(x)$, $0 \leqslant i \leqslant n$, is orthogonal to 1, giving

$$\int_0^1 \psi_i(x)\,dx = 0, \qquad 0 \leqslant i \leqslant n.$$

Let $\psi_j(y)$ be the analogous splines in the y direction so that

$$\int_c^d \psi_j(y)\,dy = 0, \qquad 0 \leqslant j \leqslant m.$$

It follows that

$$\iint_D \psi_i(x)\psi_j(y)\,dx\,dy = \int_a^b \psi_i(x)\,dx \cdot \int_c^d \psi_j(y)\,dy = 0,$$

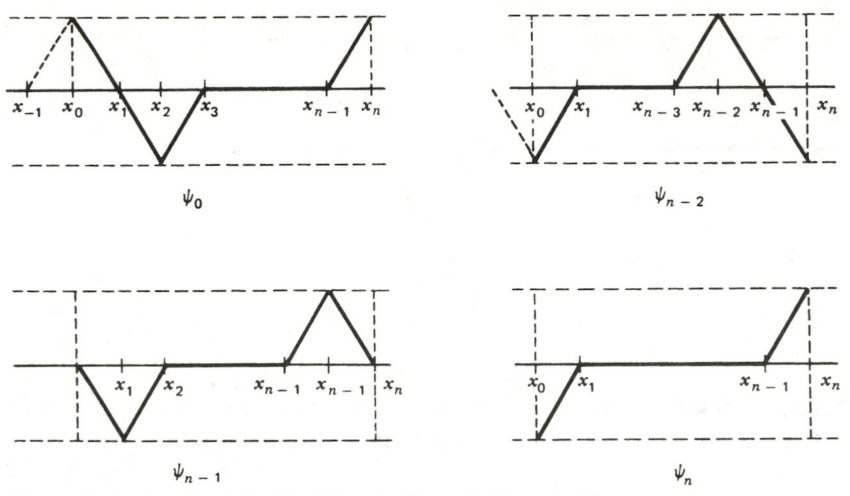

Figure 7.16 Graphs of the ψ_i's: (*a*) $i=0$, (*b*) $n-2$, (*c*) $n-1$, (*d*) n.

and therefore the two-dimensional splines

$$\phi_{ij}(x,y)=\psi_i(x)\psi_j(y), \qquad 0\leqslant i\leqslant n;\ 0\leqslant j\leqslant m$$

span an N-dimensional subspace X_N, $N=m\cdot n$, of X, each of whose members is orthogonal to 1. Moreover, we can prove that the ϕ_{ij}'s are independent on D and span the set of piecewise Lagrange polynomials of degree 1 in each of the variables x and y. As such, we know that if $u\in C^{2,2}[D]$ is the unique solution to our Neumann problem and $\bar{u}_N$ is the unique spline interpolate to u from X_N solving

$$\bar{u}_N(x_i,y_j)=u(x_i,y_j), \qquad 0\leqslant i\leqslant n;\ 0\leqslant j\leqslant m,$$

then, following our error estimates of Chapter 5,

$$\|u-\bar{u}_N\|_\infty=o(h^2).$$

To obtain an estimate of the error $\|u-u_N\|_2$, where u_N is the finite element approximate to u from X_N, we must obtain a Friedrichs-type inequality for the Neumann problem. The inequality in this case, due initially to Poincaré, is known as *Poincaré's inequality*.

POINCARÉ'S INEQUALITY

For all $u \in X$

$$\|u\|_2 \leqslant K\|u\|_A,$$

where K is a constant independent of u.

Proof. We prove this inequality on a rectangular domain $D=[0,a]\times[0,b]$. A proof in the more general case can be found in the book of Sobolev [34]. Let (x_1,y_1) and (x_2,y_2) belong to D. Then

$$u(x_2,y_2)-u(x_1,y_1)=\int_{x_1}^{x_2}\frac{\partial u(x,y_1)}{\partial x}dx+\int_{y_1}^{y_2}\frac{\partial u(x_2,y)}{\partial y}dy.$$

Squaring and invoking Cauchy's inequality, we obtain

$$u^2(x_1,y_1)+u^2(x_2,y_2)-2u(x_1,y_1)u(x_2,y_2)$$

$$\leqslant\left\{\left(2\int_{x_1}^{x_2}\frac{\partial u(x,y_1)}{\partial x}dx\right)^2+\left(\int_{y_1}^{y_2}\frac{\partial u(x_2,y)}{\partial y}dy\right)^2\right\}$$

$$\leqslant 2\left\{|x_2-x_1|\int_0^a\left(\frac{\partial u(x,y_1)}{\partial x}\right)^2dx+|y_2-y_1|\int_0^b\left[\frac{\partial u(x_2,y)}{\partial y}\right]^2dy\right\}$$

$$\leqslant 2\left\{a\int_0^a\left(\frac{\partial u(x,y_1)}{\partial x}\right)^2dx+b\int_0^b\left(\frac{\partial u(x_2,y)}{\partial y}\right)^2dy\right\}.$$

Now integrate this inequality with respect to each of the variables

x_1,y_1,x_2,y_2 over $0 \leqslant x_1,x_2 \leqslant a$ and $0 \leqslant y_1,y_2 \leqslant b$. We then obtain

$$2ab \iint_D u^2 - 2\left[\iint_D u\right]^2$$

$$\leqslant 2ab\left\{a^2 \iint_D \left(\frac{\partial u}{\partial x}\right)^2 + b^2 \iint_D \left(\frac{\partial u}{\partial y}\right)^2\right\}.$$

Dividing by $2ab$, letting

$$A = \max(a^2, b^2) \qquad \text{and} \qquad B = \frac{1}{ab},$$

and recalling that $\iint_D u = 0$, we arrive at Poincaré's inequality. ■

Following this error analysis to completion for our special choice of X_N and smoothness class to which u belongs, it follows in the usual way that

$$\|u - u_N\|_A \leqslant \|u - \bar{u}_N\|_A = o(h).$$

Since $\|u - u_N\|_2 \leqslant K\|u - u_N\|_A$, we see that there exists a constant γ independent of N such that

$$\boxed{\|u - u_N\|_2 \leqslant \gamma h,}$$

where u_N is the finite element approximate to u from X_N.

Higher order convergence can be obtained using subspaces X_N of higher degree piecewise polynomials in (x,y) when u is sufficiently smooth. For example, with products of piecewise quadratics in a rectangular grid, we can anticipate accuracy

$$\boxed{\|u - u_N\|_2 = o(h^2)},$$

when $u \in C^{3,3}[D]$, and so forth. However, each X_N must be spanned by a set of basis functions ϕ_i which are orthogonal to 1. Moreover, it is highly desirable to preserve the sparseness or bandedness of the matrix

$$A_N = ((\phi_i, \phi_j)_A),$$

as we were careful to do in our choice of X_N.

7.9 COERCIVENESS AND RATES OF CONVERGENCE

It has been observed by Nitsche (1968) and others that there exists a large class of differential operators for which the rates of convergence of the finite element approximates u_N to the solution u of a given boundary value problem are even better than indicated in Sections 7.3 through 7.6. These are the differential operators that satisfy a so-called *strong coerciveness inequality*. To see how this applies in a particular case, let us once again consider the Dirichlet problem

$$-\Delta u = f$$

$$u = 0 \qquad \text{on the boundary of } \Omega, \tag{1}$$

where Ω is a closed, bounded, connected region of the real plane. In particular, suppose we know that Δ is *strongly coercive* in that for all continuous f and all solutions u to $-\Delta u = f$, there exists a constant Γ such that

$$\|D_x^k D_y^j u\|_2 \leqslant \Gamma \|f\|_2, \qquad 0 \leqslant k + j \leqslant 2. \tag{2}$$

Let u_N be the finite element approximate to u from an N-dimensional subspace $X_N = \operatorname{span}\{l_1, l_2, \ldots, l_N\}$ of X consisting of piecewise linear Lagrange polynomials over some triangulation τ of Ω (i.e., the method of plates). Let

$$\psi = \frac{u - u_N}{\|u - u_N\|_2}$$

and let

$$-\Delta\phi = \psi.$$

If Δ is coercive, then

$$\|\Delta\phi\|_2 \leqslant (\Delta\phi, u - u_N).$$

Moreover

$$\begin{aligned}(\phi, u - u_N)_\Delta &= (\Delta\phi, u - u_N)\\ &= (\psi, u - u_N)\\ &= \|u - u_N\|_2.\end{aligned} \tag{3}$$

Let $\bar{\phi}_N$ be the unique piecewise linear interpolate to u from X_N. That is, $\bar{\phi}_N(p_i)=\phi(p_i)$ at the vertices of all triangles of τ. Then (see Section 5.4) we have

$$\|\phi-\bar{\phi}_N\|_\infty \leqslant M_2h^2,$$

where

$$M_2=\frac{1}{4}\max\left\{\|D_x^2\phi\|_\infty, 2\|D_{xy}^2\phi\|_\infty, \|D_y^2\phi\|_\infty\right\},$$

whereas

$$\|D_x(\phi-\bar{\phi}_N)\|_\infty \leqslant M_2h$$

and

$$\|D_y(\phi-\bar{\phi}_N)\|_\infty \leqslant M_2h,$$

where

$$\|g\|_\infty = \operatorname*{ess\,sup}_{p\in\Omega}|g(p)|.$$

But if Δ is strongly coercive, it follows from (2) that

$$M_2 \leqslant 2\Gamma.$$

Thus

$$\begin{cases} \|D_x(\phi-\bar{\phi}_N)\|_\infty \leqslant 2\Gamma h \\ \|D_y(\phi-\bar{\phi}_N)\|_\infty \leqslant 2\Gamma h. \end{cases} \tag{4}$$

Next note that since

$$\|u-u_N\|_\Delta = \min_{\tilde{u}\in X_N}\|u-\tilde{u}\|_\Delta,$$

$u-u_N$ is orthogonal to X_N in the energy inner product. That is,

$$(u-u_N, l_i)_\Delta = \int_\Omega [D_x(u-u_N)D_xl_i + D_y(u-u_N)D_yl_i] = 0 \tag{5}$$

for all $1 \leqslant i \leqslant N$. Thus

$$(u-u_N, \bar{\phi}_N)_\Delta = 0. \tag{6}$$

We know this immediately because the least squares fit u_N to a given function u in a Hilbert space X from an N-dimensional subspace X_N of X is that unique vector (function) for which $u-u_N$ is perpendicular to X_N (see Figure 7.17). In this case our inner product is the energy inner product $(\ ,\ )_\Delta$. Combining (5) and (6) with (3) and invoking the Schwarz, inequality, we have

$$\|u-u_N\|_2=(\phi,u-u_N)$$

$$=\int_\Omega\left[D_x(\phi-\bar{\phi}_N)D_x(u-u_N)+D_y(\phi-\bar{\phi}_N)D_y(u-u_N)\right]$$

$$\leqslant\|D_x(\phi-\phi_N)\|_2\|D_x(u-u_N)\|_2+\|D_y(\phi-\phi_N)\|_2\|D_y(u-u_N)\|_2. \tag{7}$$

But

$$\begin{cases}\|D_x(u-u_N)\|\leqslant M_2h\\ \|D_y(u-u_N)\|\leqslant M_2h.\end{cases} \tag{8}$$

Combining (7), (8), and (4), we find

$$\boxed{\|u-u_N\|_2\leqslant 4M_2\Gamma h^2.}$$

Thus the order of approximation of u_N to u in the L_2 norm is the so-called order of best approximation. This is to be compared with the estimate $\|u-u_N\|_2\leqslant\gamma h$ given in Section 7.6.

The argument easily generalizes to subspaces other than piecewise linear

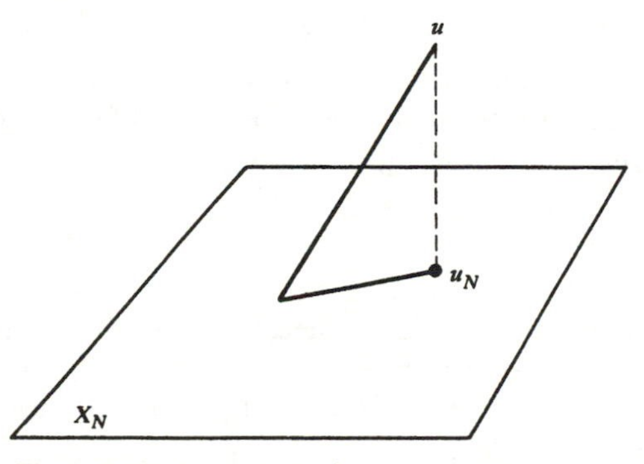

Figure 7.17 $(u-u_n,u_n)=0$.

Lagrange polynomials. We simply observe that

$$\|D^{2+k}\phi\|_\infty \leqslant \frac{2^k}{h^k}\|D^k\phi\|_\infty.$$

We can also extend this work to more general elliptic operators (see Exercises). As a rule, all one needs is strong coerciveness.

DEFINITION: STRONG COERCIVENESS

A second-order partial differential operator A is said to be *strongly coercive* on $X = X(\Omega)$ if and only if for all continuous f and all solutions u to $Au = f$, there exists a constant Γ such that

$$\|D_x^k D_y^j u\|_2 \leqslant \Gamma\|f\|_2, \qquad 0 \leqslant k+j \leqslant 2.$$

One theorem guaranteeing such coerciveness is the following, due to Birman and Skvortsov (1962).

THEOREM

Let $Au = -D_x[pu_x] - D_y[qu_y] + cu$. If $p, q \in C^1(\Omega)$, $c \in C(\Omega)$, and if there exist constants γ and μ such that

$$\gamma\|u\|_\Delta \leqslant \|u\|_A \leqslant \mu\|u\|_\Delta,$$

where

$$\|u\|_A^2 = \int_\Omega \left[p(u_x)^2 + q(u_y)^2 + cu^2\right],$$

then u is strongly coercive.

In particular, we see that Δ is strongly coercive. We leave it to the reader to generalize these ideas to higher order partial differential equations and to *ordinary differential equations* (see Exercises).

EXERCISES

1. (a) Define coerciveness for a general self-adjoint ordinary differential operator $Au = -D_x[pu_x] + cu$.
   (b) Prove a set of sufficient conditions on p and c for A to be coercive.
2. Let the ordinary differential operator A given below be strongly

coercive. Let u_N be a finite element approximate to the solution u of the boundary value problem

$$Au=-D_x[pu]+cu=f$$

$$u(a)=u(b)=0,$$

where A is strongly coercive. Estimate $\|u-u_N\|_2$ and $\|u-u_N\|_\infty$. Prove your estimates.

3. Let $Au=-D_x[pu_x]-D_y[qu_y]+cu$ be a coercive second-order partial differential operator, where $p,q\in C(\Omega)$. Define $(u,v)_A=\int_\Omega[p(u_x)^2+q(u_y)^2+cu^2]$, and suppose there exist constants μ and γ such that

$$\mu\|u\|_\Delta\leqslant\sqrt{(u,u)}_{\,A}\leqslant\gamma\|u\|_\Delta,$$

where $u=u_{xx}+u_{yy}$. Let u_N be the piecewise quadratic Lagrange polynomial finite element approximate to the solution u of the boundary value problem

$$\begin{cases} Au=f & \\ u=0 & \text{on the boundary of } \Omega. \end{cases}$$

Prove that $\|u-u_N\|_2\leqslant\beta h^3$, where β is a constant independent of the mesh of your triangulation of Ω. Find β specifically.

7.10 CURVED BOUNDARIES AND NONCONFORMING ELEMENTS

Our approximations to solutions of partial differential equations have been confined to second-order problems on a rectangle. If the domain Ω of the partial differential operator has either *polygonal* or *curved boundaries* that are more general, the approximation schemes must be revised to accommodate such boundaries. Polygonal boundaries over a domain that can be properly triangulated using any automatic mesh generation schemes present no approximation–theoretic difficulties for second-order problems using piecewise Lagrange polynomials over the triangulation. The problem of curved boundaries is more difficult. If there is a simple *isoparametric transform* T (see Chapter 5) of the unit square $\mathbb{S}$ onto Ω, there is again no difficulty, provided the Jacobian of T remains bounded uniformly as the mesh is refined. In this case the matrix for the resultant linear system is simply the finite element matrix over the rectangle $\mathbb{S}$ multiplied by the Jacobian matrix (see Exercises and Chapter 5). If there is no simple isoparametric scheme at hand, one usually approximates the domain using a triangulation or a simple isoparametric transform onto an approximation $\tilde{\Omega}$ of Ω. This represents the replacement of the original boundary value problem by another approximating boundary value problem; and this modification must be theoretically accounted for to assure convergence.

A second and equally serious difficulty arises when we have a third, fourth, or higher order problem to solve over a two- or three-dimensional domain. It is here that we feel the lack of simple, continuously differentiable approximating functions (splines) over triangular grids. Not having such smooth two-dimensional splines at hand, the engineer frequently takes recourse to continuous splines over a triangular or rectangular grid (such as piecewise Lagrange polynomials), but these fail to be even once continuously differentiable. Meanwhile, finite element schemes applied to partial differential equations of order $2k$ appear to require approximating functions that are $k-1$ times continuously differentiable. The reason for this lies in the definition of the bilinear form $a(u,v)$ giving rise to the finite element approximate to the given differential equation. For example, consider a fourth-order problem such as the biharmonic equation whose strain energy involves the bilinear form

$$a(u,v)=(\Delta u,\Delta v)=\int_{\Omega}\Delta u\Delta v,$$

where Ω is a closed, bounded, connected region of the real plane. If ϕ_i and ϕ_j were, say, basis piecewise linear Lagrange polynomials over some triangulation of Ω,

$$a(\phi_i,\phi_j)=\int_{\Omega}\Delta\phi_i\Delta\phi_j$$

would be infinite (see Exercises). On the other hand,

$$\tilde{a}(\phi_i,\phi_j)=\sum_{k=1}^{N}\int_{T_k}\Delta\phi_i\Delta\phi_j$$

is finite and well defined. When can we replace $a(\ ,\ )$ by $\tilde{a}(\ ,\ )$ and still get a reasonable approximation theory for our boundary value problem? This is the problem of *nonconforming elements*. Cases of such nonconforming elements that work and do not work have been investigated experimentally by Irons and others. The work of Irons (see Bazeley, et. al. (1965)) suggested a theory further developed by Strang (1973). Irons conjectured that a *patch test* to determine whether the modified finite element procedure using $\tilde{a}(\ ,\ )$ reproduces polynomial solutions of a given degree would decide the convergence of the modified method. Failure of convergence does not, however, mean that it is impossible to obtain fairly good coarse initial approximates using nonconforming elements.

EXERCISES

1. Let T be an isoparametric map of the unit square $\mathbb{S}$ onto a closed, bounded, connected region Ω of the real plane. Set up the equations for the finite element approximate u_N to the Dirichlet problem on Ω

$$\begin{cases} \Delta u = f \\ u = 0 \qquad \text{on } \partial\Omega \end{cases}$$

using bilinear Lagrange polynomials on $\mathbb{S}$. Write out the matrix and estimate the error $\|u-u_N\|_2$.

2. Let $\{\phi_i : 1 \leqslant i \leqslant N\}$ be basis piecewise linear Lagrange polynomials over some triangulation τ of a region Ω of the real plane. Compute

$$\int_\Omega \Delta\phi_i \, \Delta\phi_j.$$

Explain your answer.

7.11 HIGHER ORDER LINEAR ORDINARY DIFFERENTIAL EQUATIONS

The methods of the previous sections generalize to a large class of both ordinary and partial differential equations. Since our arguments so far have depended rather heavily on the *symmetry* (or *self-adjointness*) and the positive definiteness of our differential operator, it is only natural to attempt to find a large, but simply defined class of both ordinary and partial differential operators possessing these two properties (see the section 7.13 on *Galerkin methods* for the extension of the finite element method to nonsymmetric operator equations).

In the case of ordinary differential equations the natural candidate is the set of all $[2m]$th-order, linear differential operators L of the form

$$Lu(t) = \sum_{j=0}^{m} (-1)^{j+1} D^j\big[p_j(t) D^j u(t)\big], \qquad a < t < b, \tag{1}$$

where $D^j = d^j/dt^j$, the functions $p_j(t) \in C^j[a,b]$, $0 \leqslant j \leqslant n$, are given, and $p_n(t) \geqslant w > 0$ for all t in $[a,b]$ for some constant w. To discuss symmetry of such an L we need a linear space X in which to imbed the problem and an inner product on X. The choice of X is in part dictated by the solution to the homogeneous boundary value problem associated with L and will always be a linear subspace of $L_2[a,b]$. Given a function $f(t)$ continuous on

$[a,b]$, the homogeneous boundary value problem associated with L is that of finding a solution $u(t)$ to the equation

$$Lu(t)=f(t)$$

satisfying the boundary values

$$D^k u(a)=D^k u(b)=0, \qquad 0\leqslant k\leqslant n-1. \tag{2}$$

The theory of ordinary differential equations tells us when such a problem has a unique solution $u(t)$ and that $u(t)\in C^{2n}[a,b]$. In fact, if $f\in C^k[a,b]$ for some integer $k\geqslant 0$, then $u(t)\in C^{2n+k}[a,b]$. If we let u and v be any two functions satisfying our homogeneous boundary values, we obtain, after formal integration by parts, the expression

$$(Lu,v)=\int_a^b Lu(t)v(t)\,dt=\sum_{j=0}^{m}\int_a^b [p_j(t)D^j u(t)]D^j v(t)\,dt.$$

The right-hand side of this equation defines our energy inner product

$$\boxed{(u,v)_L=\sum_{j=0}^{m}\int_a^b [p_j(t)D^j u(t)]D^j v(t)\,dt.} \tag{3}$$

To ensure that this expression is well defined in as large a space as possible, we can let

$$X=\left\{u\in C^{m-1}[a,b]:\ \begin{array}{l} D^m u\in L_2[a,b], D^{m-1}u \text{ is absolutely} \\ \text{continuous on } [a,b], \text{ and } u \text{ satisfies} \\ \text{the homogeneous boundary values (2),} \end{array}\right\}$$

or we can let

$$H^m[a,b]=\{u\in L_2[a,b]: u^{(k)}\in L_2[a,b] \qquad \text{for all } 1\leqslant k\leqslant m\}$$

and let

$$X=H_0^m[a,b]=\{u\in H^m[a,b]: u^{(k)}(a)=u^{(k)}(b)=0,\ 0\leqslant k<m-1\}.$$

In this section we work in our first choice for X. The space $H_0^m[a,b]$ is also perfectly acceptable, but it requires a bit more analysis.

The reader uncomfortable with these seemingly bizarre constraints on X should reread Section 7.4 to see how to form an alternate choice for X. Roughly, all that is required other than the boundary conditions, is that $D^m u$ be piecewise continuous with $\int_a^t D^m u(s)\,ds = D^{m-1}u(t)$. This choice of X alone does *not* guarantee that $(\ ,\ )_L$ forms an inner product on X. Further constraints are needed to assure the positive definiteness of this expression. It is clear that if each of the functions $p_j(t) \geqslant 0$ and $p_n(t) \geqslant w > 0$ on $[a,b]$, then $(\ ,\ )_L$ is positive definite [i.e., $(u,u)_L \geqslant 0$ if and only if $u \equiv 0$]. In particular

$$(u,u)_L = \sum_{j=0}^{m} \int_a^b p_j(t)\left[D^j u(t)\right]^2 dt,$$

and this expression is zero only if $u \equiv 0$. If the p_j's are not all positive with $p_n(t) \geqslant w > 0$, some other constraint must be placed on L. One of the most useful is to assume that L is *strongly elliptic* and hence, satisfies the Gårding inequality

$$\boxed{\|u\|_\infty \leqslant K\|u\|_L \leqslant \gamma\|u\|_2,} \tag{4}$$

for all $u \in X$ where $\|u\|_2 = (\int_a^b [u(t)]^2 dt)^{1/2}$, $\|u\|_L = \sqrt{(u,u)_L}$, and K and γ are constants.

When L satisfies such a Gårding inequality, (3) does define an inner product, hence a norm $\|\ \|_L = \sqrt{(\ ,\ )_L}$. Proceeding along our now well-known approximation path, we let X_N be a subspace of X consisting of piecewise polynomials satisfying our boundary data and let x_N be the least squares fit from X_N in the norm $\|\ \|_L$ to the unique solution $u \in X$ of $Lu = f$. That is, x_N is the Rayleigh-Ritz or finite element approximate to x from X_N. It remains to estimate $\|x - x_N\|_\infty$. This estimate comes from the first two terms of Gårding's inequality. In particular, suppose f is continuous so that $u \in C^{2n}[a,b]$. Let X_N be a subspace of some kind of spline functions with knots at $a = t_0 < t_1 < \cdots < t_N = b$ with mesh spacing h. Let $\bar{u}_N(t)$ be the unique spline interpolate to the solution $u(t)$ from X_N. Invoke the error estimates of Chapters 1 and 2 to find

$$\begin{cases} \|u - \bar{u}_N\|_2 \leqslant M_0 h^p \\ \|u' - \bar{u}_N'\|_2 \leqslant M_1 h^{p-1} \\ \vdots \\ \|u^{(m)} - u_N^{(m)}\|_2 \leqslant M_n h^{p-m}, \end{cases} \tag{4}$$

where $M_0, M_1, \ldots, M_m$ are constants dependent on f but independent of h, and p is some positive integer, $p \geqslant 2n$. Then

$$\|u-\bar{u}_N\|_L^2=\sum_{j=0}^{m}\int_a^b p_j(t)D^j[u(t)-\bar{u}_N(t)]^2\,dt.$$

But

$$\int_a^b p_j(t)\{D^j[u(t)-u_N(t)]\}^2\,dt \leqslant \|p_j\|_\infty \int_a^b \{D^j[u(t)-\bar{u}_N(t)]\}^2\,dt$$

$$\leqslant \gamma_j\|D^j(u-\bar{u}_N)\|_2^2.$$

Thus

$$\|u-\bar{u}_N\|_L^2 \leqslant \gamma_0(\|u-\bar{u}_N\|_2)^2+\gamma_1(\|u'-\bar{u}_N'\|_2)^2+\cdots+\gamma_n(\|D^m(u-\bar{u}_N)\|_2)^2.$$

Applying the inequalities (4) we find

$$\|u-\bar{u}_N\|_2^2 \leqslant \gamma_0 M_0^2 h^{2p}+\gamma_1 M_1^2 h^{2p-1}+\cdots+\gamma_m M_m^2 h^{2(p-m)}$$

or, more simply,

$$\|u-\bar{u}_N\|_L \leqslant \gamma h^{2(p-m)}, \tag{5}$$

where $\gamma=\gamma_0 M_0^2 h^{2m}+\gamma_1 M_1^2 h^{2m-2}+\cdots+\gamma_m M_m^2$. To estimate $\|u-u_N\|_\infty$ recall that if u_N is the finite element approximate to u from X_N, then

$$\|u-u_N\|_L \leqslant \|u-\bar{u}_N\|_L,$$

since

$$\|u-u_N\|_L=\min_{\bar{u}\in X_N}\|u-\bar{u}\|_L.$$

It follows from (5) that

$$\|u-u_N\|_L \leqslant \gamma h^{2(p-m)}.$$

But if Gårding's inequality holds,

$$\boxed{\|u-u_N\|_\infty \leqslant K\|u-u_N\|_L \leqslant K\gamma h^{2(p-m)},}$$

which is our error estimate.

To see a simple instance where Gårding's inequality obtains, suppose each of the functions $p_j(t) \geqslant 0$ and $p_m(t) \geqslant w > 0$ for all t in $[a,b]$. Then letting $u \in X$, write

$$\|u\|_L^2 = \sum_{j=0}^{m} \int_a^b p_j(t)[D^j u(t)]^2 \, dt$$

$$\geqslant \int_a^b p_m(t)[D^m u(t)]^2 \, dt \geqslant w(\|D^m u\|_2)^2. \tag{6}$$

But for each $0 \leqslant k \leqslant n-1$, $D^k u(a) = 0$; thus we have

$$[D^{k-1}u(t)]^2 = \int_a^t 2D^{k-1}u(t) D^k u(t) \, dt$$

$$\leqslant 2\sqrt{b-a} \left(\int_a^b [D^k u(t)]^2 \, dt \right)^{1/2} \|D^{k-1}u\|_\infty.$$

Thus

$$\|D^{k-1}\|_\infty^2 \leqslant 2\sqrt{b-a} \, \|D^{k-1}u\|_\infty (\|D^k u\|_2).$$

It follows that

$$\|u\|_\infty \leqslant 2\|u'\|_2$$
$$\|u'\|_\infty \leqslant 2\|u''\|_2$$
$$\vdots$$
$$\|u^{(m-1)}\|_\infty \leqslant 2\|u^{(m)}\|^2.$$

Thus

$$\|u\|_\infty \leqslant 2^m \|u^{(m)}\|_2. \tag{7}$$

Applying (6) to (7), we have

$$\boxed{\|u\|_\infty \leqslant 2^m (\sqrt{b-a}\,)^m \|u^{(m)}\|_2 \leqslant \frac{2^m}{\sqrt{w}} \|u\|_L.}$$

Thus Gårding's inequality is true in this very special case. More general cases in which Gårding's inequality is true are proved in the book of Yosida (1966).

7.12 SECOND AND HIGHER ORDER ELLIPTIC PARTIAL DIFFERENTIAL EQUATIONS

It is also natural to ask for the most general class of second (and higher) order partial differential operators to which we can apply the finite element method. Although we by no means answer this question comprehensively, we can certainly indicate the usual direction of mathematical reasoning for such problems. We begin by examining second-order partial differential operators. As a simple example, consider the boundary value problem

$$\begin{cases} Lu = -D_x[pu_x] - D_y[qu_y] + cu = f \\ u = 0 \qquad \text{on } \partial\Omega, \text{ the boundary of } \Omega, \end{cases} \tag{1}$$

where p, q, c, and f are given functions belonging to $L_2(\Omega)$, and Ω is a closed, bounded, connected subset of the real plane with continuous, piecewise differentiable boundary. The most popular linear space in which to imbed this problem is the Sobolev space $H_0^1(\Omega)$. In particular, define

$$H^1(\Omega) = \{u \in L_2(\Omega) : u_x \text{ and } u_y \in L_2(\Omega)\}.$$

The space $H^1(\Omega)$ becomes a Sobolev space when we attach to it the inner product

$$(u,v)_{H^1(\Omega)} = \int_\Omega (u_x v_x + u_y v_y + uv)\blacksquare. \tag{2}$$

Since solutions to (1) satisfy zero boundary data, we are principally interested in the subspace

$$H_0^1(\Omega) = \{u \in H^1(\Omega) : u = 0 \text{ on the boundary of } \Omega\}$$

of $H^1(\Omega)$, together with the inner product (2). This inner product generates a norm in the natural way. Namely, for each $u \in H^1(\Omega)$, and, hence for each $u \in H_0^1(\Omega)$,

$$\|u\|_{H^1(\Omega)} = \sqrt{(u,u)_{H^1(\Omega)}}$$

What, if anything, has all this to do with approximating a solution to (1) by finite elements? To see the connection we need an energy inner product on $H_0^1(\Omega)$. To this end, integrate (Lu, v) formally by parts over Ω, using Green's identities in the plane to obtain the bilinear form

$$(u,v)_L = a(u,v) = \int_\Omega (pu_x v_x + qu_y v_y + cuv) + \int_{\partial\Omega} (-pu_x v + qu_y v), \tag{3}$$

where the line integral on the right vanishes because $v \in H_0^1(\Omega)$. Thus

$$(u,v)_L = a(u,v) = \int_\Omega (pu_x v_x + qu_y v_y + cuv).$$

We would like $(u,v)_L$ to form a second inner product on $H_0^1(\Omega)$, and we would be even happier if the norm $\|u\|_L$, $\|u\|_L^2 = (u,u)_L$, generated by this inner product was equivalent (see Section 6.5) to the Sobolev norm $\|u\|_{H^1(\Omega)}$ for all $u \in H_0^1(\Omega)$. As a first step in proving this, notice that $(u,v)_L$ is certainly symmetric $[(u,v)_L = (v,u)_L$ for all $u,v \in H_0^1(\Omega)]$ and is additive $[(\alpha u_1 + \beta u_2, v)_L = \alpha(u_1,v)_L + \beta(u_2,v)_L]$ on $H_0^1(\Omega)$. Thus to ascertain that $(u,v)_L$ is an inner product, it suffices to establish the positive definiteness,

$$(u,u)_L > 0 \qquad \text{unless} \quad u=0, \tag{4}$$

of $(\ ,\ )_L$ on $H_0^1(\Omega)$. If we could find positive constants μ and γ such that

$$\mu\|u\|_{H^1(\Omega)} < \sqrt{(u,u)}_L < \|u\|_{H^1(\Omega)} \tag{5}$$

for all $u \in H_0^1(\Omega)$, the positive definiteness of $(\ ,\ )_L$ would be immediate: in this case, $(u,u)_L = 0$ would imply $\mu\|u\|_{H^1(\Omega)} = 0$. Since $\|u\|_{H^1(\Omega)} > 0$ unless $u=0$, our result would follow. Thus the question facing us is what choices of p,q, and c guarantee (5). Note that (5) is tantamount to the equivalence of the norms $\|\ \|_{H^1(\Omega)}$ and $\|\ \|_L$.

One very simple set of conditions guaranteeing (5) is that p,q, and c all be strictly positive and continuous. In this case

$$(u,u)_L = \int_\Omega \left[p(u_x)^2 + q(u_y)^2 + cu^2 \right]$$

$$\geq \mu^2 \int_\Omega \left[(u_x)^2 + (u_y)^2 + u^2 \right]$$

$$= \mu^2 \|u\|_{H^1(\Omega)},$$

where $\mu^2 = \min\{p_1, q_1, c_1\}$ $p_1 = \min_{z\in\Omega} p(z)$, $q_1 = \min_{z\in\Omega} q(z)$, and $c_1 = \min_{z\in\Omega} c(z)$. The opposite inequality $(u,u)_L \leq \gamma^2 \|u\|^2_{H^1(\Omega)}$ is immediate from Schwarz's inequality and requires only p,q, and $c \in L_2(\Omega)$. A far more general set of constraints on p,q, and c is guaranteed by *Gårding's inequality* when L is *strongly elliptic*. The interested reader is referred to the books of Yosida (1966) and of Lions (1961).

Once we have (5), or simply the positive definiteness of $(\ ,\ )_L$, we can apply finite elements in the standard way. Letting u be the finite element approximate to the solution μ,

$$\|u-u_N\|_L=\min_{\tilde{u}\in H_N}\|u-\tilde{u}\|_L,$$

where $H_N=\text{span}\{\phi_1,\phi_2,\ldots,\phi_N\}$ is an N-dimensional subspace of $H_0^1(\Omega)$, we can invoke Friedrichs's first inequality to find

$$\|u-u_N\|_L\leqslant\beta\|u-u_N\|_\Delta\leqslant\beta\|u-u_N\|_{H^1(\Omega)}\leqslant\frac{\beta}{\mu}\|u-u_N\|_L$$

and proceed in the usual way.

Examples of second-order partial differential equations to which we can apply the finite element method are given by equations of the torsion of a bar, the bending of a beam by a lateral force, and the general self-adjoint second-order partial differential equation of *elliptic type*

$$Lu=-\sum_{j,k=1}^{2}\frac{\partial}{\partial x_j}\left[A_{jk}(p)\frac{\partial u}{\partial x_k}\right]+c(p)u=f(p),$$

where $p=(x,y)$ is a point in D,

$$A_{jk}=A_{kj},\text{ and}$$

$$\sum_{j,k=1}^{2}A_{jk}(p)t_jt_k\geqslant\mu_0\sum_{k=1}^{2}t_k^2,\qquad\text{for all real } t_j \text{ and } t_k,\mu_0>0.$$

Here we assume $c(p)$, $f(p)$, and the $A_{jk}(p)$'s are given functions defined in some closed region D of the plane bounded by a simple closed curve C. Moreover, the functions u belonging to the domain of L are usually subject to one of the three following types of boundary conditions:

1. $u\equiv0$ on C
2. $N(u)+\sigma u\equiv0$ on C
3. $N(u)\equiv0$ on C

where

$$N(u)=\sum_{j,k=1}^{2}A_{jk}(p)\frac{\partial u}{\partial x_k}\cos(n,x_j),$$

and n is the outward directed normal. If boundary conditions 1 or 3 apply, we define our inner product by

$$(u,v)_L=\iint_D\left(\sum_{j,k=1}^{2}A_{jk}\frac{\partial u}{\partial x_j}\frac{\partial v}{\partial x_k}+Cuv\right).$$

If condition 2 applies, we let

$$(u,v)_L = \iint_D \left(\sum A_{jk} \frac{\partial u}{\partial x_j} \frac{\partial v}{\partial x_k} + Cuv \right) + \int_C \sigma uv.$$

Both these expressions come from the formal integration of (Lu,v) by parts. It must be proved that these equations do indeed define inner products on an appropriate linear space X, and the appropriate space from the mathematician's viewpoint is again $H^1(\Omega)$. We can, of course, set the entire analysis in a much simpler linear space setting, as was illustrated in sections 7.4 through 7.7. If $\sigma(p) \geqslant C_0 > 0$ and $C(p) \geqslant C_0 > 0$ on D, it follows from the ellipticity of L that

$$\iint_D \sum_{k,j=1}^{2} A_{jk} \frac{\partial u}{\partial x_j} \frac{\partial u}{\partial x_k} \geqslant \mu_0 \iint_D \sum_{k=1}^{2} \left(\frac{\partial u}{\partial x_k} \right)^2.$$

Moreover, since a *general Friedrichs–Poincaré type inequality* can be proved in each of these cases, there exists a constant $\gamma > 0$ such that

$$\gamma \|u\|_2 \leqslant \|u\|_L = \sqrt{(u,u)_L} \tag{6}$$

for all u in X. One then chooses a computationally pleasant finite dimensional subspace X_N of X and proceeds in the now usual way.

Extending these ideas to higher order elliptic equations on n-dimensional domains follows analogous lines of reasoning. In particular, suppose Ω is a connected, closed, bounded region of R^n. The general self-adjoint equation in n variables defined on Ω is of the form

$$Lu = \sum_{|p|,|q| \leqslant k} (-1)^{|q|} D^q \left(a_{pq}(x) D^p u \right),$$

where $x = (x_1, x_2, \ldots, x_n) \in R^n$, $p = (p_1, p_2, \ldots, p_n)$, and $q = (q_1, q_2, \ldots, q_n)$ are positive multi-integers, $|p| = p_1 + p_2 + \cdots + p_n$, and

$$D^p u = \frac{\partial^{|p|} u}{\partial x_1^{p_1} \partial x_2^{p_2} \cdots \partial x_n^{p_n}}.$$

The operator L is said to be elliptic if

$$\sum_{|p|,|q| \leqslant k} a_{pq}(x) x^p x^q \geqslant C \sum_{|p| \leqslant k} |x^p|^2$$

for all $x \in \Omega$, where $x^p = x_1^{p_1} x_2^{p_2} \cdots x_n^{p_n}$, and c is a constant. One then embeds the problem in an appropriate inner product space X, usually the

Sobolev space $H^k(\Omega)=\{u\in L_2(\Omega): D^k u\in L_2(\Omega) \text{ for all } |p|\leqslant k\}$. The space so chosen admits approximating functions of a sufficiently low order of smoothness (see Exercises) and uses the ellipticity of L to guarantee the expression

$$a(u,v)=(u,v)_L=\sum_{|p|,|q|\leqslant k}\int_\Omega a_{pq}(x)D^p u D^q v+\gamma(u,v),$$

which is obtained by integrating (Lu,v) by parts and invoking the boundary values satisfied by u and v, does define an (energy) inner product on X. The expression $\gamma(u,v)$ is generally a surface integral that vanishes or does not vanish depending on the boundary conditions satisfied by u and v. Integration by parts in three or more variables requires the higher dimensional analog of Green's identities or the divergence theorem. We then choose an N-dimensional subspace X_N of X consisting of some kind of spline functions, compute the finite element approximate u_N to u in the usual way, and estimate the error $\|u-u_N\|_2$ after proving a generalized Friedrichs–Poincaré or Gårding type of inequality

$$\|v\|_2\leqslant\gamma\|v\|_L=\sqrt{(v,v)_L}$$

for all v in X.

The great difficulty here from the numerical point of view is finding appropriate approximating functions. For higher order equations in two variables we run into the difficulty of *nonconforming elements* (see Section 7.10). When dealing with partial differential equations in three or more variables, the situation is much worse. We begin with the problem of constructing spline approximates in three or more variables. This is not too hard to accomplish on "superrectangular" domains and grids (the higher dimensional analogs of rectangular grids), using higher dimensional tensor products. On tetrahedral and more general simplicial subdivisions (the higher dimensional analog of triangulations), constructing other than continuous piecewise polynomials can become quite complicated and, quite frankly, little work has been done in this area. Thus we are back to the *nonconforming element* syndrome for higher order equations. But our difficulties comprise more than simply nonconforming elements. In particular the *size of the linear system* becomes troublesome (as does the size of the nonlinear system if we are working with a nonlinear problem). For linear problems, the size of the matrices involved is roughly $n^3\times n^3$. For even modest-sized n, say $n=10$, our system is already of order 1000×1000. We are faced with storage difficulties, and we require high speed of computation. The matrices involved, however, are usually banded, and this is some help on size, although minimizing the bandwidth by appropriate numbering of knots is another problem. For these reasons, numerical computations involving approximations of even modest partial differential equations in three dimensions tend to take hours of computer time. Nevertheless, we can still talk about such approximations.

Treatment of these problems in even greater generality is afforded if one takes recourse to spaces of distributions or *generalized functions*. A "function" u is said to be a *generalized solution* to $Lu=f$ if and only if

$$(Lu-f,v)=0 \qquad \text{or} \qquad (u,v)_A-(f,v)=0$$

for all $v\in X$. A "function" w is said to be a kth *generalized derivative of u* with respect to $x_{i_1},x_{i_2},\ldots,x_{i_k}$, denoted

$$w(x)=\frac{\partial^k u}{\partial x_{i_1}\partial x_{i_2}\cdots\partial x_{i_k}},$$

if and only if, for all functions $v(x)$ that are k times continuously differentiable on the interior of Ω and vanish outside some compact set on Ω,

$$(w,v)=\left(u,\frac{\partial^k v}{\partial x_{i_1}\partial x_{i_2}\cdots\partial x_{i_k}}\right).$$

This approach to the problem entails considerable mathematical sophistication and was historically well launched by Sobolev (1950) and by L. Schwartz (1957). Choosing X here is somewhat more complicated than when we know that $D^p u$ exists for $|p|\leqslant k$. In the latter case we can simply generalize from our earlier arguments and apply our method to a great many partial differential equations. The interested reader is referred to the books of Sobolev, (1950) Courant and Hilbert (1953), Lions (1961), Lions and Magenes (1972), Aubin (1967), and of Oden and Reddy.

7.13 GALERKIN METHODS AND LEAST SQUARES METHODS

Galerkin methods subsume both the finite element method and the method of least squares, providing yet another variational method for approximating solutions to the linear (and nonlinear) operator equations

$$Au=f \tag{1}$$

from finite dimensional subspaces X_N of some inner product space X in which the operator A is defined. In particular if (,) denotes an inner product on X, and if $X_N=\operatorname{span}\{\phi_1^N,\phi_2^N,\ldots,\phi_N^N\}$ and $Y_N=\operatorname{span}\{\psi_1^N,\psi_2^N,\ldots,\psi_N^N\}$ are N-dimensional subspaces of X, the general

Galerkin method for approximating a solution u to (1) for given f seeks a function $u_N \in X_N$ satisfying the system of equations

$$(Au_N - f, \psi_i^N) = 0, \qquad 1 \leqslant i \leqslant N. \tag{2}$$

Thus if

$$u_N = c_1\phi_1^N + c_2\phi_2^N + \cdots + c_N\phi_N^N$$

and if A is linear, we find, substituting into (2), that u_N must satisfy the linear system

$$\sum_{j=1}^{N} (A\phi_i^N, \psi_j^N)c_j = (f, \psi_i^N), \qquad I \leqslant i \leqslant N. \tag{3}$$

Usually one takes $\psi_i^N = \phi_i^N$, $1 \leqslant i \leqslant N$. In this case

$$\sum_{j=1}^{N} (A\phi_i^N, \phi_j^N)c_j = (f, \phi_i^N), \qquad 1 \leqslant i \leqslant N.$$

But (3) is precisely the Rayleigh-Ritz method when A is positive definite and symmetric. The point, however, is that there are nonsymmetric and nonpositive definite operators A to which Galerkin's method applies. Thus the Galerkin method in this form subsumes the Rayleigh-Ritz method but is more general. As an example, consider the partial differential equation

$$Au = -\sum_{i,j=1}^{2} \frac{\partial}{\partial x_i}\left[a_{ij}\frac{\partial u}{\partial x_j}\right] + \sum_{j=1}^{2} b_j \frac{\partial u}{\partial x_j} + cu = f,$$

subject to $u = 0$ on the boundary $\partial\Omega$ of some closed, bounded region Ω of the real plane, where the b_j's are functions that do not vanish identically on Ω. The problem then, assuming that $Au = f$ has a solution, is to determine conditions on A guaranteeing the existence of the Galerkin approximates $u_N = c_1\phi_1^N + c_2\phi_2^N + \cdots + c_N\phi_N^N$ from subspaces X_N whose union is dense in X, and to estimate the error $\|u - u_N\|$.

Arguing heuristically, note that if u_N is the Galerkin approximate solution from X_N to (1), then

$$(Au_N, v) = (f, v) \tag{4}$$

for all $v \in X_N$. In particular, if $\{\tilde{\phi}_i^N : 1 \leqslant i \leqslant N\}$ is an orthonormal basis for X_N, it follows that

$$(Au_N, \tilde{\phi}_i^N) = (f, \tilde{\phi}_i^N), \qquad 1 \leqslant i \leqslant N. \tag{5}$$

Let $P_N f = \sum_{i=1}^{N}(f, \tilde{\phi}_i^N)\tilde{\phi}_i^N$ be the least squares fit to f from X_N. It follows from (4) and (5) that

$$P_N(Au_N) = P_N f. \tag{6}$$

But for subspaces X of spline functions arising from successive mesh refinements of the domain Ω of functions from X, $X_{N+1} \supset X_N$, and $\cup_{N=1}^{\infty} X_N$ is dense in X. It follows that

$$\lim_N P_N f = f$$

in the L_2 sense. That is, $\lim_N \|f - P_N f\|_2 = 0$. But then

$$\lim_N P_N(Au_N) = f.$$

One argues that $\lim_N Au_N$ exists, and in fact $\lim_N Au_N = f$, which means that

$$\lim_N \|Au_N - f\|_2 = 0.$$

Now suppose A satisfies a coerciveness constraint

$$\|u\|_2 \leqslant \gamma \|Au\|_2$$

for all u in the domain of A. It would follow that

$$\|u - u_N\|_2 \leqslant \gamma \|Au - Au_N\|_2 \to 0$$

as $N \to \infty$. Thus $\lim u_N = u$ under these conditions.

This highly heuristic argument does not, of course, answer the existence, uniqueness, and convergence question for us.

In another interesting case of the Galerkin method, the *method of least squares*, we let $\psi_i^N = A\phi_i^N$. Then we have

$$(Au_N - f, \psi_i^N) = 0 = (Au_N - f, A\phi_i^N),$$

giving

$$\sum_{j=1}^{N} (A\phi_j^N, \psi_i^N)c_j = \sum_{j=1}^{N} (A\phi_j^N, A\phi_i^N)c_j = (f, A\phi_i^N).$$

We easily establish the existence of u_N in this case, since such a u_N is the solution of the minimization problem

$$\|Au_N - f\|_2 = \min_{\tilde{u} \in X_N} \|A\tilde{u} - f\|_2.$$

Since $Y_N = \text{span}\{\psi_1^N, \psi_2^N, \ldots, \psi_N^N\}$ is a finite dimensional subspace of X, there exists a $y_N = Au_N$ minimizing $\|\tilde{y} - f\|_2$ over all $\tilde{y} \in Y_N$: namely, the least squares fit to f from Y_N; and this is precisely u_N solving

$$\sum_{j=1}^{N} (A\phi_j^N, A\phi_i^N) = (f, A\phi_i^N).$$

Two highly attractive features of the method of least squares applied to differential equations are:

1. It is not necessary to existence that the approximating functions ϕ_i^N satisfy the boundary conditions of the differential equation at hand.
2. From the viewpoint of computation, the matrix $(A\phi_i^N, A\phi_j^N)$ is more efficiently formed than the matrix $(A\phi_j^N, \phi_i^N)$ associated with the Ritz or finite element method.

Many open questions exist in this interesting area of approximation theory for both linear and nonlinear operator equations. The interested reader is referred to the papers of Bramble and Schatz (1971), Locker (1971), and Varga (1971).

REFERENCES

Aubin, J. P. Approximation des espaces des distributions et des operateurs differentiels. *Bull. Soc. Math. France*, Mem. 12, 1967.

Aubin, J. P. Approximation of non-homogeneous Neumann problems—Regularity of convergence and estimates of errors in terms of *n*-widths. MRC Technical Report 924, Augurs 1968.

Aziz, A. K., ed. The mathematical foundations of the finite element method with applications to partial differential equations. Proceedings of a conference held at the University of Maryland, June 26–30, 1972.

Babuska, I. The finite element method for infinite domains. *Math. Comp.*, **117** (1972), 1–11.

Babuska, I. The finite element method for elliptic differential equations. *Numerical solution of partial differntial equations*, Vol. II, B. Hubbard, ed. Academic Press, New York, 1971, pp. 69–106.

Babuska, I. and M. B. Rosenweg. A finite element scheme for domains with corners. *Numer. Math.*, **20** (1972), 1–21.

Bazeley, G. P., Y. K. Cheung, B. M. Irons, and O. C. Zienkiewicz, *Triangular elements in plate bending-comforming and nonconforming solutions*. Wright-Patterson I, (1965).

Birkhoff, G., C. de Boor, B. Swartz, and B. Wendroff. Rayleigh-Ritz-Galerkin approximations by piecewise cubic polynomials. *SIAM J. Numer. Anal.*, **3** (1966), 188–203.

Borman, M. S. and G. E. Skvortsov. On summability of the highest order derivatives of the solution of the Dirichlet problem in a domain with piecewise smooth boundary. *Izv. Vyssh. Nchebn. Zavod Mat.*, **30** (1962)8 12–21.

Bramble, J. H. and S. R. Hilbert. Estimation of linear functionals on Sobolev spaces with application to Fourier transforms and spline interpolation. *SIAM J. Numer. Anal.*, **7** (1970), 112–124.

Bramble, J. H. and A. H. Schatz. On the numerical solution of elliptic boundary value problems by least squares approximation of the data. In *Numerical solution of partial differential equations*, Vol. II, B. Hubbard, ed. Academic Press, New York, 1971, pp. 107–132.

Ciarlet, P. G., M. H. Schultz, and R. S. Varga. Numerical methods of high order accuracy for nonlinear boundary value problems. I. One-dimensional problems. *Numer. Math.*, **9** (1967), 394–430.

Ciarlet, P. G., M. H. Schultz, and R. S. Varga. Nonlinear boundary value problems. II. *Numer. Math.*, **11** (1968), 331–345.

Ciarlet, P. G., F. Natterer, and R. S. Varga. Numerical methods of high order accuracy for singular nonlinear boundary value problems. *Numer. Math.*, **15**, (1970), 87–99.

Coddington, E. and N. Levinson. *Theory of ordinary differential equations*. McGraw-Hill, New York, 1968.

Courant, R. Variational methods for the solution of problems of equilibrium and vibrations. *Bull. AMS*, **49** (1943), 1–23.

Courant, R. and D. Hilbert. *Methods of mathematical physics* vol II. Wiley-Interscience, New York, 1953.

Desai, C. S. and J. F. Abel. *Introduction to the finite element method*. Van Nostrand-Reinhold, New York, 1972.

Fairweather, Graeme. Galerkin methods for differential equations. CSIR Special Report 111, Council for Scientific and Industrial Research, Union of South Africa, Pretoria, February 1974.

Friedrichs, K. O. On the boundary value of the theory of elasticity and Korn's inequality. *Ann. Math.*, **48** (1947).

Gårding, L. Dirichlet's problem for linear elliptic partial differential equations. *Math. Scand.* **1** (1953), 55–73.

Lions, J. L. *Equations différentielles opérationelles et problemes aux limites*, vol. II. Springer-Verlag, Berlin, 1961.

Lions, J. L. and E. Magenes. *Non-homogeneous boundary value problems and applications*, vol I. Springer-Verlag, New York, 1972.

Locker, J. L. The method of least squares for boundary value problems. *Trans. AMS*, **154**, (1971), 57–68.

Mikhlin, S. G. *Variational methods in mathematical physics*. Macmillan, New York, 1964.

Mikhlin, S. G. and K. L. Smolitskiy. *Approximate methods for the solution of differential and integral equations*. American Elsevier, New York, (1967).

Nitsche, J. Ein Kriterium für die quasi-optimalität des Ritzchen verfahrens. *Numer. Math.*, **11** (1968), 346–348.

Oden, J. T. and J. N. Reddy. Variational methods in theoretical mechanics. (300 page monograph; to appear.)

Prenter, P. M. A brief survey of splines and variational methods. CSIR Special Report 134, Council for Scientific and Industrial Research of the Union of South Africa, Pretoria, February 1974.

Schultz, M. H. *Spline analysis*. Prentice-Hall, Englewood Cliffs, N.J., 1973.

Schwartz, L. Théorie des distributions à valeurs vectorielles. *Ann. Inst. Fourier*, **7** (1957), 1–141.

Schwartz, L. L_2-errror bounds for the Rayleigh-Ritz-Galerkin method. *SIAM J. Numer. Anal.*, **8**, (1971), 737–748.

Sobolev, S. L. *Partial differential equations of mathematical physics*. Gostekhizdat, Moscow. English translation published by Pergamon Press, Oxford, 1950.

Schultz, M. H. and S. C. Eisenstadt. Computational aspects of the finite element method. Research Report 72-1, Department of Computer Science, Yale University, 1973.

Strang, G. and G. Fix. *Analysis of the finite element method*. Prentice-Hall, Englewood Cliffs, N.J., 1973.

Varga, R. S. Functional analysis and approximation theory in numerical analysis. SIAM Regional Conference Series in Applied Mathematics. Philadelphia, 1971.

Yosida, K. *Functional analysis*, vols 1–3. Springer-Verlag, Berlin, 1966.

Zienkiewicz, O. C. *The finite element method in engineering science*. McGraw-Hill, London, 1971.

Zlamal, M. On the finite element method. *Numer. Math.*, **12** (1968), 394–409.

Zlamal, M. On some finite element procedures for solving second order boundary value problems. *Numer. Math.*, **14** (1969), 42–48.

Zlamal, M. A finite element procedure of second order accuracy. *Numer. Math.*, **14** (1970), 394–402.

8

THE METHOD OF COLLOCATION

8.1 INTRODUCTION

Readers with long memories will recall that in the first chapter of the book (Section 1.6) we remarked that the operator analog of finding an approximate subject to pure interpolatory constraints was the *method of collocation*. The only difference from the techniques of, say, Chapters 2 and 5, is that instead of forcing a combination of our approximating functions to interpolate the solution $x(t)$ of our operator equation $Lx=y$, we force a combination of their images under L to interpolate the right-hand side $y(t)$ of our equation. To be less vague, we assume throughout the chapter that X is a linear subspace of $L_2[D]$, the space of square integrable functions on D, where D is some subset of the real line E^1 or of the real plane E^2. Let L be a linear operator whose domain is X and whose range is also in X. Let $\{\phi_1,\phi_2,\ldots,\phi_N\}$ be a linearly independent subset of X, and let

$$X_N=\operatorname{span}\{\phi_1,\phi_2,\ldots,\phi_N\}$$

be an N-dimensional subspace of X. Suppose we are given the linear equation

$$Lx = y, \tag{1}$$

where y is given function from X. Approximating the solution $x(t)$ of (1) by the method of collocation consists of finding a function $x_N(t)$

$$x_N(t) = a_1\phi_1(t) + a_2\phi_2(t) + \cdots + a_N\phi_N(t)$$

in X_N solving the $N \times N$ system of linear equations

$$Lx_N(t_i) = \sum_{j=1}^{N} a_j L\phi_j(t_i) = y(t_i), \qquad 1 \leqslant i \leqslant N, \tag{2}$$

where $t_1, t_2, \ldots, t_N$ are N *distinct points* of D at which all the terms of (2) are defined. The function $x_N(t)$, if it exists, is said to collocate $y(t)$ at the points $t_1, \ldots, t_N$. Any function $f(t)$ so obtained is referred to as an approximate solution obtained by the *method of collocation*.

In this chapter we study some classes of operators L and subspaces X_N for which such collocation solutions exist and are unique; and we estimate the error $\|x - x_N\|$ in these cases. We see that such methods are closely related to Galerkin methods, hence to finite element methods; but collocation methods, when they work, can be much easier and more efficient for computing. Before turning to the theory of collocation methods, we look at some examples.

EXAMPLE 1

As a very simple application of the method, consider the boundary value problem

$$\begin{cases} Lx(t) = x''(t) - x(t) = 1 \\ x(0) = x(1) = 0. \end{cases}$$

Let $\phi_1(t) = t(t-1)$ and $\phi_2(t) = t^2(t-1)$. Then

$$L\phi_1(t) = 2 - t(t-1)$$

$$L\phi_2(t) = 6t - 2 - t^2(t-1)$$

and

$$L\phi_1(0) = 2$$

$$L\phi_2(0) = -2$$

$$L\phi_1(1) = 2$$

$$L\phi_2(1) = 4.$$

Let $t_1=0$ and $t_2=1$. The collocation approximate

$$\hat{x}(t)=a_1\phi_1(t)+a_2\phi_2(t),$$

collocating the right-hand side of the first equation at $t_1=0$, and $t_2=1$ is given by the solution to the linear system

$$\begin{pmatrix} 2 & 2 \\ -2 & 4 \end{pmatrix}\begin{pmatrix} a_1 \\ a_2 \end{pmatrix}=\begin{pmatrix} 1 \\ 1 \end{pmatrix},$$

where $y(0)=y(1)=1$. Solving we find $a_2=\frac{1}{3}$ and $a_2=\frac{1}{6}$, giving

$$\hat{x}(t)=\tfrac{1}{6}t(t-1)+\tfrac{1}{3}t^2(t-1).$$

The exact solution is $x(t)=1/(e+1)[e^t+e^{1-t}]-1$. We tabulate $x(t)$, $\hat{x}(t)$, and the error $e(t)=|x(t)-\hat{x}(t)|$ for several values of t in Table 8.1.

TABLE 8.1*

t	$x(t)$	$\hat{x}(t)$	$x(t)-\hat{x}(t)$
0.0	0.0000 0000	0.0000 0000	0.0000 0000
0.1	−0.0412 8461	−0.0180 0000	−0.0232 8461
0.2	−0.0729 7407	−0.0373 3333	−0.0356 4073
0.3	−0.0953 8554	−0.0560 0000	−0.0393 8554
0.4	−0.1087 4333	−0.0720 0000	−0.0367 4333
0.5	−0.1131 8112	−0.0833 3333	−0.0298 4778
0.6	−0.1087 4333	−0.0880 0000	−0.0207 4333
0.7	−0.0840 0000	−0.0840 0000	−0.0113 8554
0.8	−0.0729 7407	−0.0693 3333	−0.0036 4073
0.9	−0.0412 8461	−0.0420 0000	0.0007 1539
1.0	0.0000 0000	0.0000 0000	0.0000 0000

*The author is grateful to Mr. D. J. Laurie of the C.S.I.R., Pretoria, R.S.A. for carrying out these and other numerical computations.

Notice that we seem to have one-decimal-place accuracy using the functions $\phi_1(t)$ and $\phi_2(t)$, which we appear to have selected out of thin air, so to speak. Several immediate questions arise (just as they did in the case of straight forward interpolation problems). The first is whether we can improve the accuracy by keeping the same ϕ_1 and ϕ_2 and moving the knots t_1 and t_2. This problem is one we have seen before—the *optimal placing of knots*. The second is whether we can improve the accuracy by changing the

ϕ_i's and the knots. The answer to the second question is affirmative, and in the remainder of the chapter we want to see what happens if we choose the ϕ_i's to be a set of spline functions. This is by no means to suggest that spline functions are the best possible choice either in terms of size of the resultant linear system or in terms of accuracy. The plain truth is that we know more about what happens when we use splines than we do about best choice of ϕ_i's and best choice of knots.

Also, collocation with splines in solving differential equations leads once again to banded matrices with a small number of bands, as opposed to the full matrices one obtains using, say, polynomials, trigonometric functions, and other well-known nonpiecewise approximates. This is a particularly desirable feature when solving partial differential equations, since in this case the systems tend to become quite large. However, as we cautioned the reader at the outset, spline analysis as we present it here is not the final answer to solving differential equations.

The simple computational experiments given in the exercises at the end of this section should be done to illustrate these ideas. To see how collocation works using splines, we consider a somewhat more general boundary value problem.

EXAMPLE 2

Suppose the boundary value problem

$$Lx(t) = x''(t) + p(t)x'(t) + q(t)x(t), \qquad 0 \leqslant t \leqslant 1$$
$$x(0) + x(1) = 0$$

has a unique solution $x(t)$ and that p and q are continuous on $[0,1]$. To approximate $x(t)$ by collocation using splines, let $\pi: 0 = t_0 < t_1 < \cdots < t_n = 1$ be a partition of $[0,1]$ and let $\phi_i(t) = B_i(t)$ be cubic splines with knots at $\pi: 0 = t_0 < t_1 < \cdots < t_n = 1$. Let $X_N = \operatorname{span}\{B_{-1}, B_0, \ldots, B_{n+1}\}$. Using collocation with these approximating functions, we seek

$$x_N(t) = a_{-1}B_{-1}(t) + a_0B_0(t) + \cdots + a_{n+1}B_{n+1}(t) \tag{3}$$

such that

$$\begin{cases} x_N(0) = 0 \\ Lx_N(t_i) = f(t_i), \qquad 0 \leqslant i \leqslant n \\ x_N(1) = 0. \end{cases} \tag{4}$$

Notice that we are collocating at $n+1$ knots and that we have introduced two extra splines B_{-1} and B_{n+1}, to force $x_N(t)$ to satisfy the same boundary data as $x(t)$. The matching of boundary data is *not* essential to the method of collocation, but we do match boundary data. The reason, from the approximation point of view, is simply to make use of all information at hand. Since we know $x(0)$ and $x(1)$, why not use this information?

To compute $x_N(t)$, we first use the linearity of L. Thus

$$Lx_N(t_i)=\sum_{j=-1}^{n+1} a_j LB_j(t_i)=f(t_i), \qquad 0\leq i\leq m,$$

where

$$LB_j(t_i)=B_j''(t_i)+p(t_i)B_j'(t_i)+q(t_i)B_j(t_i), \tag{5}$$

while (see Table 8.2)

$$x_N(0)=a_{-1}+4a_0+a_1$$

and

$$x_N(1)=a_{n-1}+4a_n+a_{n+1}.$$

Recalling that $B_i(t)\equiv 0$ when $t\geq t_{i+2}$ and $t\leq t_{i-2}$ for each i, we see that (4) reduces to the linear system

$$C_N a=b, \tag{6}$$

where the coefficient matrix C_N is

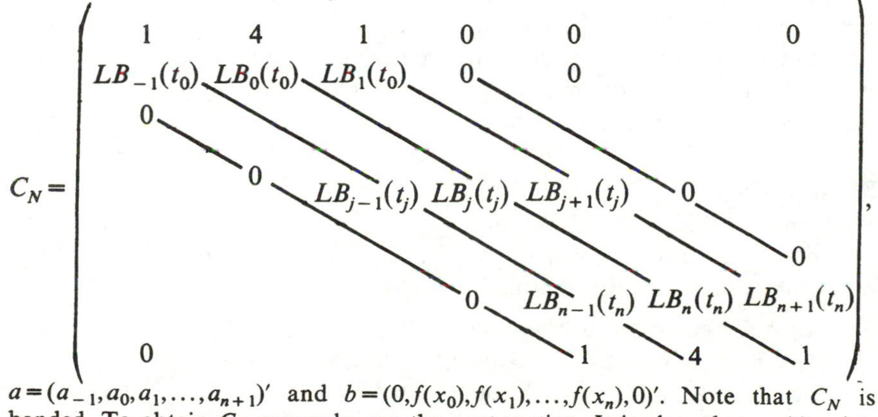

$a=(a_{-1},a_0,a_1,\ldots,a_{n+1})'$ and $b=(0,f(x_0),f(x_1),\ldots,f(x_n),0)'$. Note that C_N is banded. To obtain C_N expressly, see the next section. It is clear that $x_N(t)$ exists and is unique if and only if C_N is nonsingular. The nonsingularity of the collocation matrix C_N and estimating the error $\|x-x_N\|$ is our principal concern in the next several sections.

It is both interesting and important to compare the collocation matrix C_N with the matrix A_N (see Section 7.2) obtained in approximating $x(t)$ using the finite element or Rayleigh-Ritz method, which employs the basis splines $\{\hat{B}_i(t): 0\leq i\leq n\}$ satisfying zero boundary data. In the finite element method, the matrix $A_N=(a_{ij}^N)$ was given by

$$a_{ij}^N=\left(L\hat{B}_i,\hat{B}_j\right)$$

where

$$\left(L\hat{B}_i,\hat{B}_j\right)=\int_0^1 L\hat{B}_i(t)\hat{B}_j(t)\,dt$$

or an equivalent expression obtained by integrating $(L\hat{B}_i, \hat{B}_j)$ by parts. In this case the matrix A_N was also three banded and $3(n+1)$ integrations must be performed before the linear system can be inverted. If $a(t)=p(t)$ and $b(t)=q(t)$ are at all complicated, one must use $3(n+1)$ numerical quadratures to obtain A_N, which increases the operation count and introduces an approximation $\hat{A}_N$ to the matrix A_N, which theoretically should be checked for nonsingularity and also leads to a corresponding adjustment in the error analysis. *This is to be compared with the collocation matrix C_N for which no integrations or numerical quadratures whatsoever are involved* but only $3(n+3)$ simple point evaluations $LB_i(t_j)$ of the images of the basis functions under the operator L. This is one of the very positive aspects of the collocation method as contrasted with the finite element method. Before we can make a fair comparison of the two methods, however, we must ascertain the class of operators L for which the matrix C_N is nonsingular and we must compare the error estimates $\|x - x_N\|$ and $\|x - \bar{x}_N\|$, where $\bar{x}_N$ is the finite element approximation and x_N is the collocation approximation. We can see for ordinary differential equations with the proper choice of basis and knots that collocation certainly seems as efficient, whereas for partial differential equations it appears to be a very strong competitor. We also learn that collocation is equivalent to discrete Galerkin with a nonsingular H-matrix.

EXERCISES

1. Compute the collocation matrix $\hat{C}_N$ obtained in solving Example 2 by collocation using $X_N = \text{span}\{\hat{B}_0, \hat{B}_1, \ldots, \hat{B}_n\}$. Is either matrix C_N or $\hat{C}_N$ symmetric or positive definite? Why?
2. Solve the equation in Example 1 using
   (a) $X_N = \text{span}\{1, t, t^2, t^3\}$ collocating at $t_i = i/3$, $0 \leqslant i \leqslant 3$.
   (b) $X_N = \text{span}\{1, t, \ldots, t^9\}$ collocating at $t_i = i/7$, $0 \leqslant i \leqslant 7$ and forcing $x_N(0) = x_N(1) = 0$.
   (c) $X_N = \text{span}\{1, \cos t, \sin t, \cos 2t, \sin 2t\}$ collocating at $\frac{1}{4}$, $\frac{1}{2}$, $\frac{3}{4}$, and forcing $x_N(0) = x_N(1) = 0$.

   Compare these approximates with the actual solution. What better choice of X_N and knots might you make? What basis might you use with collocation to yield an *exact* solution?
3. Repeat Exercise 2 with

$$x''(t) - \sigma(t)x(t) = \sin 4\pi t$$

$$x(0) = x(1) = 0$$

   and $\sigma(t) = t, \sigma(t) = 3t^2 + 1, \sigma(t) = e^t$.

4. Compute the collocation matrix C_N of Example 2 when $n=4$.

8.2 A SIMPLE SPECIAL CASE: EXISTENCE VIA MATRIX ANALYSIS

One of the simplest ways of deriving existence theorems for the collocation method applied to both ordinary and partial differential equations is the direct study of the matrices C_N arising from specific choices $\{\phi_0, \phi_1, \ldots, \phi_n\}$ of basis functions ϕ_i spanning X_N and specific choices of L. This method entails invoking some of the classical theory of applied linear algebra, especially theorems concerning nonnegative matrices, diagonal dominance, graph connectedness, and Gershgorin theorems as found in the books of Varga (1959) and of Collatz (1966). It suffers the same limitations such matrix methods meet in the analysis of finite difference methods. In particular, when the differential equations or their domains become at all complicated, the extant matrix analysis is inadequate. Even though matrix methods do not give the most general results for collocation applied to ordinary differential equations, they are accessible to a broader audience than the integral equation methods of Section 8.4. Moreover, the same methods apply to the elementary theory of collocation methods for partial differential equations. For this reason we start our analysis of existence via classical matrix analysis following the result of Lee and Sinovec (1973).

The methods are best illustrated by an example. With this in mind, let us return to our old friend, the ordinary differential equation

$$Lx(t) = x''(t) - \sigma(t)x(t) = f(t), \qquad a \leq t \leq b, \tag{1}$$

$\sigma(t) > 0$, with σ continuous on $[0,1]$, subject to the boundary constraints $x(a) = x(b) = 0$. To solve this equation by collocation, we let

$$\pi_n : a = t_0^N < t_1^N < \cdots < t_n^N = b$$

be a uniform partition of the interval $[a,b]$, $t_{i+1}^N - t_i^N = h = 1/n$. We seek a function

$$x_N(t) = a_0\phi_0(t) + a_1\phi_1(t) + \cdots + a_n\phi_n(t)$$

approximating the unique solution $x(t)$ to (1) such that $Lx_N(t_i^N) = f(t_i^N)$, $0 \leq i \leq n$. Let the ϕ_i's be the cubic B splines satisfying zero boundary data, namely, the functions $\{\hat{B}_0(t), \hat{B}_1(t), \ldots, \hat{B}_n(t)\}$, where

$$\hat{B}_i(t) = B_i(t), \qquad 2 \leq i \leq n-2.$$

The functions $\hat{B}_0(t)$, $\hat{B}_2(t)$, $\hat{B}_{n-1}(t)$ and $\hat{B}_n(t)$ were given in Section 7.2.

Specifically

$$\hat{B}_0(t) = B_0(t) - 4B_{-1}(t)$$

$$\hat{B}_1(t) = B_0(t) - 4B_1(t)$$

$$\hat{B}_{n-1}(t) = B_n(t) - 4B_{n-1}(t)$$

$$\hat{B}_n(t) = B_n(t) - 4B_{n+1}(t).$$

Each of these functions has the property that $\hat{B}_i(a) = \hat{B}_i(b) = 0$. The collocation matrix $\hat{C}_N$ associated with this problem is given by $\hat{C}_N = \hat{c}_{ij}^N$, where

$$\hat{c}_{ij}^N = L\hat{B}_j(t_i^N) = \hat{B}_j''(t_i^N) - \sigma(t_i^N)\hat{B}_j(t_i^N).$$

An alternate but completely equivalent manner of setting up the problem is to let

$$x_N(t) = a_{-1}B_{-1}(t) + a_0B_0(t) + \cdots + a_nB_n(t) + a_{n+1}B_{n+1}(t). \tag{2}$$

Then force x_N to satisfy the collocation equations plus the boundary conditions (see comments on boundary values in Section 8.1). Specifically, force

$$Lx_N(t_i^N) = f(t_i^N), \qquad 0 \leqslant i \leqslant n \tag{3}$$

and

$$x_N(0) = x_N(1) = 0. \tag{4}$$

It is clear that any $x_N(t)$ solving this interpolation problem automatically solves the previous problem, and conversely. In this case we have a system of $n+3$ linear equations in $n+3$ unknowns, with a slightly modified collocation matrix $C_N = (c_{ij}^N)$, $-1 \leqslant i,j \leqslant n+1$. The first row of the matrix C_N arises from the coefficients of the boundary constraint $x_N(a) = 0$, just as the last row of C_N arises from the boundary constraint $x_N(b) = 0$. Substituting into (2) at these two points we find

$$\begin{cases} a_{-1} + 4a_0 + a_1 = 0 \\ a_{n-1} + 4a_n + a_{n+1} = 0, \end{cases} \tag{5}$$

where the values of $B_j(t_i)$ are taken from Table 8.2. To determine the rest of the matrix C_N we substitute into the collocation equations. Specifically

$$c_{ij}^N = LB_j(t_i^N), \qquad 0 \leqslant i \leqslant n; \ -1 \leqslant j \leqslant n+1.$$

Since $LB_j(t) = B_j''(t) - \sigma(t)B_j(t)$, it follows that

$$c_{ij}^N = B_j''(t_i^N) - \sigma(t_i^N)B_j(t_i^N), \qquad -1 \leqslant j \leqslant n+1;\ 0 \leqslant i \leqslant n.$$

Differentiating $B_j(t)$ and evaluating at the knots of our partition we find Table 8.2. Substituting into the collocation equations (6) and multiplying by h^2, we are led to the $n+1$ linear equations

$$(h^2\sigma_i - 6)a_{i-1} + (4h^2\sigma_i + 12)a_i + (h^2\sigma_i - 6)a_{i+1} = h^2 f(t_i), \qquad 0 \leqslant i \leqslant n, \quad (7)$$

in the $n+3$ unknowns $(x_{-1}^N, x_0^N, \ldots, x_{n+1}^N)$. Eliminating a_{-1} from the first equation of (7):

$$(\sigma_0 h^2 - 6)a_{-1} + (4h^2\sigma_0 + 12)a_0 + (\sigma_0 h^2 - 6)a_1 = h^2 f(x_0)$$

and the first equation of (5):

$$a_{-1} + 4a_0 + a_1 = 0,$$

we find

$$36a_0 = h^2 f(t_0). \qquad (8)$$

Similarly, eliminating a_{N+1} from the last equations of (7) and of (5), we find

$$36a_N = h^2 f(t_N^N). \qquad (9)$$

Coupling (8) and (9) with the second through $(n-1)$st equations of (7), we are lead to the system of $n+1$ linear equations

$$A_N x^N = d_N \qquad (10)$$

in the $n+1$ unknowns $x^N = (a_0, a_1, \ldots, a_n)^T$ with right-hand side $d_N = h^2 f^N$

TABLE 8.2 TABLE OF VALUES OF $B_j(t)$ AND ITS DERIVATIVES AT ITS KNOTS

$B_j(t)$	t_{j-2}^N	t_{j-1}^N	t_j^N	t_{j+1}^N	t_{j+2}^N
$B_j(t)$	0	1	4	1	0
$B_j'(t)$	0	$3/h$	0	$-3/h$	0
$B_j''(t)$	0	$6/h^2$	$-12/h^2$	$6/h^2$	0

$=h^2(f(t_0),f(t_1),\ldots,f(t_n))^T$ and coefficient matrix

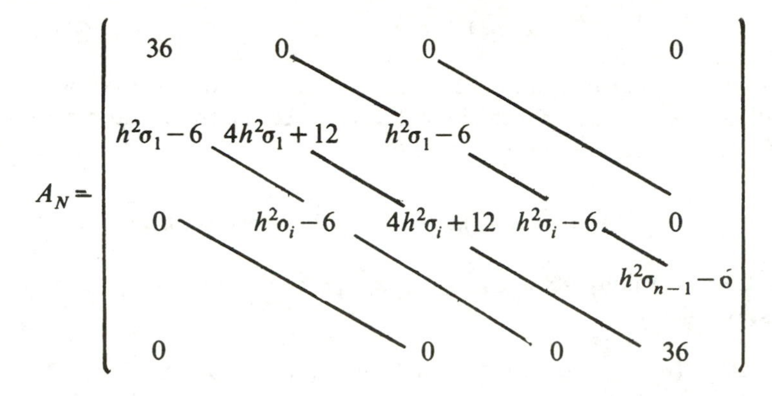

Since $\sigma(t)\geqslant 0$, it is easily seen that A_N is strictly diagonally dominant, hence nonsingular by Gershgorin's theorem (see Varga (1959). Since A_N is nonsingular, we can solve (10) for $a_0,a_1,\ldots,a_n$ and substitute into the boundary equations (5) to obtain a_{-1} and a_0. Hence the method of collocation applied to (1) using a basis of cubic splines has a unique solution $x_N(t)$ given by (2).

To estimate the error $\|x-x_N\|_\infty$, let y_N be the unique spline interpolate from X_N to the solution $x(t)$ of our boundary value problem (1). If $f\in C^2[a,b]$, then $x(t)C^4[a,b]$, and it follows from the De Boor-Hall error estimates that

$$\|D^j(x-y_N)\|_\infty \leqslant \gamma_j h^{4-j}, \qquad j=0,1,2, \tag{11}$$

where the γ_j's are constants independent of h and n. Let

$$y_N(t)=\sum_{j=-1}^{n+1} b_jB_j(t),$$

and let $\delta_j=b_j-a_j$, $-1\leqslant j\leqslant n+1$, where $x_N=\sum_{j=-1}^{n+1}a_jB_j$ is our collocation solution. It follows from the estimates (11) that

$$|Lx_N(t_i)-Ly_N(t_i)|=|f(t_i)-Ly_N(t_i)|\leqslant \beta h^2 \tag{12}$$

where $\beta=[\gamma_0h^2\|\sigma\|_\infty+\gamma_2]$. In particular

$$\begin{aligned}|Lx(t)-Ly_N(t)|&=|(x''(t)-y_N''(t))+\sigma(t)(x(t)-y_N(t))|\\&\leqslant|x''(t)-y_N''(t)|+|\sigma(t)||x(t)-y_N(t)|,\end{aligned}$$

and $Lx_N(t_i) = Lx(t_i) = f(t_i)$. Let $Ly_N = \hat{f}_N$, and $\hat{f}^N = (\hat{f}_N(t_0), \hat{f}_N(t_1), \ldots, \hat{f}_N(t_n))^T$. It is clear from (12) and (10) that the ith coordinate $[A_N(x^N - y^N)]_i$ of $A_N(x^N - y^N)$, $y^N = (b_0, b_1, \ldots, b_n)'$, satisfies the inequality

$$|[A_N(x^N - y^N)]_i| \leqslant \beta h^2, \tag{13}$$

since $(A_N x^N)_i = h^2 f(t_i)$ and $(A_N y^N)_i = h^2 \hat{f}_N(t_i)$. But the ith coordinate of $A_N(x^N - y^N)$ is simply the ith equation

$$(h^2\sigma_i - 6)\delta_{i-1} + (4h^2\sigma_i + 12)\delta_i + (h^2\sigma_i - 6)\delta_{i+1} = \tau_i, \tag{14}$$

$1 \leqslant i \leqslant n-1$, $\tau_i = h^2[f_N(t_i) - \hat{f}_N(t_i)] = h^2(Lx_N(t_i) - Ly_N(t_i))$. Thus $|\tau_i| \leqslant \beta h^4$. Let $\tau = \max_{1 \leqslant i \leqslant n} |\tau_i|$, $e_i = |\delta_i^N|$, and $\tilde{e} = \max_{1 \leqslant i \leqslant n-1} e_i$. Then $(4h^2\sigma_i + 12)\delta_i = \tau_i + (6 - \sigma_i h^2)(\delta_{i-1} + \delta_{i+1})$. Taking absolute values with sufficiently small h, we have

$$(4h^2\sigma_i + 12)e_i \leqslant \tau + 22(6 - \sigma_i h^2)e.$$

Since $0 < \sigma^* \leqslant \sigma(t)$ for all t,

$$(4h^2\sigma^* + 12)e_i \leqslant \tau + 2(6 - \sigma^* h^2)\tilde{e} \leqslant \tau + 2(6 + \sigma^* h^2)\tilde{e}$$

and, in particular,

$$(4h^2\sigma^* + 12)\tilde{e} \leqslant \tau + (12 + 2\sigma^* h^2)\tilde{e}. \tag{15}$$

Solving for $\tilde{e}$, we find $2h^2\sigma^*\tilde{e} \leqslant \tau \leqslant \beta h^4$, or

$$\tilde{e} = \max_{1 \leqslant i \leqslant n-1} e_i = \max_{1 \leqslant i \leqslant n-1} |a_i - b_i| \leqslant \frac{\beta h^2}{2\sigma^*}. \tag{16}$$

To estimate e_0, e_{-1}, e_n, and e_{n+1}, first observe that the first and last equations of the system $A_N(x^N - y^N) = h^2(f^N - \hat{f}^N)$ are $36(a_n - b_n) = h^2[f(t_n) - \hat{f}_N(t_n))]$ and $36(a_0 - b_0) = h^2[f(t_0) - \hat{f}_N(t_0)]$, respectively. Thus $e_0 \leqslant \beta h^4/36$ and $e_n \leqslant \beta h^4/3L$. Since $(a_{-1} - b_{-1})$ and $(a_{n+1} - b_{n+1})$ are given through (5) by $(a_{-1} - b_{-1}) = -(a_1 - b_1) - 4(a_0 - b_0)$ and $(a_{n+1} - b_{n+1}) = -(a_n - b_n) - 4(a_n - b_n)$, it is easy to see that there exists a constant γ such that

$$e = \max_{-1 \leqslant i \leqslant n+1} |\delta_i| = \max_{-1 \leqslant i \leqslant n+1} |a_i - b_i| \leqslant \gamma h^2, \tag{17}$$

where γ depends on x and σ^*.

The inequality (17) enables us to estimate $\|x_N - y_N\|_\infty$, hence $\|x - x_N\|_\infty$. In particular,

$$x_N(t) - y_N(t) = \sum_{i=-1}^{n+1} (a_i - b_i) B_i(t);$$

thus

$$|x_N(t) - y_N(t)| \leqslant \max |a_i - b_i| \sum_{i=-1}^{n+1} |B_i(t)|. \tag{18}$$

But (see Exercises)

$$\sum_{i=-1}^{n+1} |B_i(t)| \leqslant 10, \qquad a \leqslant t \leqslant b. \tag{19}$$

Combining (17), (18), and (19), we see that

$$\|x_N - y_N\|_\infty \leqslant 10\gamma h^2.$$

But $\|x - y_N\|_\infty \leqslant \gamma_0 h^4$. Since $\|x - x_N\|_\infty \leqslant \|x - y_N\|_\infty + \|y_N - x_N\|_\infty$, we see that

$$\boxed{\|x - x_N\|_\infty \leqslant Mh^2,} \tag{20}$$

where $M = 10\gamma + \gamma_0 h^2$. Combining the results of this section we have proved.

THEOREM 8.1

The collocation approximate $x_N(t)$ from the space of cubic splines with knots $a = t_0 < t_1 < \cdots < t_n = b$ to the solution $x(t)$ of the boundary value problem

$$-u''(t) + \sigma(t)u(t) = f(t), \qquad a \leqslant t \leqslant b$$

$$u(a) = u(b) = 0$$

exists. If $f \in C^2[a,b]$, then

$$\|x - x_N\|_\infty \leqslant Mh^2$$

for h sufficiently small.

Error estimates are also available when $f \in C^1[a,b]$ and $f \in C[a,b]$ (see Exercises). The error estimate (20) should be compared with the error estimate (Chapter 7)

$$\|x-\bar{x}_N\|_\infty \leqslant mh^4,$$

where $\bar{x}_N$ is the finite element or Rayleigh-Ritz approximate to x from X_N. From this point of view it appears that the Rayleigh-Ritz method is more accurate. However actual computational experience indicates equal accuracy. As mentioned in the previous section, the collocation matrix involves no numerical quadratures, which both increase the operation count and may result in some loss of accuracy to the matrix approximation. Therefore the collocation system is set up rather easily. One can actually attain $o(h^3)$ convergence under coerciveness when f is smooth enough (see Section 8.7). Moreover, by going to a different basis and collocating at the so-called Gaussian knots, $o(h^4)$ convergence is attainable.

EXERCISES

1. Suppose the differential equation

$$u''-pu'-qu=f$$

subject to the boundary conditions $u(0)=a$ and $u(1)=b$ has a unique solution for all t in the interval $[0,1]$. Assume that $q(t)\geqslant q^*>0$ on $[0,1]$. Prove that the method of collocation using cubic spline functions (quintic splines) has a unique solution $u_N(t)$ for all partitions π_N of $[0,1]$ whose mesh spacing is sufficiently small.

2. Program the collocation method using cubic splines (quintic splines) with uniform mesh spacing to approximate the solutions $u(t)$ to the differential equations

$$u''-(e^t+1)u=-(e^t+1+4m^2n^2)\cos 2m\pi t$$

$$u''-(e^t+1)u=\cos 2m\pi t$$

subject to the boundary constraints $u(0)=1=u(1)$ with various mesh spacings h, at least one of which gives four-decimal-place accuracy.

3. Prove that the B splines $\{B_{-1},B_0,\ldots,B_{n+1}\}$ form a so-called partition of unity with

$$\tfrac{2}{3}\sum_{i=-1}^{n+1}|B_i(t)|=4, \qquad a\leqslant t\leqslant b.$$

Also prove

$$\sum_{i=-1}^{n+1}|B_i(t)|\leqslant 10.$$

4. How is the error estimate (20) altered when $f \in C^1[a,b]$? $f \in C[a,b]$? What happens if $f(t)$ is piecewise continuous with jump discontinuities at a finite number of points of $[a,b]$?
5. Perform an error analysis in Exercise 1.

8.3 GREEN'S FUNCTIONS

It is useful, both in obtaining error estimates for the collocation method and in extending the method to a class of operators larger than those discussed in Section 8.2, to use the notion of the *Green's function* associated with a differential operator. To accomodate the greatest possible generality in our analysis, suppose we have an nth-order linear differential operator

$$Lu(t) = u^{(m)}(t) + a_{m-1}(t)u^{(m-1)}(t) + \cdots + a_0(t)u(t), \tag{1}$$

where the a_k's, $0 \leqslant k \leqslant m-1$, are continuous functions of t on the interval $[a,b]$. Let $X = H^n[a,b]$ be the domain of L where $H^n[a,b] = \{u \in C^{n-1}[a,b]: u^{(n)} \in L_2[a,b]$ and $f^{(n-1)}$ is absolutely continuous on $[a,b]\}$. The absolute continuity is tantamount to assuming $f^{(n-1)}(b) - f^{(n-1)}(a) = \int_a^b f^{(n)}(t)\,dt$. Suppose that the differential equation

$$Lu(t) = f(t), \qquad a \leqslant t \leqslant b, \tag{2}$$

subject to the m linearly independent homogeneous boundary constraints

$$U_i(u) = \sum_{k=0}^{m-1} \left[\alpha_{ik} u^{(k)}(a) + \beta_{ik} u^{(k)}(b)\right] = 0 \tag{3}$$

has a unique solution $u(t) \in X$ corresponding to each $f(t)$ belonging to $L_2[a,b]$. Then, for each f, this solution (see Naimark (1960); Coddington and Levinson (1955), Chapter 11; and Dunford and Schwartz (1963)) is given by an integral equation

$$u(t) = \int_a^b G(t,s) f(s)\,ds = Gf(t),$$

where $G(t,s)$ is the *Green's function* associated with the operator (1) together with the boundary conditions (3). (There is no loss of generality in assuming homogeneous boundary values because any operator (2) with nonzero boundary values can always be converted to such a problem.) The function $G(s,t)$ is continuous on the rectangle $[a,b] \times [a,b]$ and, as such, forms a compact operator from the space $X=C[a,b]$ with the uniform norm into itself. An operator G from X to X *compact* if the image $(Gv_n(t):n=1,2,\ldots)$ of any bounded sequence of functions $(v_n(s):n=1,2,\ldots)$ in X has a subsequence $(v_{n_k}(s):k=1,2,\ldots)$ whose image $(Gv_{n_k}(t):k=1,2,\ldots)$ converges to a function $v(t)$ belonging to X. That is, the images of bounded sequences must have convergent subsequences.

To illustrate the notion of a Green's function by a simple example consider the second-order differential equation $u''(t)=D^2u(t)=v(t)$ subject to the boundary conditions $u(a)=u(b)=0$. Integrating u'' we find

$$u'(s)-u'(a)=\int_a^s u''(\tau)\,d\tau=\int_a^s v(\tau)\,d\tau. \tag{5}$$

Integrating once again, we have

$$u(t)-u(a)=\int_a^t u'(s)\,ds=\int_a^t u'(a)\,ds+\int_a^t ds\int_a^s v(\tau)\,d\tau. \tag{6}$$

Since $u(a)=0$

$$u(t)=u'(a)(t-a)+\int_a^t ds\int_a^s v(\tau)\,d\tau. \tag{7}$$

Interchanging the order of integration in the double integral, we have

$$\int_{s=a}^t ds\int_{\tau=a}^s v(\tau)\,d\tau=\int_{\tau=a}^t d\tau\int_{s=\tau}^t v(\tau)\,d\tau=\int_a^t (t-\tau)v(\tau)\,d\tau.$$

Letting $\tau=s$ in the right-hand integral and substituting into (7), we find

$$u(t)=u'(a)(t-a)+\int_a^t (t-s)v(s)\,ds. \tag{8}$$

In particular when $t=b$

$$u(b)=0=u'(a)(b-a)+\int_a^b (b-s)v(s)\,ds.$$

Solving for $u'(a)$ and substituting into (8) gives

$$u(t)=\int_a^b G(t,s)v(s)\,ds,$$

where the Green's function $G(s,t)$ is given by

$$G(t,s)=\begin{cases}(t-s)-\dfrac{(t-a)(b-s)}{(b-a)}, & \text{for } a\leqslant s\leqslant t\\ -\dfrac{(t-a)(b-s)}{(b-a)}, & \text{for } t<s\leqslant b.\end{cases}$$

It is easy to see that G is continuous on the square $[a,b]\times[a,b]$. Note, however, that

$$\frac{\partial G}{\partial t}(t,s)=\begin{cases}1-\dfrac{(b-s)}{(b-a)}, & a\leqslant s<t\\ -\dfrac{(b-s)}{(b-a)}, & t<s\leqslant b.\end{cases}$$

Thus

$$\lim_{s\to t+} G(t,s)-\lim_{s\to t-} G(t,s)=-1,$$

and we see the jump discontinuity of $G(s,t)$ along the line $s=t$ typical of Green's functions associated with second-order ordinary differential operators.

It is also important to our analysis that every boundary value problem (2) coupled with (3) can be converted to a Fredholm integral equation of the second kind. In particular, let $v(t)=u^{(m)}(t)$, where $u(t)$ is the unique solution to (2) plus (3). It follows that $u(t)=K_0v(t)=\int_a^b k_0(t,s)v(s)\,ds$, where $k_0(t,s)$ is the Green's function associated with the boundary value problem

$$\begin{cases} lu(t)=u^{(m)}(t)\\ U_iu=0, \qquad 1\leqslant i\leqslant n.\end{cases} \tag{9}$$

Moreover, we also know (see Naimark (1960) that

$$u'(t)=K_1v(t)=\int_a^b \frac{\partial k_0}{\partial t}(t,s)v(s)\,ds$$

$$u''(t)=K_2v(t)=\int_a^b \frac{\partial^2 k_0}{\partial t^2}(t,s)v(s)\,ds$$

$$\vdots$$

$$u^{(m-1)}(t)=K_{m-1}v(t)=\int_a^b \frac{\partial^{m-1} k_0}{\partial t^{m-1}}(t,s)v(s)\,ds,$$

where the operators K_i, $1 \leqslant i \leqslant m-1$, are also compact. Substituting these quantities into (2) and (3) we find

$$v(t) + K_{m-1}v(t) + \cdots + K_0 v(t) = f(t);$$

or simply

$$v(t) + \int_a^b k(t,s)v(s)\,ds = f(s), \tag{10}$$

where the kernal

$$k(t,s) = \sum_{j=0}^{m-1} \frac{\partial^j k_0}{\partial t^j}(t,s)$$

once again determines a compact operator on X into X. Note that any solution $v(t)$ determines a unique solution to (2) and (3) via the boundary value problem (9). Thus problems (10) and (2) plus (3) are equivalent. Both the integral equations (4) and (10) are important to the analysis of the next two sections.

EXERCISE

1. Find the Green's function associated with the boundary value problem

$$x''(t) - x(t) = 0$$

$$x(0) = x(1) = 1.$$

8.4 COLLOCATION EXISTENCE VIA GREEN'S FUNCTIONS

(Ordinary Differential Equations)

We want to use the analysis of the previous section to prove a quite general existence theory for the method of collocation applied to the numerical solution of ordinary differential equations. This method of proof was early adopted by the Russians (see Vainikko (1966), Karpelovskaya (1970), and Shindler (1969) and later by Russell and Shampine (1972) to collocation with piecewise polynomials. We start with a single second-order equation and then indicate how the results are easily extended to mth-order equations. With this in mind, we consider the second-order boundary value problem

$$\begin{cases} Lu = x''(t) + p(t)x'(t) + q(t)x(t) = f(t), & a \leqslant t \leqslant b, \\ x(a) = 0 \quad \text{and} \quad x(b) = 0, \end{cases} \tag{1}$$

where p and q are continuous on $[a,b]$. We assume that (1) has a unique solution for each continuous function $f(t)$. Let $\pi: a=t_0 \leqslant t_1 \leqslant \cdots \leqslant t_n = b$, let $n=pd$ be a partition of the interval $[a,b]$, and let X_N be the set of piecewise polynomials of degree $d+2$ on each of the intervals $[t_{kd}, t_{(k+1)d}]$, $0 \leqslant k \leqslant p-1$, which are twice continuously differentiable on $[a,b]$. For example, when $d=1$, X_N is the space of cubic splines with knots at π, but if $d=2$, X_N is a space of twice continuously differentiable piecewise polynomials that reduce to quartics on each of the intervals $[t_{kd}, t_{(k+1)d}]$. In any case X_N is a subset of $C^2[a,b]$.

The first step in our method is to notice that for any function $u_N(t)$ belonging to X_N, $u_N''(t)$ is a piecewise Lagrange polynomial of degree d on $[a,b]$. Applying the method of collocation to (1) with this choice of X_N, we seek a unique function $x_N(t)$ from X_N solving the collocation equations

$$Lx_N(t_i) = f(t_i), \qquad 0 \leqslant i \leqslant n. \tag{2}$$

Note that $x_N(t)$ reduces to a polynomial

$$a_{i,d+2}t^{d+2} + \cdots + a_{i1}t + a_{i0} \tag{3}$$

on each of the intervals $I_k = [t_{kd}, t_{(k+1)d}]$, $0 \leqslant k \leqslant p-1$. As such $x_N(t)$ contains $p(d+1)$ unknown coefficients a_{ij}, $0 \leqslant i \leqslant p-1$, $0 \leqslant j \leqslant d+2$. On the other hand, (2) is a system of $n+1$ linear equations in our unknowns, whereas the boundary conditions $x_N(a)=\alpha$ and $x_N(b)=\beta$ provide two additional linear constraints. The remaining $3(p-1)$ linear constraints came from forcing

$$\lim_{t \to t_{kd}+} x_N^{(j)}(t) = \lim_{t \to t_{kd}-} x_N^{(j)}(t), \qquad 0 \leqslant j \leqslant 2; \tag{4}$$

thus x_N is a $C^2[a,b]$ function, and the total number of constraints does equal the total number of unknowns.

To prove that the collocation problem has a solution using this choice of X_N, we work through the Green's function. Let $k_0(s,t)$ be the Green's function associated with the boundary value problem

$$\begin{cases} u''(t) = y(t) \\ u(a) = \alpha \quad \text{and} \quad u(b) = \beta. \end{cases} \tag{5}$$

It follows from the arguments of the previous section that solving (1) is equivalent to solving the integral equation

$$v(t) + \int_a^b k(t,s)v(s)\,ds = f(s). \tag{6}$$

The solution $x(t)$ to (1) is then simply the unique solution to $x''(t)=v(t)$, $x(a)=a$ and $x(b)=b$, where $v(t)$ is the unique solution to (6). We usually write (6) in operator form

$$(I+K)v=f. \tag{6'}$$

The key to our method is noting that finding x_N in X_N solving the collocation equations (2) is equivalent to finding $v_N(t)=x_N''(t)$, $x_N(a)=\alpha$ and $x_N(b)=\beta$, solving the integral collocation equations

$$v_N(t_i)+Kv_N(t_i)=f(t_i), \qquad 0\leqslant i\leqslant n. \tag{7}$$

For each function $g(t)$ defined in $[a,b]$, let P_N be the unique piecewise Lagrange polynomial of degree d interpolating f at the knots t_i. In particular P_N is determined by the linear system $P_N g(t_i)=g(t_i)$, $0\leqslant i\leqslant n$. Next note that two functions g and f have the same piecewise Lagrange interpolate if and only if $P_N f(t_i)=P_N g(t_i)$, $0\leqslant i\leqslant n$. Applying this reasoning to (7) we see that this equation has a solution v_N if and only if

$$P_N v_N+P_N K v_N=P_N f. \tag{8}$$

But (8) is closely related to

$$v_N+P_N K v_N=P_N f. \tag{9}$$

In fact, if v_N is a solution to (9), $v_N=P_N f-P_N K v_N=P_N(f-Kv_N)$. Then $v_N=P_N v_N$, and so v_N automatically satisfies (8). Thus it suffices to prove that (9) has a unique solution lying in $C[a,b]$ for all sufficiently fine partitions π_n of $[a,b]$.

With this in mind, let $h_n=\max\{t_{i+1}-t_i:0\leqslant i\leqslant n-1\}$ be the mesh spacing of π_n, and suppose we choose a sequence of partitions (π_n) such that $h_n\to 0$ as $n\to\infty$. We can then prove the following theorem.

THEOREM 8.2

The sequence $\{(I+P_N K_N)^{-1}:N\geqslant N_0\}$ exists for all N_0 sufficiently large and forms a bounded sequence of linear operators. In particular there exists a constant γ independent of N_0 and of y in $C[a,b]$ such that

$$\|(I+P_N K)^{-1}y\|\leqslant\gamma\|y\|$$

for all $N\geqslant N_0$ and all y in $C[a,b]$.

The proof of this theorem depends on a series of lemmas combined with Neumann's theorem on the invertibility of linear operators. *Throughout our discussion we view the integral operator* (6′) *as a linear operator on the space* $V = C[a,b]$ *into itself with the Tchebycheff norm*. This is clearly possible if we take $C^2[a,b] = X$ as the domain of L. We also assume throughout that partitions π_n are proper.

DEFINITION: PROPER PARTITIONS

A sequence of partition $(\pi_n : n = d, 2d, \ldots)$ of the interval $[a,b]$ is said to be proper if

1. $\lim_{n\to\infty} h_n = 0$.
2. There exists a constant $\mu_0 > 0$ such that

$$\frac{h_n}{\bar{h}_n} \leqslant \mu_0$$

for all n where $\bar{h}_n = \min\{x_{i+1} - x_i : 0 \leqslant i \leqslant n-1\}$.

The lemmas we need are the following.

LEMMA 1

The projections P_N converge strongly to the identity and are uniformly bounded on V.

Proof. Recall the error estimate for Lagrange and piecewise Lagrange interpolates. (See Section 2.6.) In particular, for all continuous functions f, $\|P_N f - f\| \leqslant 6\omega(f, h_n)$. Thus $P_N f \to f$ as $N \to \infty$ for all f in V, and P_N converges strongly to the identity. The projection P_N is uniformly bounded if and only if there exists a constant $M > 0$ such that $\|P_N f\| \leqslant M$ for all $f \in V$ with $\|f\| \leqslant 1$. But on the interval $I_k = [t_{kd}, t_{(k+1)d}]$,

$$P_N f(t) = \sum_{i=kd}^{(k+1)d} f(t_i) l_i(t)$$

where the l_i's are the basis Lagrange polynomials of degree d on I_k. That is, $l_i(t_j) = \delta_{ij}$, $kd \leqslant i, j \leqslant (k+1)d$. But recall the "useful estimate" (1) of Section 2.6,

$$\sum_{i=kd}^{(k+1)d} |l_i(t)| \leqslant 2d\left(\frac{h_n}{\bar{h}_n}\right)^d,$$

where $\bar{h}_n = \min\{x_{i+1} - x_i : 0 \leqslant i \leqslant n\}$ and $t \in I_k$. We see from this that if

$f \in V$ and $\|f\| \leqslant 1$, then

$$
\begin{aligned}
|P_N f(t)| &\leqslant \sum_{i=kd}^{(k+1)d} |f(t_i)||l_i(t)| \\
&= \sum_{i=kd}^{(k+1)d} |l_i(t)| \\
&\leqslant 2^d \left(\frac{h_n}{\bar{h}_n} \right)^d \\
&\leqslant \mu_0 2^d \qquad \text{for all} \quad t \in I_k.
\end{aligned}
$$

But this estimate is independent of k, thus

$$
\|P_N f\| \leqslant 2^d \left(\frac{h_n}{\bar{h}_n} \right)^d \leqslant \mu_0 2^d.
$$

This completes the proof of the lemma. ■

LEMMA 2

$\|K - P_N K\| \to 0$ as $N \to \infty$.

Proof. Let β be the unit ball in V. In particular $\beta = \{v \in V : \|v\| \leqslant 1\}$. Since the operator K is compact, $K[\beta]$ is totally bounded. In particular, given $\varepsilon > 0$, there exists a finite set $N_\varepsilon = \{y_1, y_2, \ldots, y_q\}$ of points from V such that for each $v \in \beta$ there exists a $y_i \in N_q$ such that $\|Kv - y_i\| < \epsilon$. But then

$$
\begin{aligned}
\|Kv - P_N Kv\| &= \|(I - P_N)Kv\| \\
&\leqslant \|(I - P_N)(Kv - y_i)\| + \|(I - P_N)y_i\| \\
&\leqslant \|I - P_N\| \varepsilon + \|(I - P_N)y_i\|.
\end{aligned}
$$

Choose N_0 so large that $\|(I - P_N)y_i\| < \epsilon$ for all $y_i \in N_\epsilon$ and all $N \geqslant N_0$. Since the P_N's are uniformly bounded, $\|I - P_N\| \leqslant 1 + \mu_0 2^d$. Thus $\|Kv - P_N Kv\| \leqslant (2 + \mu_0 2^d)\epsilon$ whenever $N \geqslant N_0$. Thus $\|K - P_N K\| \to 0$ as $N \to \infty$, as was to be proved. ■

Let $[V]$ denote the set of all bounded linear operators from V into V and recall Neumann's theorem.

THEOREM A (NEUMANN'S THEOREM)

Let S and $T \in [V]$ and let $T^{-1} \in [V]$. If $\Delta = \|T^{-1}\|\,\|S-T\| < 1$, then S^{-1} exists, belongs to $[V]$, and

$$\|S^{-1} - T^{-1}\| \leqslant \frac{\|T^{-1}\|^2 \|T-S\|}{1-\Delta}.$$

This theorem combined with Lemma 2 suffices to prove Theorem 1.

Proof (Theorem 1).

We know from Lemma 2 that $\|K - P_N\| \to 0$ as $N \to \infty$. Moreover, since our differential equation (1) always has a unique solution, the integral equation (6) always has a unique solution, which means that $(I+K)^{-1}$ exists and is bounded. Let $T = I + K$ and $S = I + P_N K$ in the Neumann theorem. Then

$$\|S - T\| = \|K - P_N K\| \to 0$$

as $N \to \infty$. Thus from Theorem A, $(I + P_N K)^{-1}$ exists for all $N \geqslant N_0$. The second part of Theorem A suffices to prove that the sequence $\{(I + P_N K)^{-1} : N \geqslant N_0\}$ is uniformly bounded. ■

The equivalence of equations (9) and (2) coupled with Theorem 8.2 prove the existence of a unique solution $x_N(t)$ to the collocation equations (2). The reader has undoubtedly observed, perhaps with distaste, that the method of proof adopted here seems to be far more complicated than the Gershgorin type of matrix analysis of Section 8.2. However, Theorem 8.1 is *easily generalized* to mth-order linear boundary value problems using a variety of subspaces X_N of $X = C^m[a,b]$. The Gershgorin and nonnegative matrix arguments, however, are not at all easily generalized (see Exercises) to higher order equations, as anyone who has worked with finite differences will easily recognize. For this reason, the arguments of this section are much more powerful than those in Section 8.2.

Also note that we have chosen our approximating functions from $C^2[a,b]$ and f from $C[a,b]$. This is not essential. It suffices to choose our approximates from $X_0 = \{u \in H^2[a,b] : u(a) = u(b) = 0\}$ and to choose any f from $L_2[a,b]$ which is defined at the knots since L maps X_0 one to one onto $L_2[a,b]$. For example we could use piecewise cubic Hermites (Section 8.7) and collocate at the Gaussian knots. Finite element approximates need only come from $C[a,b]$. It is important whether the collocation method or some modification thereof can be used with trial functions (approximates) chosen from $C[a,b]$. If it cannot, such a smoothness inequity is a disadvantage for collocation as compared with finite elements. This would be especially so in higher dimensions and in higher order operator analogous of the problem.

Before passing on to an error analysis of the collocation method, a word must be said about *computation* using the piecewise polynomials introduced in this section. We have seen that in the case of our second-order problem (1), X_N consisted of piecewise polynomials of degree $d+2$ on each of the subintervals $I_k=[t_{kd},t_{(k+1)d}]$; thus they could be viewed from (3) as involving $p(d+2)$ unknown coefficients a_{ij}. However, the old problems of computation and choice of basis are again with us, a proper choice of basis considerably reducing the number of unknowns, hence the computational effort involved. For example, when $d=1$, X_N is simply the space of cubic splines with knots at π_n spanned by the B splines $\{B_{-1},B_0,\ldots,B_{n+1}\}$. Thus as we have seen in Section 8.2,

$$x_N(t)=c_{-1}B_{-1}(t)+c_0B_0(t)+\cdots+c_{n+1}B_{n+1}(t);$$

therefore there are only $n+3$ unknowns to determine, and not $4p=4n$. A similar observation pertains to the cases $d>1$.

EXERCISES

1. Consider the third-order linear boundary value problem

$$\left[\begin{array}{ll} Lu(t)=u'''(t)+a(t)u''(t)+b(t)u'(t)+c(t)u(t)=f(t), & a\leqslant t\leqslant b \\ a_{i1}u(a)+a_{i2}u'(a)+a_{i3}u(b)+a_{i4}u'(b)=\gamma_i, & i=1,2,3, \end{array}\right.$$

where the γ_i's and a_{ij}'s, $1\leqslant i,j\leqslant 3$, are given constants. Assume that the problem has a unique solution for all $f(t)$ continuous on $[a,b]$. Let $X_N\subset C^3[a,b]=X$ be the space of three times continuously differentiable piecewise polynomials of degree 5 in each of the subintervals $I_k=[t_k,t_{(k+1)}]$ of a partition $\pi_n: a=t_0<t_1<\cdots<t_n$ of $[a,b]$. Describe how you would set up the method of collocation applied to this problem, and prove that it works.

2. Using the boundary value problem stated in Exercise 1, let X_N be the set of quintic splines with knots at $a=t_0<t_1<\cdots<t_n=b$. That is, any function belonging to X_N is four times continuously differentiable on $[a,b]$ and reduces to a fifth-degree polynomial on each subinterval $[t_i,t_{i+1}]$ of π_n. Does there exist a unique function $x_N\in X_N$ solving the

collocation equations

$$Lx_N(t_i)=f(t_i), \qquad 0\leqslant i\leqslant n?$$

Explain.

8.5 ERROR ANALYSIS VIA GREEN'S FUNCTIONS

(ordinary differential equations)

Continuing with our analyses of the previous section, let $x_N(t)$ be the collocation approximate to the unique solution to

$$\left[\begin{array}{ll} Lx(t)=x''(t)+p(t)x'(t)+q(t)x(t)=f(t), & a\leqslant t\leqslant b \\ x(a)=\alpha, \qquad x(b)=\beta, & \end{array}\right. \tag{1}$$

p and q continuous, from the set X_N of twice continuously differentiable piecewise polynomials of degree $d+2$ between the knots $a=t_0<t_1<\cdots<t_{pd}=b$ of $[a,b]$ collocating $f(t)$ at each of the points t_i. We saw (Section 8.4) that

$$Lx_N(t_i)=f(t_i), \qquad 0\leqslant i\leqslant pd=n, \tag{2}$$

if and only if $v_N(t)=x_N''(t)$ solved the integral equation

$$v_N+P_NKv_N=P_Nf, \tag{3}$$

where P_Ng is the unique piecewise Lagrange polynomial of degree d on each of the intervals $I_k=[t_{kd},t_{(k+1)d}]$, $0\leqslant k\leqslant p-1$, interpolating g at the knots of our partition. Recalling the Tchebycheff error estimates for Lagrange interpolates (see Chapter 2), we know that

$$\|g-P_Ng\|_\infty\leqslant\frac{\|g^{(d+1)}\|_\infty}{2(d+1)}h^{d+1} \tag{4}$$

if $y\in C^{d+1}[a,b]$ and h is the mesh of our partition, and

$$\|g-P_Ng\|_\infty\leqslant\left[\begin{array}{lll} \gamma_k\|g^{(k)}\|_\infty h^k & \text{if } f\in C^ka,b, & 1\leqslant k\leqslant d, \\ \gamma_0\omega(g,h) & \text{if } f\in C[a,b], & \end{array}\right. \tag{5}$$

where the γ_i's are constants independent of g and of h.

Next let $v=x''$, where $x(t)$ is the solution to (1). Then $v(t)$ solves the integral equation

$$v+Kv=f. \tag{6}$$

Projecting both sides of this equation, we find

$$P_N v+P_N Kv=P_N f,$$

and adding v to both sides and subtracting gives

$$v+P_N Kv=P_N f+(v-P_N v). \tag{7}$$

Subtracting (3) and (7), we find

$$(v-v_N)+P_N K(v-v_N)=(v-P_N v)$$

or

$$v-v_N=(I+P_N K)^{-1}(v-P_N v). \tag{8}$$

Letting $G(t,s)$ be the Green's function associated with the boundary value operator $lx=x''(t)$, $x(a)=\alpha$ and $x(b)=\beta$, we see that $x=Gv$, while $x_N=Gv_N$. Applying G to both sides of (8) we find

$$x-x_N=G(I+P_N K)^{-1}(x''-P_N x'').$$

Recalling that G is compact and bounded on $C[a,b]$ with the Tchebycheff norm, we have

$$\|x-x_N\|_\infty \leqslant \|G\|\,\|(I+P_N K)^{-1}\|\,\|x''-P_N x''\|_\infty.$$

But the operators $(I+P_N K)^{-1}$, $N \geqslant N_0$ were uniformly bounded by Theorem 1 of Section 8.4 hence there exists a constant $M>0$ such that $\|(I+P_N K)^{-1}\| \leqslant M$ for all $N \geqslant N_0$. Moreover, if $x \in C^{2+k}$, $1 \leqslant k \leqslant d+1$, inequalities (4) and (5) imply

$$\|x-x_N\|_\infty \leqslant M\gamma_k \|G\|\,\|u^{(k)}\|_\infty,$$

$$\|x-x_N\|_\infty \leqslant B_k \|x^{(k+2)}\|_\infty h^k, \qquad \text{if } x \in C^{k+2}[a,b]; \tag{9}$$

and

$$\|x-x_N\|_\infty \leqslant B_0\omega(x'',h), \qquad \text{if} \quad x \in C^2[a,b]. \tag{10}$$

The constants $B_k=M\gamma_k\|G\|$, $1 \leqslant k \leqslant d+1$ and $B_0=\|G\|M\gamma_0$. That is, we have proved the following

THEOREM 8.3 ERROR ESTIMATE FOR THE COLLOCATION METHOD

Let x_N be the approximate solution from X_N to (1) obtained by the method of collocation and let x be the solution to (1). Then

$$\|x-x_N\|_\infty \leqslant \begin{bmatrix} B_k\|x^{(k+2)}\|h^k, & \text{if} \quad x\in C^{k+2}[a,b]; \quad 1\leqslant k\leqslant d+1 \\ B_0\omega(x'',h), & \text{if} \quad x\in C^2[a,b], \end{bmatrix} \tag{11}$$

where $B_k = M\|G\|\gamma_k$ and $B_0 = \|G\|M\gamma_0$ are constants and $N \geqslant N_0$.

Looking at this estimate in the particular case $d=1$, we are collocating with cubic splines, and we see once again that our initial estimate only guarantees order h^2 accuracy using collocation, whereas the finite element method applied to the same problem guarantees order h^4 accuracy with evenly spaced knots under strong coercivity. The same disparity between rates of convergence using collocation versus Rayleigh-Ritz applies to using other bases as well as to higher order equations. However, as indicated at the end of Section 8.2, the problem is that we have neither collocated at *optimal knots* nor invoked *strong coercivity*. As Section 8.8 demonstrates, collocating at strategic knots other than the natural knots of our spline can do a lot to improve the situation, making collocation highly competitive with finite elements.

8.6 COLLOCATION AND PARTIAL DIFFERENTIAL EQUATIONS*

Collocation methods also apply to partial differential equations, although our knowledge in this area is nowhere near as advanced as is our knowledge of the finite element method. Existence theorems are somewhat harder to come by for collocation with partial differential equations than for collocation with ordinary differential equations. This happens because one does not usually recognize the image of the approximate u_N under L even in the simplest of cases such as $L=-\Delta$. Thus the interpolatory capacities of Δu_N are not readily apparent. This is in constrast to the ordinary differential operator $Lu=u^{(m)}$. In the latter case L applied to a piecewise polynomial is just another piecewise polynomial whose interpolatory properties are well known. There are of course, notable exceptions

*The analysis of this section can easily be omitted, that of section 8.7 being easier. We have included it to show that matrix arguments are applicable and yield Tchebycheff error estimates.

such as the work of Karpilovskaya (1970) using truncated Fourier series. In general, however, existence theory for collocation, applied to partial differential equations, is not as transparent as the one dimensional theory.

In this section we apply the methods of Ito (1972) and Cavendish (1972) to a simple Dirichlet problem. The analysis, while direct, is rather lengthy and suffers the same limitations born by matrix analysis of finite difference methods. It does, however, provide Tchebycheff error estimates without introducing Green's functions. The approximates of the next section are not only more accurate but the existence analysis is also considerably more simple.

Consider the simple boundary value problem on the unit square $R = [0,1]\times[0,1]$

$$\left[\begin{array}{ll} Lu = u_{xx} + u_{yy} - \sigma u = f & \text{on } R \\ u \equiv 0 & \text{on the boundary } \partial R \text{ of } R. \end{array}\right. \tag{1}$$

We assume that $\sigma(x,y)$ is continuous R, and that $\sigma(x,y) \geqslant \sigma^* > 0$. To solve (1) by collocation we let $\pi_x: 0 = x_0 < x_1 < \cdots < x_n = 1$ and $\pi_y: 0 = y_0 < y_1 < \cdots < y_n = 1$ be partitions in the x and y directions, respectively, where for simplicity we choose the same number of knots in π_x and π_y. Let $\{\hat{B}_0(x), \hat{B}_1(x), \ldots, \hat{B}_n(x)\}$ and $\{\hat{B}_0(y), \hat{B}_1(y), \ldots, \hat{B}_n(y)\}$ be the B splines satisfying zero boundary data in the x and y directions, respectively (see Section 8.2); thus $\hat{B}_j(0) = \hat{B}_j(1) = 0$. Then letting

$$\hat{B}_{ij}(x,y) = \hat{B}_i(x)\hat{B}_j(y),$$

we take our N-dimensional subspace $X_N = \operatorname{span}\{\hat{B}_{ij}: 0 \leqslant i,j \leqslant n\}$ and seek an approximate solution

$$u_N = \sum_{i,j=0}^{n} z_{ij} B_{ij} \tag{2}$$

to (1) which collocates the right-hand side of (1). However, in this case we must do more than collocate at the corner knots (0,0), (1,0), (0,1), and (1,1). The reason for this is apparent if we assume $f \in C^2[R]$. The unique solution u to (1) belongs to $C^4[R]$. In particular since $u \equiv 0$ on the boundary of R, $u_{xx} = 0 = u_{yy}$ at the corner knots. Thus $Lu = 0$ at the corner knots, and no additional information is gained at these knots (i.e., the collocation matrix would be singular in this case). To overcome this difficulty we replace collocation at the corners by interpolation of mixed partials at the corners. Specifically, we force

$$D_{xxyy}u_N(x_l,y_m)=D_{xxyy}u(x_l,y_m), \qquad l,m=0,n. \tag{3}$$

Substituting (2) into (3) we find

$$z_{lm}B_l''(x_l)B_m''(x_m)=\frac{81}{h^4}z_{lm}=u_{xxyy}(x_l,y_m), \tag{4}$$

when $l,n=0,m$; thus the corner boundary equations *uncouple*. Next note that $u_{xxyy}=f_{xx}-u_{xxxx}-\sigma_{xx}u-\sigma u_{xx}$. But since u is C^4 and $u\equiv 0$ along $\{(x,0):0\leqslant x\leqslant 1\}$ and along $\{(x,1):0\leqslant x\leqslant 1\}$, it follows that $u_{xxxx}\equiv 0$ along these sides, as does u_{xx}. Thus u, u_{xx}, and u_{xxxx} vanish at the four corners, and we have

$$u_{xxyy}(x_l,y_m)=f_{xx}(x_l,y_m) \qquad \text{when } l,m=0,n.$$

Substituting into (4) we find

$$z_{lm}=\frac{h^4}{81}f_{xx}(x_l,y_m), \qquad l,m=0,n. \tag{5}$$

The collocation equations that u_N, given by (2), must satisfy* are

$$\left[\begin{array}{ll} u_N(x_l,y_m)=f(x_l,y_m), & 0\leqslant l,m\leqslant n \\ D_{xxyy}u_N(x_l,y_m)=f_{xx}(x_l,y_m), & l,m=0,n. \end{array}\right. \tag{6}$$

The other boundary collocation equations (i.e., the noncorner equations) also *uncouple*. Specifically

$$\begin{aligned} Lu_N(0,y_m)&=\sum_{i,j=0}^{n} z_{ij}B_i''(0)B_j(y_m) \\ &=-\frac{g}{h^2}\sum_{j=0}^{n} z_{0j}B_j(y_m)=f(0,y_m), \end{aligned}$$

or

$$\tfrac{1}{4}z_{0,m-1}+z_{0,m}+\tfrac{1}{4}z_{0,m+1}=-\frac{h^2}{g}f(0,y_m), \qquad 1\leqslant m\leqslant n-1. \tag{7}$$

*One very valid objection to this method is to collocation on the boundary which is somewhat unnatural to the existence theory of such boundary value problems. From this point of view, the methods of the next section not only give better results but also are closer to the theory of these problems.

Using (5), we can write (7) as the matrix equation

$$B_N z_0^N = b_0^N,$$

where $z_0^N = (z_{01}, z_{02}, \ldots, z_{0,n-1})^T$ and

$$b_0^N = -\frac{h^2}{9}\Big(f(0,y_1) + \frac{h^2}{36} f_{xx}(0,0), f(0,y_1), f(0,y_2), \ldots,$$

$$f(0,y_{n-1}), f(0,y_n) + \frac{h^2}{36} f_{xx}(0,y_n)\Big)^T.$$

It is easy to see that the matrices B_N are diagonally dominant with uniformly bounded inverse. The remaining equations of (6) involving collocation at all nonboundary knots reduce, after transposing known "boundary" coefficients z_{0j}, z_{nj}, z_{j0}, and z_{jn}, $0 \leqslant j \leqslant n$, to a system

$$A_N z^N = b^N \tag{8}$$

of $(n-1)^2$ linear equations in the remaining unknown coefficients z_{ij}, $1 \leqslant i,j \leqslant n-1$. If we order these unknowns from left to right in each row, starting with the bottom row, and work from the bottom row to the top row (see Figure 8.1) across the rows, we see that the matrix A_N is block-tridiagonal. Specifically

$$A_N = \frac{3}{4h^2}\begin{bmatrix} D_1 & C_1 & & & & 0 \\ B_2 & D_2 & C_2 & & & \\ & B_3 & D_3 & C_3 & & \\ & & & & & \\ & & & & & C_{n-2} \\ 0 & & & & & \\ & & & & B_{n-1} & D_{n-1} \end{bmatrix}, \tag{9}$$

where B_j, D_j, and C_j are also tridiagonal matrices with entries

$$(C_j)_{i,i-1} = -1 + \frac{h^2}{12}\sigma_{ij}, \qquad (C_j)_{i,i} = -1 + \frac{h^2}{3}\sigma_{ij}, \qquad (C_j)_{i,i+1} = -1 + \frac{h^2}{12}\sigma_{ij},$$

$$(D_j)_{i,i-1} = -1 + \frac{h^2}{3}\sigma_{ij}, \qquad (D_j)_{i,i} = 8 + \frac{4}{3}h^2\sigma_{ij}, \qquad (D_j)_{i,i+1} = -1 + \frac{h^2}{3}\sigma_{ij}$$

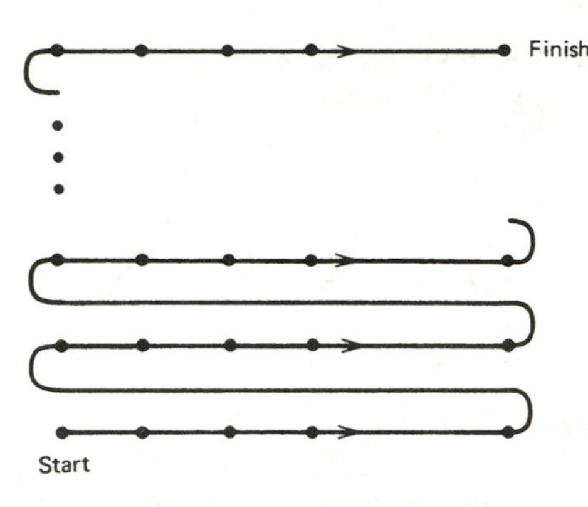

Figure 8.1 Ordered path through the knots.

and

$$B_j = C_j, \tag{10}$$

where $\sigma_{ij} = \sigma(x_i, y_j)$. From (10) and the assumption $\sigma \in C^0[R]$, it is clear that for h sufficiently small, the matrix A_N is irreducibly diagonally dominant. That is, system (8) has a unique solution and consequently, the collocation approximate (2) satisfying (6) is uniquely defined. Moreover, note that in the special case where $\sigma(x,y)$ is a nonnegative constant, A_N is symmetric, irreducibly diagonally dominant with positive diagonal entries. Thus A_N is positive definite and the system (8) can be solved by line S.O.R. (see Varga (1959), Chapter 4). Finally, one can prove (see Cavendish (1972), Chapter 3)

THEOREM 8.4

For all h sufficiently small, $\|A_N^{-1}\| \, \| \leqslant 2/9$.

Proof. It is clear from (7) that the matrices B_N^{-1} are also uniformly bounded with $\|B_N^{-1}\| \leqslant 2$. In particular, let $w = (1, 1, \ldots)$ be the unit vector in E^{n-1} and suppose $B_N^{-1}w = \hat{g}$ where $\hat{g} = (\hat{g}_0, \hat{g}_1, \ldots, \hat{g}_n)$. Then $w = B_N g$ and

$$\tfrac{1}{4}\hat{g}_{i-1} + \hat{g}_i + \tfrac{1}{4}\hat{g}_{i+1} = 1, \qquad 1 \leqslant i \leqslant n-1.$$

Let $\|g\| = \max_{0 \leqslant i \leqslant n} |\hat{g}_i|$. It follows that

$$\hat{g}_i = 1 - \tfrac{1}{4}(\hat{g}_{i-1} + \hat{g}_{i+1});$$

thus $|g_i| \leqslant 1 + \frac{1}{2}\|g\|_\infty$ for all $0 \leqslant i \leqslant n$. Thus $\|g\|_\infty \leqslant 2$, giving $\|B_N^{-1}w\| \leqslant 2$ for all w with $|w_i| \leqslant 1$ and $\|B_N^{-1}\| \leqslant 2$. ■

These uniform bounds on the norms of the inverses of A_N and B_N enable us to estimate the Tchebycheff norm of the error $\|u-u_N\|$. Specifically, let $\hat{u}_N$ be the natural cubic spline interpolate to u on R. Then

$$\hat{u}_N=\sum_{i,j=0}^{n}\hat{z}_{ij}\hat{B}_i(x)\hat{B}_j(y), \tag{11}$$

where

$$\begin{aligned}\hat{u}_N(x_i,y_j)&=u(x_i,y_j), && 0\leqslant i,j\leqslant n\\ D_{xxyy}\hat{u}_N(x_i,y_j)&=D_{xxyy}u(x_i,y_j), && i,j=0,n.\end{aligned} \tag{12}$$

We know from our error estimates for spline approximates (see Chapter 5) that $\|L\hat{u}_N-Lu\|=o(h^2)$, where h is the mesh of our partition $\pi_x\pi_y$. In particular $|Lu(x_i,y_j)-L\hat{u}_N(x_i,y_j)|=o(h^2)=|Lu_N(x_i,y_j)-L\hat{u}_N(x_i,y_j)|$. Moreover, since $D_{xxyy}\hat{u}_N(x_i,y_j)=D_{xxyy}u_N(x_i,y_j)$, we see that $B_N(z_0^N-\hat{z}_0^N)=o(h^2)$, from which it follows that $|z_0^N-\hat{z}_0^N|=\max_{1\leqslant j\leqslant n-1}|z_{oj}-\hat{z}_{oj}|=o(h^2)$. In particular, if $B_N(z_0^N-\hat{z}_0^N)=\epsilon_0^N$, then $z_0^N-\hat{z}_0^N=B_N^{-1}\epsilon_0^N$ and $\|z_0^N-\hat{z}_0^N\|\leqslant\|B_N^{-1}\|_\infty\|\epsilon_0^N\|_\infty\leqslant 20(h^2)=o(h^2)$. An analogous argument using the uniform boundedness of the A_N^{-1} suffices to prove $|z_{ij}-\hat{z}_{ij}|=o(h^2)$ when $1\leqslant i,j\leqslant n-1$. It follows that $|z_{ij}-\hat{z}_{ij}|=o(h^2)$ for all $0\leqslant i,j\leqslant n$. But $u_N(x,y)-\hat{u}_N(x,y)=\sum_{i,j=0}^{n}(z_{ij}-\hat{z}_{ij})\hat{B}_i(x)B_j(y)$. Since $B_i(x)B_j(y)=0$ when $x\geqslant x_{i+2}$, $x\leqslant x_{i-2}, y\geqslant y_{j+2}$, and $y\leqslant y_{j-2}$, we see that

$$|u_N(x,y)-\hat{u}_N(x,y)|\leqslant\max_{0\leqslant i,j\leqslant n}|z_{ij}-\hat{z}_{ij}|\sum_{m=j-2,l=i-2}^{m=j+2,l=i+2}|B_l(x)B_m(y)|=o(h^2).$$

But then $\|u_N-\hat{u}_N\|=o(h^2)$ while $\|u-\hat{u}_N\|=o(h^4)$. It follows that

$$\boxed{\|u-u_N\|_\infty\leqslant\gamma h^2.} \tag{13}$$

■

That is we have proved.

THEOREM 8.5 (ERROR ESTIMATE)

Let $u_N(x,y)$ be the approximate solution to $Lu=u_{xx}+u_{yy}+\sigma u=f$, $u=0$ on ∂R, obtained by the method of collocation. If $\sigma(x,y)$ is continuous on R and $f\in C^2[R]$, $R=[a,b]\times[a,b]$, there exists a constant γ dependent on u

but independent of h such that $\|u-u_N\|_\infty \leqslant \gamma h^2$, where u is the unique solution of the given boundary value problem.

We note that collocation using cubic splines with collocation occurring at the knots of the splines guarantees at least order h^2 accuracy, whereas Rayleigh-Ritz or finite elements using exactly the same splines guarantees order h^3 accuracy. Thus two-dimensional collocation appears to have the same difficulties we witnessed with one-dimensional collocation, compared with the finite element method. In the next section we discuss means of speeding up the convergence by collocating at the so-called Gaussian knots of the partition rather than at the natural knots of the splines.

8.7 ORTHOGONAL COLLOCATION

One means of speeding the rate of convergence of collocation methods to make them competitive with Galerkin methods is to collocate at the Gaussian knots of the partition rather than at the natural knots. Collocation in this context is known in the engineering literature as the method of *orthogonal collocation* and dates back at least to Lanczos (1938, 1956) where Tchebycheff polynomials collocating at Gaussian knots were used to approximate solutions to initial value problems. Chemists have used the method extensively to solve one and two dimensional initial and boundary value problems arising in reactor dynamics and other systems (see Horvay and Spiess, 1954, Wright, 1964, Villadsen and Stewart, 1967, Ferguson and Finlayson, 1970, Finlayson, 1971, 1972, and Ferguson, 1971). Existence, uniqueness, and error estimates for one dimensional orthogonal collocation using splines rather than full polynomials was first presented in the paper of de Boor and Schwartz (1973). The idea of orthogonal collocation is a simple one and is based on the theory of Gaussian quadratures with weight function $w(x)\equiv 1$. A Gaussian quadrature rule is a rule $\Sigma_{k=1}^{m} w_k f(\bar{t}_k)$ for approximating $\int_a^b w(x)f(x)\,dx$ which integrates polynomials of degree less than $2m$ exactly. It is well known that $\Sigma_{k=1}^{m} w_k f(\bar{t}_k)$ integrates such polynomials exactly if the $\bar{t}_k$'s are the zeros of the mth-degree Legendre polynomial $Q_m(t)$ and if $w_k=\int_{-1}^{1} l_k(t)\,dt$, where $l_k(t)$ is the basis Lagrange polynomial of degree m solving the interpolation problem $l_k(\bar{t}_i)=\delta_{ik}$ (see Davis (1965) and Bellman (1966)). Moreover, these zeros all lie in $[a,b]$ and

$$\int_a^b Q_m(t)p(t)\,dt=0 \tag{1}$$

whenever $p(t)$ is a polynomial of exact degree $m-1$ or less. The knots $\bar{t}_1,\bar{t}_2,\ldots,\bar{t}_m$ are known as the *Gaussian knots* of the interval $[a,b]$, and (1) will be the key to our convergence acceleration just as it is the key to rapid convergence of Gaussian quadrature rules.

We show that *collocation at Gaussian knots using a basis of piecewise cubic Hermite polynomials gives an order h^4 algorithm for approximating the unique solution $x(t)$* to a second-order boundary value problem such as Example 1 given below. This is to be compared with order h^2 or better (see Exercises) convergence using cubic splines and collocating at the natural knots of a partition of $[a,b]$. We then indicate how the method can be generalized to higher order equations and to partial differential equations.

To see how the method works, we continue with the simple second-order boundary value problem

$$\left[\begin{array}{ll} Lx(t) = -x''(t) + \sigma(t)x(t) = f(t), & a \leqslant t \leqslant b \\ x(a) = x(b) = 0, & \end{array}\right. \tag{2}$$

where $\sigma(t) \geqslant 0$ is continuous on $[a,b]$. Let $\pi: a = t_0 < t_1 < \cdots < t_n = b$ be a partition of $[a,b]$ with mesh spacing h. The two-point Gaussian quadrature rule for approximating $\int_{t_{i-1}}^{t_i} g(t)\,dt$ is

$$\frac{h_i}{2}\left[g(\bar{t}_{i1}) + g(\bar{t}_{i2}) \right] \approx \int_{t_{i-1}}^{t_i} g(t)\,dt, \tag{3}$$

where

$$\left[\begin{array}{l} \bar{t}_{i1} = t_{i-1} + h_i\left(\dfrac{1}{2} - \dfrac{1}{2\sqrt{3}}\right) \\ \bar{t}_{i2} = t_{i-1} + h_i\left(\dfrac{1}{2} - \dfrac{1}{2\sqrt{3}}\right) \\ h_i = t_i - t_{i-1}. \end{array}\right. \tag{4}$$

Noting that $t_{i-1} < \bar{t}_{i1} < \bar{t}_{i2} < t_i$, we refer to the set $\bar{\pi}: \bar{t}_{11} < \bar{t}_{12} < \bar{t}_{21} < \bar{t}_{22} < \cdots < \bar{t}_{n1} < \bar{t}_{n2} = \bar{t}_1 < \bar{t}_2 < \cdots < \bar{t}_{2n}$ as the two-point Gaussian knots of π or simply as the *Gaussian knots of* π.

We are going to collocate at these $2n$ *Gaussian knots*. This requires $2n$ linearly independent spline functions $\{\hat{\phi}_j(t): 1 \leqslant j \leqslant 2n\}$ for which the matrix $(L\hat{\phi}_j(\bar{t}_i)) = c_{ij}^N$ $1 \leqslant i,j \leqslant 2n$ is nonsingular. The natural choice is the set of *piecewise cubic Hermite polynomials with knots at* π (not at $\bar{\pi}$) satisfying zero boundary data. In particular (see Chapter 2) we use the basis, piecewise cubic Hermite polynomials,

$$\phi_i(t)=\begin{bmatrix} -\dfrac{2(t-t_{i-1})^3}{h^3}+\dfrac{3(t-t_{i-1})^2}{h^2}, & t_{i-1}\leqslant t\leqslant t_i \\ \dfrac{2(t-t_i)}{h^3}-\dfrac{3(t-t_i)^2}{h^2}, & t_i\leqslant t\leqslant t_{i+1} \\ 0, & \text{otherwise} \end{bmatrix}, \tag{5}$$

$1\leqslant i\leqslant n-1$, and

$$\xi_i(t)=\begin{bmatrix} \dfrac{(t-t_{i-1})^2(t-t_i)}{h^2}, & t_{i-1}\leqslant t\leqslant t_i \\ \dfrac{(t-t_i)(t_{i+1}-t)^2}{h^2}, & t_i\leqslant t\leqslant t_i-1 \\ 0, & \text{otherwise} \end{bmatrix}, \tag{6}$$

$0\leqslant i\leqslant n+1$. Note that each of these functions does indeed satisfy zero boundary data $\xi_i(a)=\xi_i(b)=\phi_i(a)=\phi_i(b)=0$. Renumbering, let

$$\hat{\phi}_i(t)=\begin{bmatrix} \xi_{\frac{i-1}{2}} & \text{when } i \text{ is odd} \\ \phi_{\frac{i}{2}}-1 & \text{when } i \text{ is even} \end{bmatrix}. \tag{7}$$

That is, $\{\xi_0,\phi_1,\xi_1,\phi_2,\xi_2,\ldots,\phi_{n-1},\xi_{n-1},\xi_n\}=\{\hat{\phi}_1,\hat{\phi}_2,\ldots,\hat{\phi}_{2n}\}$. Let X_N be the span $\{\phi_i\}$. We seek

$$x_N(t)=x_1\hat{\phi}_1(t)+x_2\hat{\phi}_2(t)+\cdots+x_{2n}\hat{\phi}_{2n}(t)$$

solving the collocation equations

$$Lx_N(\bar{t}_i)=\sum_{j=1}^{2n}L\hat{\phi}_j(t_i)x_j=f(t_i), \qquad 1\leqslant i\leqslant 2n, \tag{8}$$

where $\bar{t}_1<\bar{t}_2<\cdots\bar{t}_{2n}$ are the Gaussian knots of our partition.

Existence and error analysis follows either the line of arguement given in Sections 8.4 and 8.5 or that of Theorem 8.6. Specifically, note that $v_N(t) = x_N''(t)$ reduces to a polynomial of degree 1 on each of the intervals $[t_{i-1}, t_i]$, $1 \leqslant i \leqslant n$. As such, there is a unique $v_N(t)$ interpolating $x''(t) = v(t)$ at the knots $\bar{t}_i$, $1 \leqslant i \leqslant 2n$, which satisfies the boundary conditions $v_N(a) = v_N(b) = 0$. Here $x(t)$ is the unique solution to (1). Moreover if $v \in C^2[a,b]$, then $\|v - v_N\| \leqslant \gamma h^2$ by our Lagrange error estimates. It follows by arguments almost analogous to those of Sections 8.4 and 8.5 that $x_N(t)$ exists, is unique, and

$$x(s) - x_N(s) = \int_a^b G(s,t)[Lx(t) - Lx_N(t)]\,dt. \tag{9}$$

To estimate the error $\|x - x_N\|$, where we are working in the Tchebycheff norm, we go through the Green's function (9), and we use property (1). Specifically, let $Lx_N(t) = f_N(t)$. Then substituting into (9) we write

$$x(s) - x_N(s) = \int_a^b G(s,t)[f(t) - f_N(t)]\,dt, \tag{10}$$

since $Lx = f$. But $f(\bar{t}_i) = f_N(\bar{t}_i)$. Replacing $f(t) - f_N(t)$ by its third-order divided difference on $[t_{i-1}, t_i]$, we see that

$$f(t) - f_N(t) = g_i^N(t)(t - t_{i1})(t - t_{i2}),$$

where $g_i^N(t) = (f - f_N)[t_{i1}, t_{i2}, t]$, and t_{i1}, t_{i2} are the Gaussian knots lying in the interval $[t_{i-1}, t_i]$. Thus

$$\int_{t_{i-1}}^{t_i} G(s,t)(f - f_N)(t)\,dt = \int_{t_{i-1}}^{t_i} G(s,t) g_i^N(t)(t - t_{i1})(t - t_{i2})\,dt. \tag{11}$$

Note that it follows immediately from the arguments of Section 8.5 and the inequality $\|v - v_N\| \leqslant \gamma h^2$ that $\|x - x_N\| \leqslant \beta_0 h^2$, where β_0 is constant. However, if G is sufficiently smooth (and it is) we can actually obtain order h^4 convergence. Specifically, holding s fixed, we expand $G(s,t)g_i^N(t)$ in a Taylor's expansion about t_{i1}. That is, letting $\Omega_{Ni}(s,t) = G(s,t)g_i^N(t)$, we find

$$\Omega_{Ni}(s,t) = \Omega_{Ni}(s,t_{i1}) + \frac{\partial \Omega_{Ni}}{\partial t}(s,t_{i1})(t - t_{i1}) + \frac{1}{2}\frac{\partial^2 \Omega_{Ni}(s,\tau)}{\partial t^2}(t - t_{i1})^2. \tag{12}$$

Invoking (1) with $Q_{2i}(t)=(t-t_{i1})(t-t_{i2})$, we see that

$$\int_{t_{i-1}}^{t_i} \Omega_{Ni}(s,t_{i1})Q_{2i}(t)\,dt = \int_{t_{i-1}}^{t_i} \frac{\partial \Omega_{Ni}}{\partial t}(s,t_{i1})(t-t_{i1})Q_{2i}(t)\,dt = 0.$$

Thus substituting (12) into (10) and (11) we find

$$x(s)-x_N(s) = \frac{1}{2}\int_{t_{i-1}}^{t_i} \frac{\partial^2\Omega_{Ni}}{\partial t^2}(s,\tau)(t-t_{i1})^3(t-t_{i2})\,dt.$$

Invoking Schwarz's inequality and summing, we find

$$\boxed{|x(s)-x_N(s)| \leqslant \beta h^4,} \tag{13}$$

provided

$$\int_{t_{i-1}}^{t_i}\left[\frac{\partial^2\Omega_{Ni}}{\partial t^2}(s,\tau)\right]^2 dt \leqslant \beta \qquad \text{for all } i \text{ and } n,$$

where β is a constant independent of i and of n. One can prove the existence of such a uniform bounded when $f''(t)$ is piecewise continuous on $[a,b]$ and where discontinuities of f'' occur at most at the natural knots t_i, $0 \leqslant i \leqslant n$, of our splines.

The method generalizes easily to mth-order ordinary differential equations. In this case one introduces m Gaussian knots $\bar{t}_{i1} < \bar{t}_{i2} < \cdots < \bar{t}_{im}$ into each subinterval $[t_{i-1},t_i]$ of the initial partition of $[a,b]$ and uses a basis of piecewise Hermite polynomials belonging to $C^{m-1}[a,b]$ which are of degree $2m-1$ on each subinterval and satisfy the boundary conditions associated with the differential operator. When $f^{(m)}$ is piecewise continuous on $[a,b]$, one can anticipate order h^{2m} convergence, whereas order h^{2m-j} convergence can be realized when $f^{(m-j)}$, $1 \leqslant j \leqslant m$ is piecewise continuous.

Applying these ideas to partial differential equations is no more difficult than applying collocation at the natural knots. However, the existence and uniqueness theory for the method of collocation in a setting of two or more dimensions becomes increasingly complex with increasing degree of the spline approximates, whatever set of knots is chosen as collocation points. A simple existence proof in the case $Lu = -\Delta u + cu$ where $c(x,y) \geqslant 0$ on the rectangle $\Omega = [a,b]\times[c,d]$ using bicubic Hermites collocating at the 2 point Gaussian knots of each subrectangle $\Omega_{ij}=[x_i,x_{i+1}]\times[y_j,y_{j+1}]$ is given in Theorem 8.6. It has been shown by Prenter and Russell without recourse to either Green's functions or matrix analysis that collocation at Gaussian knots using bicubic Hermites on a rectangular grid provides a fourth-order scheme for approximating the solution $u(x,y)$ to the elliptic partial differential equation

$$Lu=-D_x[pu_x]-Dy[qu_y]+cu=f$$

$$u\equiv 0 \qquad \text{on } \partial R,\ R \text{ a rectangle.}$$

In this case the Gaussian knots on each subrectangle $[x_{i-1},x_i]\times[y_{j-1},y_j]$ are the points $(\bar{x}_{i1},\bar{y}_{j1})$, $(\bar{x}_{i1},\bar{y}_{j2})$, $(\bar{x}_{i2},\bar{y}_{j1})$ and $(\bar{x}_{i2},\bar{y}_{j2})$, where $\bar{x}_{ik}$ and $\bar{y}_{jk}$ are given by (4) with $\bar{t}_{ik}$ replaced by $\bar{x}_{ik}$ and $\bar{y}_{ik}$ (see Figure 8.2). We now prove Theorem 8.6.

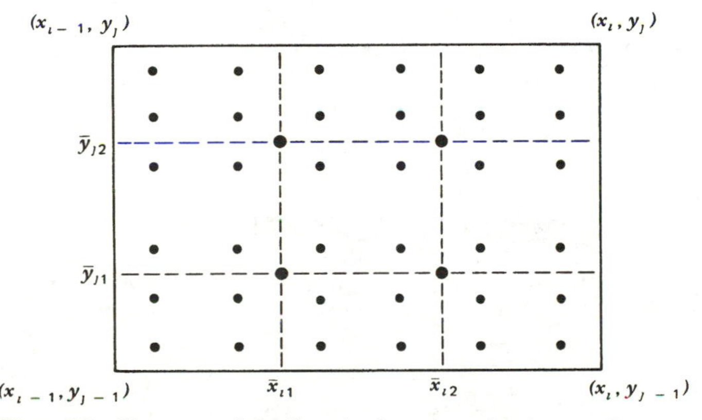

Figure 8.2 Placement of the Gaussian knots in each subrectangle.

THEOREM 8.6

The collocation matrix C_N, arising from orthogonal collocation at the four Gaussian points of each subrectangle Ω_{ij} of a partition of $\Omega=[a,b]\times[c,d]$ using bicubic Hermite polynomials over that partition applied to the boundary value problem,

$$\begin{cases} Lu=-\Delta u+cu, \\ u=0 \qquad \text{on the boundary of } \Omega, \\ c(x,y)\geqslant 0 \qquad \text{on } \Omega \text{ and continuous,} \end{cases}$$

is nonsingular, provided $u \in C^4[\Omega_{ij}]$ for each i,j.

Proof. Given any $C^4[a,b]$ function $f(x)$, the two-point Gaussian quadrature of f on the interval $[x_j, x_{j+1}]$ is given by

$$\int_{x_j}^{x_{j+1}} f(x)\,dx = \frac{1}{2}h_j[f(x_{j1}) + f(x_{j2})] + \frac{f^{(4)}(\zeta_j^x)}{4320}h_j^5$$

where $x_j < \zeta_j^x < x_{j+1}$ (see Davis and Rabinowitz, p. 36, 1967). A similar formula holds in the y-direction with $\eta_i = y_{i+1} - y_i$. Let $E_x^n f = \sum_{j=1}^n \frac{1}{2}h_j[f(x_{j1}) + f(x_{j2})]$ be the full two point Gaussian quadrature of $f(x)$ over $[a,b]$. If $f(x) = p''(x)p(x)$ where $p(x)$ is a cubic polynomial, $f^{(4)}(\zeta_i^x) = 4[p'''(\zeta_i^x)]^2$. With this in mind, let $u[x,y)$ be a bicubic Hermite polynomial satisfying zero boundary data. Then

$$\begin{aligned}
\|u_y(x_{j1},\cdot)\|_2^2 &= \int_c^d [u_y(x_{j1},t)]^2\,dt \\
&= -\int_c^d u(x_{j1},t)u_{yy}(x_{j1},t)\,dt \\
&= E_x^n[-u(x_{j1},\cdot)u_{yy}(x_{j1},\cdot)] \\
&\quad - \sum_{i=1}^n \frac{[u_{yyy}(x_{j1},\zeta_{i1}^y)]^2}{1080}\eta_i^5 \\
&= \langle u(x_{j1},\cdot), -u_{yy}(x_{j1},\cdot)\rangle - \sum_{i=1}^n \frac{[u_{yyy}(x_{j1},\zeta_{i1}^y)]^2}{1080}\eta_i^5.
\end{aligned}$$

Similarily

$$\|u_y(x_{j2},\cdot)\|_2^2 = \langle u(x_{j2},\cdot), -u_y(x_{j2},\cdot)\rangle - \sum_{i=1}^n \frac{[u_{yyy}(x_{j2},\zeta_{i2}^y)]^2}{1080}\eta_i^5,$$

$$\|u_x(\cdot,y_{i1})\|_2^2 = \langle u(\cdot,y_{i1}), -u_{xx}(\cdot,y_{i1})\rangle - \sum_{j=1}^n \frac{[u_{xxx}(\zeta_{j1}^x,y_{i1})]^2}{1080}h_j^5,$$

and

$$\|u_x(\cdot,y_{i2})\|_2^2=\langle u(\cdot,y_{i2}),-u_{xx}(\cdot,y_{i2})\rangle-\sum_{j=1}^{n}\frac{[u_{xxx}(\zeta_{j2}^x,y_{i2})]^2}{1080}h_j^5.$$

Let

$$\|u\|_{H_0^1(\Omega)}^2=\sum_{j=1}^{n}\tfrac{1}{2}h_j\{\|u_y(x_{j1},\cdot)\|_2^2+\|u_y(x_{j2},\cdot)\|_2^2\}$$

$$+\sum_{i=1}^{n}\tfrac{1}{2}\eta_i\{\|u_x(\cdot,y_{i1})\|_2^2+\|u_x(\cdot,y_{i2})\|_2^2\}.$$

It follows that

$$\|u\|_{H_0^1(\Omega)}^2=\langle-\Delta u,u\rangle-\sum_{j=1}^{n}\frac{h_j\eta_i^5}{1080}\{[u_{yyy}(x_{j1},\zeta_{i1}^y)]^2+[u_{yyy}(x_{j2},\zeta_{i2}^y)]^2\}$$

$$-\sum_{i=1}^{n}\frac{\eta_i h_j^5}{1080}\{[u_{xxx}(\zeta_{j1}^x,x_{i1})]^2+[u_{xxx}(\zeta_{j2}^x,x_{i2})]^2\}$$

$$\leqslant\langle-\Delta u,u\rangle \tag{1}$$

where $\langle-\Delta u,u\rangle$ is the full 4-point quadrature of $-\int_\Omega u\Delta u$. Let $u=\sum_{k=1}^{n^2}b_kB_k(x,y)$, where the B_k's are the canonical basis bicubic Hermites (shape functions) associated with our partition. Suppose $C_Nb=0$ where C_N is the collocation matrix associated with our boundary value problem and partition of Ω and where $b=(b_1,b_2,\ldots,b_n)'$. Then $\langle u,-\Delta u+cu\rangle=0$. But

$$\begin{aligned}\langle-\Delta u+cu,u\rangle&=\langle-\Delta u,u\rangle+\langle cu,u\rangle\\&\geqslant\langle-\Delta u,u\rangle\\&\geqslant\|u\|_{H_0^1(\Omega)}^2.\end{aligned}$$

Thus $\|u\|^2_{H_0^1(\Omega)}=0$ and

$$\|u_y(x_{j1},\cdot)\|_2=\|u_y(x_{j2},\cdot)\|_2=\|u_y(\cdot,y_{i1})\|_2$$
$$=\|u_y(\cdot,y_{i2})\|_2=0. \tag{2}$$

Since u is a bicubic Hermite satisfying zero boundary data, (2) holds if and only if $u\equiv 0$. Thus $C_N b=0$ if and only if $b=0$. That is, C_N is nonsingular.■

Some data comparing the method of collocation at Gaussian knots to the finite element method using various basis are given in the examples below.

EXAMPLE 1

We solve the differential equation

$$Lu=u_{xx}+u_{yy}$$
$$u=0 \qquad \text{on } \partial\Omega$$
$$u=3e^x e^y(x-x^2)(y-y^2)$$

approximately using collocation on the rectangle $[0,1]\times[0,1]$ with step size P_n and average error e_n based on 25 points per interval using P_n (see Table 8.3).

TABLE 8.3 COLLOCATION

Bicubic Splines		Bicubic Hermites	
P_n	e_n	P_n	e_n
$\frac{1}{7}$	0.310(−2)	$\frac{1}{2}$	0.885(−2)
$\frac{1}{8}$	0.243(−2)	$\frac{1}{3}$	0.170(−2)
$\frac{1}{9}$	0.196(−2)	$\frac{1}{4}$	0.569(−3)
$\frac{1}{10}$	0.160(−2)	$\frac{1}{5}$	0.235(−3)
$\frac{1}{11}$	0.133(−2)	$\frac{1}{6}$	0.113(−3)
		$\frac{1}{7}$	0.616(−4)
		$\frac{1}{8}$	0.358(−4)
		$\frac{1}{9}$	0.225(−4)
		$\frac{1}{10}$	0.148(−4)

EXAMPLE 2

We solve the differential equation

$$Lu = -D_x[pu_x] - D_y[pu_y] + cu$$

$$p = e^{xy} \qquad q = e^{x+y}$$

$$c = \frac{1}{1+x+y}$$

$$u = \sin \pi x \sin \pi y$$

approximately on the rectangle $[0,1]\times[0,1]$ using collocation and finite elements with bicubic Hermites, step size P_n and average error e_n (see Table 8.4).

TABLE 8.4 COMPARISON OF COLLOCATION AND FINITE ELEMENTS

	Collocation with Bicubic Hermites	Finite Elements with Bicubic Hermites
P_n	e_n	e_n
$\frac{1}{2}$	0.114(−1)	0.827(−2)
$\frac{1}{3}$	0.413(−2)	0.215(−2)
$\frac{1}{4}$	0.112(−2)	0.841(−3)
$\frac{1}{5}$	0.526(−3)	0.354(−3)
$\frac{1}{6}$	0.239(−3)	0.184(−3)
$\frac{1}{7}$	0.136(−3)	0.993(−4)
$\frac{1}{8}$	0.773(−4)	0.601(−4)
$\frac{1}{9}$	0.496(−4)	0.374(−4)
$\frac{1}{10}$	0.320(−4)	0.250(−4)

EXAMPLE 3

Let

$$\Delta u = 1$$

$$u \equiv 0 \qquad \text{on } \partial\Omega.$$

This problem has a logarithmic singularity in its second derivatives at the corners of the unit square. For collocation with bicubic splines we use the estimate

$u(\frac{1}{2},\frac{1}{2})\approx 0.07366$, and for collocation with bicubic Hermites we have used $u(\frac{1}{2},\frac{1}{2}) \approx 0.07367116$, which is the computed value for a uniform mesh and $p=\frac{1}{10}$ (see Table 8.5).

TABLE 8.5 COLLOCATION

Cubic Hermites		Cubic splines	
h	e_n	h	e_n
$\frac{1}{3}$	0.260(−3)	$\frac{1}{4}$	0.11(−1)
$\frac{1}{5}$	0.338(−4)	$\frac{1}{8}$	0.79(−2)
$\frac{1}{7}$	0.906(−5)	$\frac{1}{12}$	0.56(−2)
$\frac{1}{9}$	0.358(−5)	$\frac{1}{16}$	0.43(−2)
$\frac{1}{2}$	0.103(−4)		
$\frac{1}{4}$	0.652(−5)		
$\frac{1}{6}$	0.107(−5)		
$\frac{1}{8}$	0.119(−6)		
$\frac{1}{10}$	...		

EXERCISES

1. Assuming the operator (2) is strongly coercive, try to establish $o(h^3)$ convergence of the cubic spline collocation approximate to u.
2. Let $u_N(x,y)$ be the bicubic Hermite approximate of Theorem 8.6. Try to prove $\|u-u_N\|_2=o(h^4)$.

8.8 A CONNECTION BETWEEN COLLOCATION AND GALERKIN METHODS

It is both interesting and important to see what theoretical connections, if any, exist between the method of collocation and the Galerkin methods described in Section 7.10. We have already noted that Galerkin methods reduce to the Rayleigh-Ritz method (hence to the finite element method) when A is positive definite and symmetric; this one connection between the two methods is well known. One way of connecting Galerkin to collocation is through numerical quadrature schemes. In particular, let us once again assume that $X=L_2(D)$ for some region D in E^1 or E^2 and that A is a linear operator from some subspace of X into X. Letting $X_N = \text{span}\{\phi_1^N,\phi_2^N,\ldots,\phi_N^N\}$ be an N-dimensional subspace of X, the Galerkin approximate x_N (should it exist) to the unique solution of our operator equation

$$Ax=y \tag{1}$$

is given by

$$x^N = x_1^N\phi_1^N + x_2^N\phi_2^N + \cdots + x_N^N\phi_N^N. \tag{2}$$

The coefficients x_j^N, $1 \leqslant j \leqslant N$, are the (unique) solution to the linear system

$$\sum_{j=1}^{N} (A\phi_j^N, \phi_i^N)x_j^N = (y, \phi_i^N), \qquad 1 \leqslant i \leqslant N. \tag{3}$$

Since we are working in $L_2[D]$, the parenthetical expressions are given as integrals. In particular

$$(A\phi_j^N, \phi_i^N) = \int_D A\phi_j^N(p)\phi_i^N(p)\,dp \tag{4}$$

$$(y, \phi_i^N) = \int_D y(p)\phi_i^N(p)\,dp, \tag{5}$$

where $p \in D$. When A is a differential operator we usually integrate the first expression by parts replacing the integral (4) by the bilinear form $a(\phi_j^N, \phi_i^N)$ (see Chapter 7). Suppose, however, we do not integrate (4) by parts but instead replace both the integrals (4) and (5), just as they appear, by the numerical quadrature

$$\int_P f(p)\,dp \approx f(p_1)w_1 + f(p_2)w_2 + \cdots + f(p_n)w_n,$$

where $p_1, p_2, \ldots, p_N$ are N distinct points of D and the w_i's are constants determined by the approximation scheme giving rise to the rule. It follows that

$$(A\phi_j^N, \phi_i^N) \approx \sum_{k=1}^{N} A\phi_j^N(p_k)\phi_i^N(p_k)w_k$$

and

$$(f, \phi_i^N) \approx \sum_{k=1}^{N} \phi_i^N(p_k)w_k f(p_k).$$

If the matrix $(\phi_i^N(p_k)) = H_N$ *is nonsingular*, then

$$B_N = D_N H_N C_N^T,$$

where $D_N = \operatorname{diag}(w_1, w_2, \ldots, w_N)$, and C_N^T *is precisely the collocation matrix* $C_N^T = (\hat{c}_{ij}^N)$

$$\hat{c}_{ij}^N = A\phi_j^N(p_i).$$

Let $y_N = (y_1^N, y_2^N, \ldots, y_N^N)$, where $y_i^N = (y, \phi_i^N)$ and $\hat{y}_N = (y(p_1), y(p_2), \ldots, y(p_N))$. Then the approximate system obtained replacing (4) and (5) by our quadratures in (3) is

$$H_N D_N C_N^T \hat{x}_N = H_N D_N \hat{y}_N,$$

where $\hat{x}_N = (\hat{x}_1^N, \hat{x}_2^N, \ldots, \hat{x}_N^N)$. Thus when H_N is nonsingular, our fully discretized Galerkin system is equivalent to

$$C_N^T \hat{x}_N = \hat{y}_N. \tag{6}$$

If quadrature errors were small enough, $H_N D_N C_N^T$ would be nonsingular, and the errors for this discrete Galerkin solution would be the same as for the Galerkin solution $x_N(p)$. Moreover, since the matrices are square, nonsingularity of $H_N D_N C_N^T$ implies nonsingularity of C_N^T, hence of C_N. If $\hat{x}_N(p) = \hat{x}_1^N \phi_1^N(p) + \hat{x}_2^N \phi_2^N(p) + \cdots + \hat{x}_N^N \phi_N^N(p)$, (6) would be equivalent to

$$A\hat{x}_N(p_i) = y(p_i);$$

thus $\hat{x}_N(p)$ is the collocation solution.

Similar observations connecting the method of least squares and the method of collocation have also been observed (see Ciarlet and Raviart (1973) and Russell and Varah) and can be derived easily by the interested reader.

REFERENCES

Bellman, R., R. Kalaba, and J. Lockett. *Numerical inversion of the Laplace transform.* American Elsevier, New York, 11966.

Cavendish, J. C. Collocation methods for elliptic boundary value problems. Ph.D. thesis, University of Pittsburg, 1972.

Ciarlet, R. G. and P. A. Raviart. The combined effect of curved bounderies and numerical integration in isoperimetric finite element methods. *Proceedings of the conference on the mathematical foundations of the finite element method with applications to partial differential equations.* Academic Press, New York, 1973, pp. 404–474.

Coddington, E. A. and N. Levinson. *Theory of ordinary differential equations.* McGraw Hill, New York, 1955.

Collatz, L. *Functional analysis and numerical mathematics.* Academic Press, New York, 1966.

Davis, P. J., *Interpolation and Approximation.* Blaisdell, New York, 1965.

Davis, P. J. and P. Rabinowitz. *Numerical integration,* Ginn-Blaisdell, 1967.

De Boor, C. and B. Schwartz. Collocation at Gaussian points. *SIAM J. Numer. Anal.,* **10** (1973), pp. 582–606.

Dunford, N. and J. T. Schwartz. *Linear operators part II.* Wiley, New York, 1967.

Ferguson, H. B. *Orthogonal collocation as a method of analysis in chemical engineering.* Ph.D. Thesis, University of Washington, Seattle (1971).

Ferguson, N. B. and B. A. Finlayson. Transient chemical reaction analysis by orthogonal collocation, *Chem. Eng. J.*, **1**, (1970), pp. 327–336.

Finlayson, B. A. Packed bed reactor analysis by orthogonal collocation, *Chem. Eng. Sci.*, **26**, (1971), pp. 1081–1091.

Finlayson, B. A. *The method of weighted residuals and variational principles*, Academic Press, 1972.

Horvay, G. and F. N. Spiess. Orthogonal edge polynomial in the solution of boundary value problems, *Quant. Appl. Math.*, **12**, (1954), pp. 57–65.

Hulme, B. L. Discrete Galerkin and related one-step methods for ordinary differential equations. *Math. Comp.*, **120** (1972), pp. 881–891.

Ito, F. A collocation method for boundary value problems using spline functions. Ph.D. thesis, Brown University, 1972.

Karpilovskaya, E. B. A method of collocation for integro-differential equations with biharmonic principal part, *U.S.S.R. Comp. Math. Phys.*, **10**, No. 6, (1970), pp. 240–260.

Lanczos, C. *Applied analysis*. Prentice Hall, New Jersey, 1956.

Lanczos, C. Trigonometric interpolation of empirical and analytical functions, *J. Math. Phys.*, **17**, (1938), pp. 123–199.

Lee, E. T. Y. and R. F. Sincovec. Spline function collocation methods for linear boundary value problems. *Bull. Instit. Math.* Sinica, **1**, No. 1 (1973)8 pp. 41–55.

Naimark, M. A. *Linear differential operators*, Vol I. Ungar, New York, 1960.

Prenter, P. M. and R. D. Russell. Collocation for elliptic partial differential equations. Submitted to *SIAM J. Numer. Anal.*

Russell, R. D. and J. M. Varah. A comparison of global methods for two-point boundary value problems. Submitted to *SIAM J. Numer. Anal.*

Russell, R. D. and L. F. Shampine. A collocation method for boundary value problems, *Numer. Math.*, **19**, (1972), pp. 1–28.

Shindler, A. A. Rate of convergence of the enriched collocation method for ordinary differential equations. *Sib. Math. Zh.*, **10**, (1969), pp. 229–233.

Vainikko, G. The convergence of the collocation method for nonlinear differential equations. *USSR Comp. Math Phys.*, **6** (1966), pp. 35–42.

Varga, R. J. *Matrix iterative analysis*. Prentice-Hall, Englewood Cliffs, N.J., 1959.

Villadsen, J. V. *Selected Approximation Methods for Chemical Engineering Problems* Instit. for Kemiteknik Numer. Instit. Danmarks Tekniske Hojskole, 1970.

Villadsen, J. V. and W. E. Stewart. Solution of boundary value problems by orthogonal collocation, *Chem. Eng. Sci.*, **22**, (1967), pp. 1483–1501.

Wright, K. Chebychev collocation methods for ordinary differential equations, *Comp. J.*, **16**, (1964), pp. 358–365.

GLOSSARY OF SYMBOLS

$a(u, v)$	202
$B_i(t)$	79, 90
$C[a, b]$	3
$C^n[a, b]$	9
$C[\Omega]$	118
$C^\circ[\Omega]$	118
$C^k[\Omega]$	118
$C^{m,n}[\Omega]$	118
$H_3(\pi)$	59
$H_{2m-1}(\pi) = H_{2m-1}$	69
$H^1(\Omega)$	262
$H_0^1(\Omega)$	262
(u, v)	182
$(u, v)_A$	184, 202, 218, 231
$L_d(\pi)$	48
$L_1(\tau)$	154
$L_2(\tau)$	154
$L_k(\tau)$	154
$L_2[a, b]$	12
$\Vert\ \Vert_2$	14
$\Vert\ \Vert_A$	202, 219
$\Vert\ \Vert_\infty$	6, 118, 187
$\Vert\ \Vert_{H^1(\Omega)}$	262
$P_n[a, b]$	4
$P_{m,n}$	121
$PC[a, b]$	5, 11, 15
$S_3(\pi)$	78
$S_d(\pi, k)$	95

INDEX